High Speed Serdes Devices and Applications

David R. Stauffer • Jeanne Trinko Mechler
Michael Sorna • Kent Dramstad
Clarence R. Ogilvie • Amanullah Mohammad
James Rockrohr

High Speed Serdes Devices and Applications

 Springer

David R. Stauffer
IBM Corporation
Essex Junction, VT
USA

Jeanne T. Mechler
IBM Corporation
Essex Junction, VT
USA

Kent Dramstad
IBM Corporation
Essex Junction, VT
USA

Clarence R. Ogilvie
IBM Corporation
Essex Junction, VT
USA

Amanullah Mohammad
IBM Corporation
Research Triangle Park, NC
USA

James D. Rockrohr
IBM Microelectronics
Hopewell Junction, NY
USA

Michael A. Sorna
IBM Microelectronics
Hopewell Junction, NY
USA

ISBN 978-0-387-79833-2   e-ISBN 978-0-387-79834-9

Library of Congress Control Number: 2008925643

Printed on acid-free paper.

9 8 7 6 5 4 3 2 1

springer.com

Preface

The simplest method of transferring data through the inputs or outputs of a silicon chip is to directly connect each bit of the datapath from one chip to the next chip. Once upon a time this was an acceptable approach. However, one aspect (and perhaps the only aspect) of chip design which has not changed during the career of the authors is Moore's Law, which has dictated substantial increases in the number of circuits that can be manufactured on a chip. The pin densities of chip packaging technologies have not increased at the same pace as has silicon density, and this has led to a prevalence of High Speed Serdes (HSS) devices as an inherent part of almost any chip design.

HSS devices are the dominant form of input/output for many (if not most) high-integration chips, moving serial data between chips at speeds up to 10 Gbps and beyond. Chip designers with a background in digital logic design tend to view HSS devices as simply complex digital input/output cells. This view ignores the complexity associated with serially moving billions of bits of data per second. At these data rates, the assumptions associated with digital signals break down and analog factors demand consideration. The chip designer who oversimplifies the problem does so at his or her own peril.

Despite this, many chip designers who undertake using HSS cores in their design do not have a sufficient background to make informed decisions on the use of HSS features in their application, and to appreciate the potential pitfalls that result from ignoring the analog nature of the application. Databooks describe the detailed features of specific HSS devices, but usually assume that the reader already understands the fundamentals. This is the equivalent of providing detailed descriptions of the trees, but leaving the reader struggling to get an overview of the forest.

This text is intended to bridge this gap, and provide the reader with a broad understanding of HSS device usage. Topics typically taught in a variety of courses using multiple texts are consolidated in this text to provide sufficient background for the chip designer that is using HSS devices on his or her chip. This text may be viewed as consisting of four sections as outlined below.

The first three chapters relate to the features, functions, and design of HSS devices. Chapter 1 introduces the reader to the basic concepts and the resulting features and functions typical of HSS devices. Chapter 2 builds upon these concepts by describing an example of an HSS core, thereby giving the reader a concrete implementation to use as a framework for topics throughout the remainder of the text. Although loosely based on the HSS designs offered in IBM ASIC products, this HSS EX10 is a simplified tutorial example and shares many features/functions with product offerings from other vendors. Finally, Chap. 3 introduces interested readers to the architecture and design of HSS cores using the HSS EX10 as an example.

The next two chapters describe the features and functions of protocol logic used to implement various network protocol interface standards. Chapter 4 introduces concepts related to interface standards, as well as design architectures for various protocol logic functions. Chapter 5 provides an overview of various protocol standards in which HSS cores are used.

The next four chapters cover specialized topics related to HSS cores. Chapter 6 describes clock architectures for the reference clock network which supplies clocks to the HSS core, as well as floorplanning and signal integrity analysis of these networks. Chapter 7 covers various topics related to testing HSS cores and diagnostics using HSS cores. Chapter 8 covers basic concepts regarding signal integrity, and signal integrity analysis methods. Chapter 9 covers power dissipation concepts and how these relate to HSS cores.

Finally, any HSS core is not complete without a set of design kit models to facilitate integration within the chip design. Chapter 10 discusses various topics regarding the design kit models that require special consideration when applied to HSS cores.

Acknowledgments

The authors wish to thank the following IBM colleagues without whose contributions and reviews this text would not be possible: William Clark, Nanju Na, Stephen Kessler, Ed Pillai, M. Chandrika, Peter Jenkins, Douglas Massey, Suzanne Granato, Della Budell, and Jack Smith.

In addition, the authors would like to thank Thucydides Xanthopoulos of Cavium Networks for his detailed and insightful review of this text, and Andrea Kosich for making it possible to utilize material from Optical Internetworking Forum Interoperability Agreements.

Table of Contents

Preface   v
Acknowledgments   vii

Chapter 1: Serdes Concepts. 1
  1.1 The Parallel Data Bus   1
  1.2 Source Synchronous Interfaces   2
    Reducing the Number of I/O Pins   2
    Clock Forwarding   3
    Higher Speed Source Synchronous Interfaces   4
  1.3 High-Speed Serdes   8
    Serializer / Deserializer Blocks   9
    Equalizers   10
    Clock and Data Recovery (CDR)   14
    Differential Driver   15
    Differential Receiver   17
    Diagnostic Functions   17
    Phase-Locked Loop   19
  1.4 Signal Integrity   19
    The Channel   19
    Package Models   21
    Jitter   21
    Channel Analysis Tools   23
  1.5 Signaling Methods   24
  1.6 Exercises   27

Chapter 2: HSS Features and Functions 31
  2.1 HSS Core Example: HSS EX10 10-Gbps Core   31
    HSS EX10 Input/Output Pin Descriptions   32
    HSS EX10 Register Descriptions   41
  2.2 HSS EX10 Transmitter Slice Functions   53
    Transmitter Parallel Data   54
    Transmitter Signal Characteristics   56
    Transmitter FFE Programming   58
    Transmitter Power Control   59
    Half-Rate/Quarter-Rate/Eighth-Rate Operation   60
    JTAG 1149.1 and Bypass Mode Operation   62
    PRBS / Loopback Diagnostic Features   64
    Out of Band Signalling Mode (OBS)   65
    Features to Support PCI Express   65
  2.3 HSS EX10 Receiver Slice Functions   66
    Receiver Data Interface   68
    DFE and Non-DFE Receiver Modes   70

Serial Data Termination and AC/DC Coupling   71
Signal Detect   71
Receiver Power Control   72
JTAG 1149.1/1149.6 and Bypass Mode Operation   73
Half-Rate/Quarter-Rate/Eight-Rate Operation   76
PRBS / Loopback Diagnostic Features   77
Phase Rotator Control/Observation   78
Support for Spread Spectrum Clocking   78
Eye Quality   79
SONET Clock Output   80
Features to Support PCI Express   80
2.4 Phase-Locked Loop (PLL) Slice   80
Reference Clock   81
Clock Dividers   82
Power On Reset   82
VCO Coarse Calibration   83
PLL Lock Detection   83
Reset Sequencer   84
HSS Resynchronization   84
PCI Express Power States   87
2.5 Reset and Reconfiguration Sequences   87
Reset and Configuration   87
Changing the Transmitter Configuration   90
Changing the Receiver Configuration   92
2.6 References and Additional Reading   93
2.7 Exercises   94

Chapter 3: HSS Architecture and Design. 99
3.1 Phase Locked Loop (PLL) Slice   100
PLL Macro   101
Clock Distribution Macro   102
Reference Circuits   103
PLL Logic Overview   105
3.2 Transmitter Slice   107
Feed Forward Equalizer (FFE) Operation   109
Serializer Operation   112
3.3 Receiver Slice   114
Clock and Data Recovery (CDR) Operation   116
Decision Feedback Equalizer (DFE) Architectures   118
Data Alignment and Deserialization   121
3.4 References and Additional Reading   122
3.5 Exercises   123

Chapter 4: Protocol Logic and Specifications 125

4.1 Protocol Specifications 125
    Protocol Layers 125
    Serial Data Specifications 126
    Basic Concepts 132
4.2 Protocol Logic Functions 134
    Bit/Byte Order and Striping/Interleaving 134
    Data Encoding and Scrambling 136
    Error Detection and Correction 143
    Parallel Data Interface 147
    Bit Alignment 152
    Deskewing Multiple Serial Data Links 153
4.3 References and Additional Reading 158
4.4 Exercises 159

Chapter 5: Overview of Protocol Standards 165

5.1 SONET/SDH Networks 168
    System Reference Model 169
    STS-1 Frame Format 170
    STS-N Frame Format 174
    Clock Distribution and Stratum Clocks 176
5.2 OIF Protocols 177
    System Reference Model 177
    SFI-5.2 Implementation Agreement 180
    SPI-S Implementation Agreement 184
    CEI-P Implementation Agreement 188
    Electrical Layer Implementation Agreements 190
5.3 Ethernet Protocols 197
    Physical Layer Reference Model 198
    Media Access Control (MAC) Layer 201
    XGMII Extender Sublayer (XGXS) 204
    10-Gb Serial Electrical Interface (XFI) 207
    Backplane Ethernet 213
    PMD Sublayers for Electrical Variants 218
5.4 Fibre Channel (FC) Storage Area Networks 220
    Storage Area Networks (SANs) 220
    Fibre Channel Protocol Layers 222
    Framing and Signaling 222
    Physical Interfaces 229
    10-Gbps Fibre Channel 236
5.5 PCI Express 237
    PCI Express Architecture 238
    Physical Layer Logic 241
    Electrical Physical Layer 246
    Power States 249
    PCI Express Implementation Example 250

5.6 References and Additional Reading   251
5.7 Exercises   254

Chapter 6: Reference Clocks 263
6.1 Clock Distribution Network   263
    Single-Ended vs. Differential Reference Clocks   263
    Reference Clock Sources   265
    Special Timing Requirements   268
    Special Test Requirements   270
6.2 Clock Jitter   270
    Jitter Definitions   271
    Jitter Effects   276
    PLL Jitter   277
6.3 Clock Floorplanning   281
    Clock Tree Architecture   281
    Clock Tree Wiring   282
6.4 Signal Integrity of the Clock Network   283
    Analog Signal Levels and Slew Rates   283
    Duty Cycle Distortion   286
    Differential Clock Analysis Methodology   288
6.5 References and Additional Reading   293
6.6 Exercises   293

Chapter 7: Test and Diagnostics 297
7.1 IEEE JTAG 1149.1 and 1149.6   298
    JTAG 1149.1 Overview   299
    HSS Core Support for JTAG 1149.1   302
    HSS Core Support for JTAG 1149.6   303
7.2 PRBS Testing and Loopback Paths   306
    Loopback Paths   306
    PRBS Circuits and Data Patterns   309
    PRBS Test Sequence   314
7.3 Logic Built-In-Self-Test (LBIST)   317
    LBIST Architecture   317
    LBIST Considerations for HSS Cores   319
7.4 Manufacturing Test   320
    Chip Level Test   320
    HSS Macro Test   324
7.5 Characterization and Qualification Testing   327
    Transmitter Tests   328
    Receiver Tests   335
    General Tests   338
7.6 References and Additional Reading   340
7.7 Exercises   340

Chapter 8: Signal Integrity 345

8.1 Probability Density Functions 345
    Gaussian Distribution 345
    Dual-Dirac Distribution 348
8.2 Jitter 349
    Jitter Components 349
    Deterministic Jitter 352
    Random Jitter 356
    Total Jitter and Mathematical Models 358
    Jitter Budgets 362
    Jitter Tolerance 364
8.3 Spice Models 365
    Traditional Spice Models 365
    Hybrid Spice/Behavioral Models 367
    Spice Simulation Matrices 369
8.4 Statistical Approach to Signal Integrity 372
    Analysis Approach 373
    HSSCDR Software 388
8.5 References and Additional Reading 393
8.6 Exercises 394

Chapter 9: Power Analysis. 397

9.1 Digital Logic Circuits 397
    Digital Logic Active or AC Power 397
    Digital Logic Leakage or DC Power 402
9.2 Non Digital Logic Circuits 410
    AC (Active) Power 410
    DC (Leakage) Power 410
    Quiescent Power 410
9.3 HSS Power 411
    HSS Power Equation 411
    Multiple Power Supplies 412
    Chip Fabrication Process 413
    Mode-Dependent Power 414
    Power Dissipation Breakdown 416
9.4 Reducing Power Dissipation 417
    Power Concerns for the HSS Core Design 417
    Power Dissipation Concerns for the Chip Designer 420
9.5 References and Additional Reading 421
9.6 Exercises 421

Chapter 10: Chip Integration 425

10.1 Simulation Models   427
   Reset and Initialization Short Cuts   427
   Simulation 'X' States   429
   Modeled and Unmodeled Behavior   432
10.2 Test Synthesis   434
   Scan Test Support   435
   Macro Test Support   436
   JTAG Logic Connections   440
   Automation of Test Requirements   442
   Running Macro Test using the JTAG Interface   444
10.3 Static Timing Analysis   445
   Clock Timing   445
   Receiver Parallel Data Outputs   450
   Register Interface   452
   Transmitter Synchronization   454
   Serial Data Timing   456
   Skew Management   457
   Timing Backannotation for Simulation   458
10.4 Chip Floorplan and Package Considerations   459
   Packages   459
   Chip Physical Design   466
10.5 References   471
10.6 Exercises   472

Index. **475**

Chapter 1
Serdes Concepts

This chapter describes basic methods of transferring data from one chip to another chip, either on the same circuit board or across a cable or backplane to another circuit board. After reading this chapter, the reader should have a basic understanding of the rationale for using high-speed serializer/deserializer (Serdes) devices, and the inherent problems introduced by the high-speed operation of such devices.

1.1 The Parallel Data Bus

The simplest method of transferring data through the inputs or outputs of a silicon chip is to directly connect the datapath from one chip to the next chip (see Fig. 1.1). Since data often consists of more than one bit of information, the datapath is more than one-bit wide. In the figure, an *n*-bit datapath inside *Chip #1* is driven through chip outputs, across an *n* bit interconnect, through inputs of *Chip #2*, to an *n*-bit datapath inside the receiving chip. Synchronous data is transferred between the two chips since both chips are clocked by the same clock source.

There are two inherent problems of the parallel data bus shown in Fig. 1.1. The first problem is that *n* input/output (I/O) pins are required on each chip to transfer the data. At one point in history this was acceptable. However, Moore's Law has driven substantial increases in the number of circuits that can be manufactured on a chip compared to a few decades ago. The pin densities of chip packaging technologies have not increased at the same pace as silicon density. Therefore, I/O pins are substantially more expensive than silicon circuits, and dedicating *n* I/O pins for the above data bus is not acceptable for most chip applications.

The second inherent problem involves meeting timing requirements. The data is launched synchronously by *Chip #1* and is captured synchronously in *Chip #2* using the same clock. The data at the inputs of *Chip #2* must meet setup and hold times relative to the clock input of the chip. These setup and hold times must be calculated with sufficient margin to allow for differences in delay of the clock distribution path to the two chips, and through the chips to the launch and capture flip-flops. Delays may vary based on chip process, voltage, and temperature (PVT) conditions, and margin must be added to account for worst case variations. For higher clock frequencies, it may be necessary to use phase-locked loops (PLLs) in the chips to adjust the clock phase in order to compensate for the clock distribution delay within the chip and adapt to changing process, voltage, and temperature conditions. If the clock frequency is high enough, it will not be possible to build a system that will reliably transfer the data across this data bus.

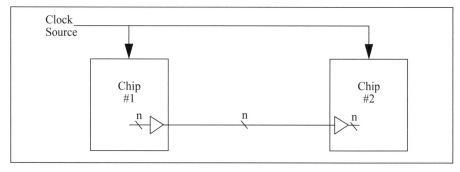

Fig. 1.1 Parallel data bus between two chips

1.2 Source Synchronous Interfaces

The two problems with the parallel data bus in Sect. 1.1 can be eliminated with the modifications to the system which are discussed in this section. These approaches are extensions of the parallel data bus. The parallel data bus and all of the extensions described in this section are considered to be *source synchronous interface* architectures. Such architectures include any interface where a clock input exists that can be used to capture the received data. This may be either a reference clock used by both the transmitting and receiving chip or the transmitting chip drives a clock to the receiving chip. In either case, clock recovery circuits are *not* required for source synchronous interfaces.

1.2.1 Reducing the Number of I/O Pins

The first issue to be addressed is reducing the number of I/O pins required to transfer the data between the chips. This is accomplished by multiplexing the n bits of data at the output of *Chip #1* onto k bits of interconnect ($k < n$), and then demultiplexing the k bits of interconnect at the input of *Chip #2* onto an n bit internal datapath. This is shown in Fig. 1.2. The resulting system only requires k I/O pins on each chip rather than the n pins previously required.

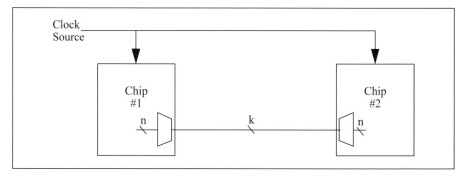

Fig. 1.2 Serializing the data to reduce pin counts

Of course, while the pin count requirements have been reduced by the ratio of $k:n$, the required frequency of the reference clock has increased by the inverse of this ratio. System designers generally do not like to distribute high-speed reference clocks within the system due to noise, electromagnetic interference (EMI), and power dissipation concerns. Often, a lower frequency clock is distributed, and PLLs in the chips are used to multiply this reference clock to a usable frequency. Variability of the phase of the resulting clock, along with the higher frequency of data transfer, tends to exacerbate the timing issues of the parallel data bus approach.

1.2.2 Clock Forwarding

In Fig. 1.3, a high-speed clock has been added to the datapath between the two chips. This clock source is assumed to supply a clock frequency somewhat lower than the frequency required to clock the data flip-flops on the chip interconnect. PLLs are used in each chip to generate clocks at a multiple of this frequency. The resulting clocks are used to launch and capture data in the respective chips. The output clock of the PLL in *Chip #1*, which is used to launch the data from this chip, is also an output of this chip. This clock is used by *Chip #2* to capture the data. This approach is called *clock forwarding*.

The advantage of this approach is that the high-speed clock used to launch the data at *Chip #1* is available to *Chip #2* as a reference to capture the data. Any variations in delays through clock distribution network driving the two chips does not need to be taken into account in timing analysis. Only delay variations between the clock path and the data bits are relevant. Variations between these paths due to process, voltage, and temperature track each other to some extent. The result is that timing analysis of the interface requires less margin and setup and hold times are therefore easier to meet.

So far we have not made any distinction or recommendations regarding the frequency of the high-speed clock relative to the bit rate of the interface. In general, the high-speed clock shown in the figure could be *single data rate* (SDR) or *double data rate* (DDR) (Fig. 1.4). The receiving chip captures data on every rising (or every falling) edge of an SDR clock; while the receiving chip captures data on every edge (both rising and falling edges) of a DDR clock.

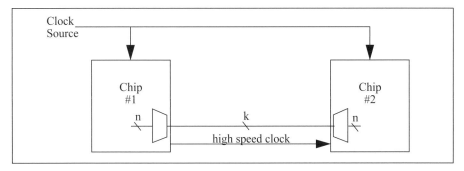

Fig. 1.3 High-speed clock forwarded with the data

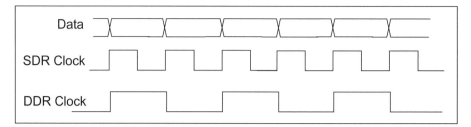

Fig. 1.4 Single data rate and double data rate clocks

The advantage of DDR clocks over SDR clocks lies in the bandwidth requirements for the corresponding I/O drivers and receivers. An I/O cell being used at a bit rate of b Mbits per second requires a bandwidth sufficient to transmit a 101010... data pattern. This corresponds to a frequency spectrum with an upper fundamental frequency limit of $b/2$ MHz. The corresponding frequency of an SDR clock is b MHz, twice the spectral limit of the data. However, the frequency of a DDR clock for the same interface is only $b/2$ MHz, consistent with the frequency spectrum of the data. Therefore, the same I/O drivers and receivers can be used to drive and receive both the data and the DDR clock.

Regardless of whether the high-speed clock is a SDR or a DDR clock, the receiving chip uses this clock to directly capture the data. This chip also uses the reference clock to generate an internal system clock at the same frequency. These clocks are mesochronous. While the frequency is the same (given that they share a common frequency reference), the phase relationship between the clocks is unknown and may vary due to PVT variations. Therefore, the receiving chip usually retimes the received data from the interface clock domain to the clock domain of the internal chip clock. FIFOs are used to perform this retiming function. It is desirable to minimize the number of flip-flops being clocked by the interface clock in order to minimize delay in the clock distribution network; otherwise timing issues will be exacerbated.

1.2.3 Higher Speed Source Synchronous Interfaces

The window of time during which data bits can be assumed to be valid is called the *eye*. This name originates from the shape of the waveform when the data signal is monitored on an oscilloscope that is continuously triggered. An example of a serial data eye is shown in Fig. 1.5a. *Eye closure* results from process, voltage, and temperature effects, as well as differences between signal rise and fall times, slew rates, etc. The more the eye is closed, the more difficult it is to find a point at which the signal can be reliability sampled to receive the data. The serial data eye shown in Fig. 1.5b is completely closed.

The largest possible *eye opening* is desirable. The width and height of an open eye can be measured as shown in Fig. 1.5c. The expected bit error rate (BER) of the link directly correlates to the amount of eye opening (both width and height). This section briefly describes some approaches to minimize eye closure of the data signal. Eye waveforms are discussed further in later chapters.

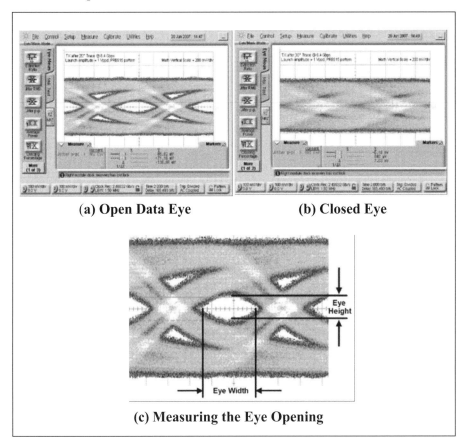

(a) Open Data Eye   **(b) Closed Eye**

(c) Measuring the Eye Opening

Fig. 1.5 Example of a serial data eye

1.2.3.1 Differential Signals

Unequal signal rise and fall times of nondifferential signals contribute to eye closure. Signals switching on the chip also create current variations on the power distribution grid of the chip, which in turn cause variations in voltage drop (noise) that can cause variation in delays of surrounding circuits. One method of reducing the effects of these phenomenon on the eye width is to drive differential signals between chips.

Differential signals represent the data bit using two electrical signals (*true* and *complement* signals). A logic "0" is represented by the *true* signal driven to its lower voltage limit, and the *complement* signal driven to its upper voltage limit; a logic "1" is represented by the *true* signal driven to its upper voltage limit, and the *complement* signal driven to its lower voltage limit. A differential receiver device interprets the logic bit value based on the difference between the two signals, and not based on the level of either signal individually.

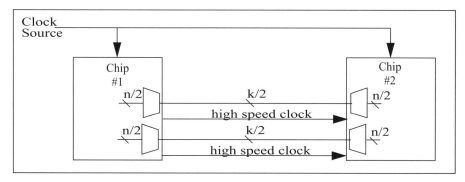

Fig. 1.6 Multiple sets of data with separate high-speed clocks

Differential driver circuits tend to have linear current draw and generate less noise on the power supply than equivalent single-ended drivers. Most noise sources induce voltage variation equally on both the true and complement signals; such *common mode noise* is ignored by the receiver. Also, since one leg of the differential signal is rising while the other is falling, or vice versa, unequal rise and fall time effects cancel.

The drawback of differential signals is that two chip pins are required for each data bit. However, this is offset by the higher speeds possible with differential signals that are not possible with single-ended signals.

1.2.3.2 Multiple Interface Clocks

The interface clock in Fig. 1.3 is the same clock as is used to launch the data, and in general is driven from a point in the clock distribution network as close to the actual flip-flops that launch the data as possible. Phase variation is introduced by any circuits which are not common to both the data path and clock path. Silicon process variables do vary from circuit to circuit on the same chip, the power distribution network may have unequal voltage drops to different circuits which may vary based on switching currents, and the temperature may vary from point to point on the chip. Tolerances and limits for all of these parameters must be taken into account when calculating delays, setup times, and hold times necessary for correct capture of the received data. At higher bit/baud rates, these parameters may significantly reduce the eye opening, and become the dominant mechanism for limiting the speed of the interface.

To maximize the eye width, the path through the clock tree to each of the data flip-flops and to the clock output should share as many circuits as possible, and the output driver for the clock should be similar to the output drivers for the data. Ideally, the same clock buffer should drive the clock to the output driver and should drive the clock input to all of the data flip-flops. The larger the number of bits in the data bus, the more difficult this becomes to implement. I/O drivers must be physically distributed based on the groundrules for connections to package pins. The greater the distance between circuits, the more process, voltage, and temperature variation, and the more circuits in the clock distribution network which cannot be shared due to lack of proximity.

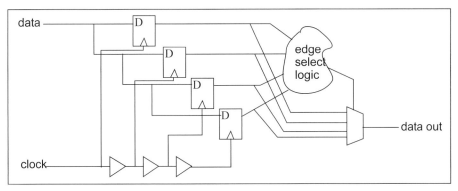

Fig. 1.7 Adapting the sampling clock phase in the receiver

One technique used to improve eye width is to limit the number of data bits associated with a given interface clock line. Wider data busses are built by using multiple interface clocks, each clock associated with a subset of the data bits. An example of this is shown in Fig. 1.6, where the k bit interconnect has been subdivided into two groups, each with its own high-speed interface clock. Note that the receiving chip must capture each group of data bits in separate clock domains, and needs to retime this data to the common clock domain internal to the chip.

1.2.3.3 Sample Edge Adaptation

Another technique used to permit higher speed operation of source synchronous interfaces is to process the data signal at the receiver and adapt the sampling phase of the clock on a per-bit basis. This is done by connecting the received interface clock signal to the input of a multitap delay line, and capturing the data signal in multiple flip-flops clocked by different clock phases. Logic can then be used to determine the clock phases between which data transitions are occurring, and select the optimal clock phase to be used to capture the data. This scheme is shown in Fig. 1.7.

Schemes, such as shown in Fig. 1.7, may require a training pattern either upon initialization of the interface or at regular intervals. If a training pattern is used, phase selections remain static between training periods. More complex implementations adjust dynamically based on the received data or based on training patterns embedded in the data stream. Alternative architectures which apply the data to the delay line are also possible. Note, however, that an inherent characteristic of most of these schemes is that the phase adjustment is less than plus/minus one bit time, and there must be a sufficient eye opening such that an optimal sampling phase exists.

Given the advanced schemes discussed above, data rates for source synchronous interfaces can be extended to several Gigabits per second (Gbps) per interconnect bit. However, PVT variations make further increases in interface speeds prohibitively complex. Beyond these speeds, High-Speed Serdes devices that extract the clock from edge transitions in the data stream become the preferred solution.

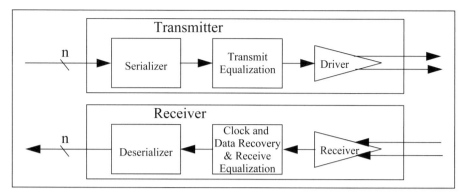

Fig. 1.8 Basic block diagram of typical high-speed serdes

1.3 High-Speed Serdes

High-Speed Serializer/Deserializer (HSS) devices are the dominant implementation of I/O interfaces at speeds of 2.5 Gbps and higher. Such devices are differentiated from source–synchronous interfaces in that the receiver device contains a clock and data recovery (CDR) circuit which dynamically determines the optimal sampling point of the data signal based upon the transition edges of the signal. In other words, clock information is extracted directly from the data rather than relying on a separate clock.

Figure 1.8 illustrates the basic block diagram of the transmit and receive channels of an HSS device. The transmitter serializes parallel data, equalizes it for reasons that will be explained shortly, and then drives the serial data onto a differential signal pair of interconnect wires. Feed forward equalizers (FFE) are commonly used in High-Speed Serdes devices, as discussed in Sect. 1.3.2. The receiver consists of a differential receiver, a CDR circuit which may also integrate an equalizer, and deserializes the data based upon the sample point established by the CDR. Peaking amplifiers and/or decision feedback equalizers (DFE) are commonly used for equalization in High-Speed Serdes receiver devices.

Note that Serdes cores are often designed to group multiple transmit and/or receive channels into a single device. The individual channels generally operate independently. Grouping channels allow some circuits to be shared across channels (for example the PLL noted below), and therefore the resulting block is more efficient in terms of chip area, cost, and power.

Serdes cores which contain only transmit or only receive channels are called *simplex* cores; Serdes cores which contain both transmit and receive channels are called *full duplex* cores. Note that the terminology "full duplex" does not imply that the electrical interface is bidirectional. Any given electrical interconnect channel has a fixed direction of data transmission. If a protocol application requires "full duplex" communication, then independent transmit and receive channels with independent interconnections are used to implement the interface. Rationale for using simplex vs. full duplex cores may include

(1) chip floorplan to minimize wiring crossings in the package design or circuit board design; (2) signal integrity concerns due to near-end crosstalk from transmit signals onto receive signals; (3) or applications where the number of transmit and receive channels is not equal.

The remainder of this section generically describes various circuits mentioned above in more detail, as well as providing generic descriptions of other circuits and functions commonly found in High-Speed Serdes cores.

1.3.1 Serializer/Deserializer Blocks

Conceptually, the input to the serializer transmit stage is an n-bit datapath which is serialized to a one-bit serial data signal for application to the FFE and Driver stages. Generally the value of n is a multiple of 8 or 10, and may be programmable on some implementations. Values of n which are multiples of 8 are useful for sending unencoded and/or scrambled data bytes; values of n which are multiples of 10 are useful for protocols which use 8B/10B coding, as discussed further in Sect. 4.2.2.1. (The 8B/10B encoder is generally implemented by logic outside the Serdes core.)

For simplicity, the block diagram in Fig. 1.8 illustrates the serializer feeding one-bit data into the transmit equalization block. Actual implementations may vary, and this datapath may be one or more bits wide. A wider datapath through the equalizer block results in a more complex design, but requires a lower operating frequency. Some implementations may initially multiplex the n-bit input to an m-bit datapath ($m < n$) prior to the equalizer, and perform the remainder of the serialization at the driver stage.

The serializer stage latches data on the n-bit input at the frequency of *baud rate/n*. The high-speed clock in the Serdes is divided down to generate a sample clock for the parallel data. Because the phase of this clock is determined by the internal state of the serializer, the Serdes channel generally provides this clock as an output for use by logic driving data to the transmit channel.

Conceptually, the deserializer receive block performs the inverse function of the serializer block. Serial data is deserialized onto an n-bit databus of similar width to the serializer. A sample clock is generated by dividing down the internal high-speed clock, and this clock is supplied as an output for use by logic latching the parallel data. In a similar manner to the serializer, actual implementations may perform partial deserialization in a prior stage.

Many Serdes receivers also include a feature to assist with data alignment of the output. Most applications organize data into bytes or words (groups of bytes). For 8B/10B encoded applications, data is organized into 10-bit encoded symbols. The initialization of the clock divider in the deserializer is arbitrary, and the data received on the parallel data bus will have an arbitrary alignment that is unlikely to match the byte or symbol boundaries of the protocol. This can be corrected by downstream logic to steer data onto the appropriate byte, symbol, or word boundary. Alternatively, many Serdes receivers provide an input which forces the deserializer to "slip" one bit. Downstream logic detects that data is not aligned to the appropriate boundary, and repeatedly pulses the deserializer control until the data "slips" to the desired alignment.

1.3.2 Equalizers

The interconnect between the transmitter and receiver device (known as the *channel*) acts as a filter at typical baud rates, and distorts the serial data signal to varying extents. Figure 1.9 illustrates this distortion: The input waveform is a clean digital signal, but the output waveform is significantly distorted. The illustrated frequency response function for the channel is characteristic of a low-pass filter. Signal distortion occurs because the signal baud rate is above the cut-off frequency for this filter.

Signal integrity concerns frequently dictate that the data signal be equalized at the transmitter and/or receiver in order to counter the effects of the channel and decode the signal properly. Many variations on filter architectures are possible, all of which accomplish this. Fig. 1.10 illustrates the addition of an equalizer at the transmitter with a transfer function that is roughly the inverse of the channel's frequency response. This equalizer distorts the signal at the transmitter output such that the resulting signal at the receiver input is a clean waveform.

Most Serdes transmitter implementations include a FFE. The block diagram for a three-tap FFE is shown in Fig. 1.11. The serial data signal is delayed by several flip-flops which implement the *taps* for the filter. Each tap is multiplied by a *tap weight* value (also called a *filter coefficient*), and the results are summed and driven to the serial data output. FFE operation is described further in Sect. 3.2.1.

The number of FFE taps on the filter, the spacing of these taps relative to the baud rate, and the granularity of these tap weight values vary based on implementation. The terminology *preemphasis* or *deemphasis* refer to the FFE architecture, and indicate whether the data signal amplitude is increased or decreased as compared to the nonemphasized value by the FFE tap. The terminology *precursor taps* and *postcursor taps* refer to whether the FFE filter taps operate on an advanced or delayed signal (respectively) relative to the $t = 0$ tap. *Baud-spaced* taps are defined as taps where the delay from one filter tap to the adjacent tap is one-bit time interval; fractional spacing of the taps is also possible.

The FFE tap weights are selected to generate a filter with the inverse transfer function of the channel transfer function. Various algorithms exist for determining optimal FFE coefficient values; some select filter coefficients to maximize signal amplitude at the receiver, while others optimize eye width (i.e., minimizing jitter). More complex algorithms may search for an optimal trade-off between amplitude and jitter in order to optimize a more complex parameter (such as projected BER). FFE tap weights are determined for many applications by design and coded as fixed values within system software, however, there are some applications where FFE tap weights are adjusted dynamically by the protocol based on signal characteristics at the receiver.

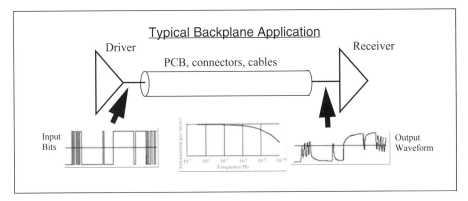

Fig. 1.9 Signal distortion for a typical channel application

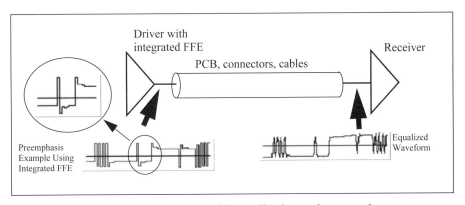

Fig. 1.10 Typical channel application with equalization at the transmitter

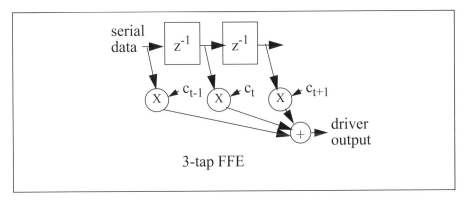

Fig. 1.11 Three-tap feed forward equalizer operation

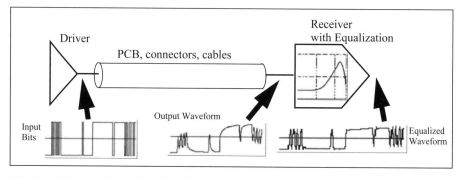

Fig. 1.12 Typical channel application with equalization at the receiver

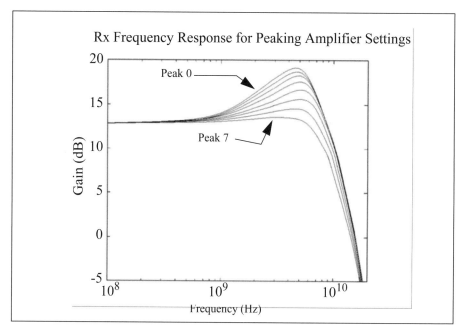

Fig. 1.13 Receiver frequency response for peaking amplifier settings

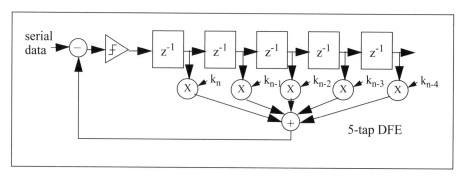

Fig. 1.14 Decision feedback equalizer architecture

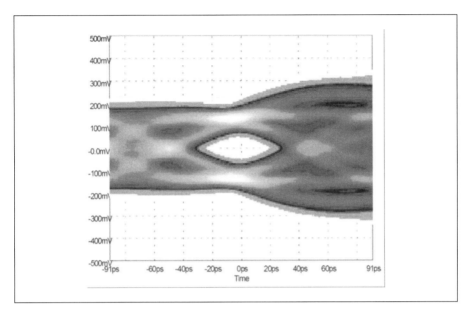

Fig. 1.15 Virtual eye after equalization

Equalization may also be performed at the receiver as illustrated in Fig. 1.12. Despite the signal distortion at the input of the receiver, this equalizer corrects for the distortion and produces a clean waveform. For lower speed or lower loss links, the most prevalent approach is to use some variant of a peaking amplifier. Peaking amplifier circuits amplify the higher frequency signal components more than the lower frequency components. If the peaking amount is matched to the high frequency loss (difference between high frequency and low frequency), then the channel is equalized and the eye is opened up. Some Serdes devices allow programmable peaking levels; the frequency response of such a peaking amplifier for various provisioned settings is shown in Fig. 1.13.

For higher baud rates, the transfer function of the channel can cause jitter exceeding the bit width of the data and significant loss of signal amplitude at higher frequencies. A DFE stage is often included in receivers for these baud rates in order to recover data despite the otherwise "closed" eye.

A conceptual block diagram of a DFE circuit is shown in Fig. 1.14. The serial data signal is applied to a slicer circuit which makes decisions as to whether the incoming signal is a "0" or a "1". The received serial data is then delayed by a number of flip-flops which implement the filter taps. Each tap is multiplied by a corresponding tap weight value, and the results are summed. This sum is then used to correct the amplitude of the incoming signal, affecting the decisions made by the slicer circuit. Slicer decisions are thus affected by feedback based on prior data received. Although not shown in Fig. 1.14, some DFE architectures use feedback to affect both the amplitude and the sample time of the slicer circuit. Such architectures adjust the CDR sample time from bit to bit based on feedback regarding the last several bits received.

As was the case for FFE circuits, the number of DFE taps on the filter, the spacing of these taps relative to the baud rate, and the granularity of these tap weight values vary. DFE implementations usually also contain logic which trains the DFE and sets DFE tap weights to optimal values dynamically.

Using a DFE, the closed eye shown in Fig. 1.5b is cleaned up to produce the virtual eye shown in Fig. 1.15. Note that the eye in this illustration is produced by a DFE architecture which corrects the CDR sample time from bit to bit. The DFE correction is valid for only one instance in time (based on the history of the previous bits). As such, once the DFE makes a decision as to whether the bit is "0" or "1", the DFE then proceeds to make adjustments for the next bit time which are different for the various signal traces of the composite waveform. For this reason, the signal eye shown in the figure is open for the bit of interest, but does not appear open for adjacent bits.

Many variations on equalizer architectures exist. As the baud rate increases, equalizer architectures become increasing complex. In some cases, protocol standards specify a base level of required equalizer functionality.

1.3.3 Clock and Data Recovery (CDR)

Conceptually, CDR circuits monitor transitions of the data signal and select an optimal sampling phase for the data at the mid-point between edges. Since the timing of data transitions includes a jitter component, the CDR must perform some averaging to provide stability of this sampling point from one bit to the next. Intersymbol interference (ISI) and other components of deterministic jitter (DJ) are dependent on the spectral content of the data signal, and this frequency spectrum does change based on the data content. Shifts in this frequency spectrum sustained for hundreds of bits or more cause the CDR to adjust the optimal sampling phase dynamically.

CDR architecture is discussed further in Sect. 3.3.1. Features of the CDR may be of some significance to the Serdes user are discussed below.

1.3.3.1 Maximum Run Length

A significant parameter for the Serdes which is primarily the result of the CDR design is the maximum number of consecutive "0" or "1" bits which can be received before the sampling point of the CDR risks incorrectly sampling the bits. An excessively long run of consecutive bits of the same value means that the CDR is not detecting any data transitions, and therefore cannot recover any clock information to ensure the data continues to be sampled in the center of the eye. A small drift in the sampling point relative to the baud rate of the data may cause the CDR to sample more "0" or more "1" bits than were actually transmitted. Also, the sampling point may require recentering when data transitions resume, and additional bits may be sampled incorrectly as this adjustment occurs. Some CDR implementations drive the receive data to a PLL and use the output of the PLL as the sample clock; clock outputs of the receiver may change frequency or stop when such CDRs do not receive data transitions for a sustained length of time.

The maximum run length of consecutive "0" or "1" bits which must be tolerated depends on the protocol application and the data encoding defined for a given protocol. For example, protocols which use 8B/10B encoding are guaranteed to have no more than 5 bit times between data transitions. Protocols using another common encoding, 64B/66B, are guaranteed to see run lengths no longer than 66 bit times. Scrambled protocols may encounter much longer run lengths, and must determine requirements using statistical analysis. For example, Sonet/SDH is a scrambled protocol which specifies systems must meet a BER of 1×10^{-12}. It is generally accepted that run lengths of scrambled Sonet/SDH data longer than 80 bits statistically occur less frequently than the specified BER. Therefore, a Serdes used to receive Sonet/SDH data must tolerate a run length of 80 bits.

The run length which can be tolerated by a CDR design is related to the frequency tolerance between the two clock sources. In a system using plesiosynchronous clocks, the reference clock used by the receiver (and the CDR circuit) may be running at a slightly different frequency from the reference clock used by the transmitter, as is described further in Sect. 4.1.3.1. The frequency tolerance between the two clock sources is generally specified in parts per million (ppm). In a plesiosynchronous system, the CDR must continually correct the phase of the sample clock to remain in the center of the data eye. During periods where no data transitions are being received, the error in phase position builds up. Therefore, as the frequency tolerance of the system is increased (corresponding to larger allowed frequency difference between the clock sources), the run length which can be tolerated by the CDR design is reduced for a given performance (BER) target.

1.3.3.2 Clock Operation During System Initialization

In the above discussion, it was noted that some CDR architectures derive the sample clock from the received data using a PLL. During system initialization or during system operation when cables are unplugged, etc., no data transitions are received for a substantial period of time. For some Serdes, this results in clock outputs of the receiver changing frequency or stopping. Any downstream logic clocked by these clock outputs must be designed to be tolerant of this frequency change or to assume logic is not clocked during these periods.

1.3.4 Differential Driver

The differential driver stage is an analog circuit which drives the *true* and *complement* legs of the differential signal. Output data must be driven such that jitter is minimized. In some architectures, data is latched in a flip-flop clocked at the baud rate, and the output of this flop is driven onto the differential output. Such implementations require an internal high-speed clock running at the baud rate. This is illustrated in Fig. 1.16.

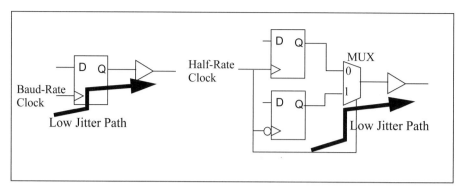

Fig. 1.16 Driver stage architectures

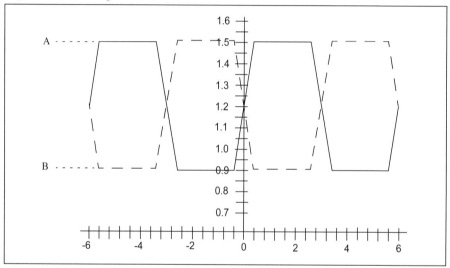

Fig. 1.17 Single-ended complementary signals

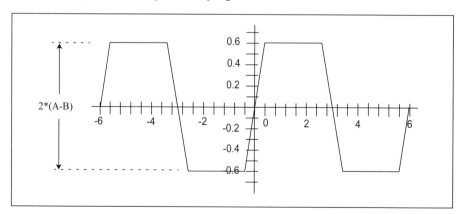

Fig. 1.18 Differential peak-to-peak signal

An alternative architecture, also shown in Fig. 1.16, uses an internal high-speed clock running at a frequency equal to half of the baud rate. Data is latched in two flip-flops on alternate edges of the high-speed clock. The high-speed clock also controls a multiplexor which alternately selects which of the flops drives the differential driver. Depending on the characteristics of the silicon technology, this architecture may result in lower jitter than the full-rate architecture.

Figure 1.17 illustrates typical voltage swings for the two legs of the differential signal, assuming a termination voltage of approximately 1.8 V. The average voltage on the signal is the *common mode voltage* (V_{cm}). For this example:

$$V_{cm} = (1.5\,V + 0.9\,V) / 2 = 1.2\,V.$$

The *differential voltage* (V_{diff}) is calculated by taking the voltage of the true leg and subtracting the voltage of the complement leg. Figure 1.18 illustrates the differential waveform corresponding to the single-ended signals from Fig. 1.17.

This differential voltage swings between the following limits:

$$V_{diff} = 1.5\,V - \quad 0.9\,V = +0.6\,V$$
$$V_{diff} = 0.9\,V - \quad 1.5\,V = \quad 0.6\,V$$

This waveform has a total *peak-to-peak differential voltage* of $1.2\,V_{ppd}$. Note that the peak-to-peak voltage of the differential signal is twice the peak-to-peak voltage of either single-ended signal considered individually.

1.3.5 Differential Receiver

The differential receiver stage is an analog comparator circuit which compares the *true* and *complement* legs of the differential signal and outputs a "0" or "1" logic level based on the relative signal voltages. Differential receiver stages used with DFEs are linear amplifiers; the comparator circuit is incorporated into the DFE.

1.3.6 Diagnostic Functions

Additional logic is often incorporated into the transmitter and receiver designs to provide diagnostic capabilities for chip manufacturing test, circuit board manufacturing test, and system diagnostic tests. Typical functions include:

1. Pseudo random Bit Sequence (PRBS) Checker. PRBS sequences can be checked by comparing received data to the output of a local linear feedback shift register implementing the corresponding characteristic polynomial. Receiver devices often include a PRBS checker capable of checking one or more PRBS test patterns.

2. Loopback or Wrap Paths. Full duplex Serdes devices often provide the capability to wrap transmitter outputs to receiver inputs in order to self-check the functionality of the Serdes. Simplex cores do not have this capability, although some simplex transmitters include a test receiver, and some simplex receivers include a test transmitter to perform self-test.

3. JTAG 1149.1 and JTAG 1149.6. These JTAG standards are used for manufacturing test of circuit boards, and require insertion of boundary scan cells on all chip I/O to support this testing. Since such logic cannot be inserted on high-speed I/O without impacting signal integrity, the Serdes core must provide appropriate hooks to drive differential outputs from boundary scan cells at the transmitter device, and sample inputs in boundary scan cells at the receiver device. JTAG 1149.6 expands the capabilities of JTAG 1149.1 to permit testing through decoupling capacitors and support independent testing of the true and complement legs of differential signals. JTAG 1149.1 and 1149.6 are covered in detail in Sect. 7.1.

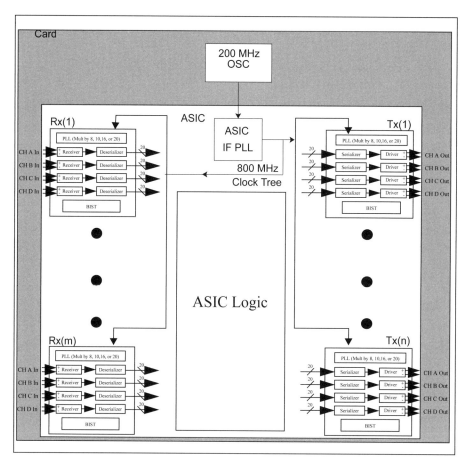

Fig. 1.19 Clock distribution example using an ASIC IF PLL

1.3.7 Phase-Locked Loop

The Serdes core requires an internal clock running at either the baud or half-baud rate depending on the architecture of the driver and receiver stages. Rather than distribute a high-speed clock throughout the chip, a lower frequency reference clock is distributed within the chip, and a PLL is used in the Serdes to multiply this clock to the appropriate frequency. A given Serdes implementation may contain multiple instances of transmitter and/or receiver channels. In such cases, it is common for a single PLL in the core to generate clocks for all channels within the core.

The off-chip clock source often operates at an even lower frequency than the on-chip reference clock. An additional PLL may be used in the chip to multiply the frequency of the off-chip reference clock to meet the desired on-chip reference clock frequency. Because the frequency of the on-chip reference clock is usually higher than the off-chip reference clock (but less than the internal clock in the Serdes core) the PLL which produces this clock is sometimes called an *intermediate frequency* (IF) *PLL*.

An example of clock distribution using such an IF PLL is shown in Fig. 1.19. An IF PLL is used to multiply the 200-MHz clock from an off-chip oscillator by four. The resulting 800-MHz reference clock is distributed on-chip to various Serdes cores. These Serdes cores each contain a PLL which additionally steps up the frequency of the 800-MHz reference clock to the desired baud rate.

1.4 Signal Integrity

This section provides an overview of the importance of signal integrity analysis to the design of successful systems using High-Speed Serdes.

1.4.1 The Channel

The *channel* is defined as the electrical path between the transmitter and the receiver, including printed circuit board traces, vias, cables, connectors, decoupling capacitors, etc. The channel may traverse the printed circuit board between two chips on the same card, or may traverse a system backplane connecting two printed circuit boards.

At frequencies of interest, the printed circuit board is not a perfect connection. Major channel impairments include insertion loss, reflections, and crosstalk. Channel frequency response, including these impairments, is typically measured using a vector network analyzer (VNA), and captured in a *Scattering Parameter* matrix format (commonly called *S-Parameters*) as is described in "Channel Response" under Sect. 8.4.1.2 in Chap. 8. Each of these impairments impacts the BER of the link. Interface standards typically require link BER performance in the range of 10^{-12} to 10^{-15}.

The equalization scheme used by Serdes devices must compensate for the channel loss and other impairments in order to achieve the desired BER. A common figure of merit is based on the evaluation of the insertion loss of the channel at the Nyquist rate (two times the highest fundamental frequency of the signal). From this metric, a qualitative assessment as to the difficulty of signal propagation (and the necessary complexity of the equalization scheme) can be made.

Figure 1.20 illustrates the measured insertion loss curve for a number of examples of channels intended to support baud rates of 5 Gbps and higher. The frequency range of 5-Gbps data has a Nyquist rate of 5 GHz, and the loss of these channels at 5 GHz is substantially higher than at lower frequencies. Higher frequency components are therefore attenuated more than lower frequencies, resulting in varied signal amplitudes at the receiver. (In addition, the signal contains harmonic frequencies above 5 GHz, but these are typically filtered and are not critical to receiver operation.)

The channel transfer function is not strictly resistive, but also contains capacitive and inductive components. This results in frequency-dependent phase shift of the propagated signal. Such phase shift in effect causes the propagation delay of the signal to vary based on frequency, appearing as data-pattern-dependent jitter (discussed later in this section).

Although the above effects already contribute to significant signal degradation, the insertion loss and phase shift associated with the channel is usually not a straightforward linear function. The channel is an electrical transmission line which is terminated at the receiver, and has impedance discontinuities at each circuit board via, connector pin and abrupt bend in the circuit board trace. Each impedance discontinuity results in reflections of the electrical energy. As with any transmission line, reflected energy adds to or subtracts from the signal amplitude at various points along the transmission line, and results in resonances in the transfer function. This signal degradation is generally worse for shorter channel lengths; loss characteristics of longer channels tend to dampen reflections whereas signals may reflect between the transmitter and receiver multiple times on a short channel.

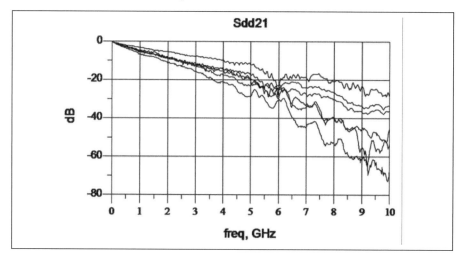

Fig. 1.20 Insertion loss for various channel examples

Crosstalk is energy coupled from an aggressor signal as noise on a victim signal, and is another significant characteristic of the channel which causes significant degradation of high-speed serial signals. While it would be nice to isolate each differential signal pair such that crosstalk was not significant, this is not practical in many real systems. The economics of most systems demand signal densities through package pins, connectors, and backplanes which result in neighboring signals causing crosstalk on the differential pair. Common-mode noise is ignored, but any differences in the noise on the two legs of the differential pair results in signal degradation.

1.4.2 Package Models

In addition to the channel as described above, the differential signal must also propagate through the chip package for the transmitting chip and the chip package for the receiving chip. While it would be convenient from an analysis viewpoint to consider chip packages as part of the channel interconnect, it is not practical to access the connections inside the package in order to measure channel response in the lab. For this reason, channel measurements are performed from package pads as described previously, and the transfer function of the package is considered separately. *Package models* are supplied by the chip manufacturer which model the transfer function of the package. Analysis of the overall interconnect is performed by cascading the transmitter package model, the channel transfer function, and the receiver package model.

Since trace lengths in the package substrate tend to be very short, insertion loss and phase shift are usually not the dominant source of signal degradation due to the package. Impedance mismatches and discontinuities tend to be a far greater concern. A measurement of the returned energy from a signal launched into the package ball at either the transmitter or the receiver is called *return loss*. Better impedance matching of connections to the silicon device within the package results in better return loss and less degradation of the signal.

1.4.3 Jitter

In an ideal world data bits would always transition at a fixed point relative to the clock which launches each bit. In the real world there are variations in the clock cycle time, data propagation delays, and signal slew times, etc.Variation in the timing of the bit transition relative to an ideal clock is called *jitter*. Jitter reduces the width of the data eye.

Any transmitter device has some amount of jitter in the launched data bit. *Jitter generation* is a measure of the timing variation in the transmitted data stream. Applications may include specifications for the maximum allowable jitter generation by a transmitter device, as measured under specified test conditions. Test conditions may include specification of test patterns to be used, and the load to which the transmitter device is connected.

The jitter generated by the transmitter device is amplified by the channel. The channel distorts the signal and introduces frequency-dependent phase shift. As the data pattern of 1's and 0's shifts, the spectral content of the data changes. This results in varying delay which becomes additional jitter at the receiver. Crosstalk and reflections due to impedance discontinuities and return loss also contribute to shift the transition points of signal edges at the receiver.

The receiver must be able to tolerate the amount of jitter that occurs on its input. *Jitter tolerance* is a specification for the receiver device which defines the "worst case" signal that the receiver is expected to receive correctly with errors no greater than allowed by the specified BER. Applications may include specifications for how to test jitter tolerance compliance, including test patterns to be used and methods for generating a datastream with specified amounts of jitter.

Jitter and jitter tolerance may be specified in picoseconds, but are more often specified in terms of unit intervals (UI). Parameters are usually specified as either peak-to-peak or root-mean-square, depending on the type of jitter. One *Unit Interval* is the cycle time for the transmission of one bit on the interface at the speed at which the link is running. A jitter specification of 0.30 UIpp indicates that the total peak-to-peak jitter cannot exceed 30% of the bit time, leaving an eye opening of 70% of the bit time. A jitter tolerance specification of 0.70 UIpp indicates that the receiver cannot expect more than an eye opening of 30% of the bit time at its input.

Jitter transfer is the amount of the input jitter on the receiver input which is passed through to the output (also called *jitter gain*). This parameter is significant in applications which must retransmit data with the same timing as the receive data. Since each stage of retransmission contributes additional jitter to the overall system, a jitter transfer specification is required in such applications to allocate how much jitter can be added within any one stage.

Many standards subdivide jitter into various types, and may use varying terminology for the types of jitter. The following discussion describes a commonly used terminology, and is sufficient for the purposes of general understanding. More comprehensive definitions are found in Sect. 8.2.1:

1. Total Jitter (TJ). This is the total jitter of the signal as seen at the point of measurement. TJ can be measured directly on hardware and is the ideal bit time minus the actual eye width, specified as either a peak or peak-to-peak value.

2. Deterministic Jitter (DJ). This is the amount of the total jitter for which the jitter distribution is non-Gaussian. Several components of DJ are dependent on the data pattern being sent. The pattern of 1's and 0's which precedes the bit transition affects when the transition occurs. Jitter caused by variations in rise, fall, and slew times are also mostly deterministic. DJ is specified as either a peak or peak-to-peak value.

3. Random Jitter (RJ). This is the amount of the total jitter for which the jitter distribution is Gaussian. RJ does *not* correlate with the data pattern being sent. Some amount of jitter has nothing to do with the data pattern, and is simply the result of random processes. Because RJ is statistical in nature, it may be specified either as a peak, peak-to-peak , or as a root-mean-square value.

Generally interface standards specify two or three of the above jitter types. If two of the above types are specified (typically TJ and DJ, or TJ and RJ), then requirements for other type is implied. This approach allows trade-offs to be made by system and component designers, while still complying with the standard.

The rationale for specifying different types of jitter is that some of the jitter can be corrected. For instance, the FFE in the transmitter device alters the transitions of the data edges based on the pattern of 1's and 0's being sent. In effect, the FFE is used to inject DJ on the transmitter output. If the equalizer coefficients are set properly, this injected DJ is the opposite of the DJ that is created by the channel's transfer function. The result is that the two DJ contributors cancel and there is less jitter at the receiver.

The DFE in the receiver also operates to cancel the effects of DJ and thereby improve the jitter tolerance of the receiver. The DFE has the advantage of adapting to changing conditions of the channel, which may alter the characteristics of the DJ over time. On the other hand, RJ is not predictable, and therefore cannot be corrected by equalization. It is important to limit the amount of RJ on the serial link.

1.4.4 Channel Analysis Tools

At serial link speeds up to 4 Gbps it is generally possible to follow reasonable design practices for the channel design. Spice ^{TM} or other equivalent circuit simulators are used to perform simulations at these baud rates. (In this text the term Spice is used generically as a designation for any of these simulators.) (Note: Spice is a trademark of Synopsys, Inc.) Spice simulations are used to verify signal integrity, and demonstrate that transmitter and receiver devices will interoperate across the channel. This approach is generally not sufficient at serial link speeds of 5 Gbps and higher.

At higher serial links speeds, some care is required to design a channel which does not unduly degrade the signal. There are cases where small design changes can have substantial unexpected effects. Spice simulation can be used to simulate the channel design. However, remember that signal degradation is dependent on the data pattern and the crosstalk. Also, the more sophisticated DFE-based cores use complex algorithms to determine tap coefficients, etc., real time. This behavior must be accurately simulated to account for algorithmic errors and compensation techniques performed in logic. Spice simulation cannot capture this behavior and is therefore not generally used for the higher speed links. In addition, it is not sufficient to simulate a few thousand bits; simulations must be long enough and contain enough randomly generated data patterns on both the primary channel and the crosstalk channel to ensure the system operates within the specified BER limits. This may require a substantial amount of simulation.

Because of these simulation run-time requirements, statistical analysis of channel designs has become prevalent at higher speeds. First, the channel design is prototyped and lab measurements are made to determine the transfer function characteristics of the channel. Transfer functions are measured for both the channel of interest, and for noise coupling between the primary channel and crosstalk channels. The resulting transfer functions are coded in a form called *S-Parameters*. Alternatively, S-Parameters for subcomponents of the channel (i.e., printed circuit boards, connectors, backplane, etc.) can be measured individually and cascaded.

Next, statistical simulation of the channel is performed using a transmitter model, the channel S-Parameters, package models for both transmitter and receiver chips, and a receiver model. The transmitter and receiver models may be models for specific vendor Serdes devices, or they may be ideal reference models specified by an interface standard. The resulting "eye" on the output of the receiver model is characterized by an eye width and height and is a function of BER. If the "eye" is sufficiently "open" for the desired BER, then the system will operate with no more than the specified number of bit errors.

The terminology "eye" is used loosely in the above paragraph, hence the quotation marks. The receiver device often contains a DFE. By convention, the "eye" is the signal at the sampling latch of a classical DFE design. However for some DFE designs this may not be a single node, or may be beyond the point where analog-to-digital conversion of the signal has been performed. The notion of an analog "eye" at this point is a virtual mathematical concept; this eye is not a measurable analog signal.

Note that signal integrity engineers often prefer to view analysis results in the form of a *bathtub curve* which graphs eye opening as a function of Q of the system (which relates to BER), as is described further in Sect. 8.2.4. The geometry of this curve results from the *dual-Dirac* probability density function, which models pattern-dependent jitter sources.

There are a number of implementations of tools that can be used to perform statistical signal integrity analysis. A number of EDA vendors supply software tools. Various silicon vendors have proprietary software to perform this analysis (i.e., IBM's HSSCDR software) which have built-in transmitter and receiver models for the vendor's Serdes. "StatEye" is an open source software tool which performs this analysis (see http://www.stateye.org). HSSCDR software is discussed in Sect. 8.4. Note that many of these software tools use MatLab^{TM} as the underlying calculation engine.

1.5 Signaling Methods

This chapter has thus far assumed differential signals which use two signal levels that convey whether the bit being transmitted is a "0" or a "1". This is called *non-return-to-zero* (NRZ) signaling. (This name originally differentiated NRZ signaling from signals which always returned to their zero level between each transmitted bit.) The signal eye of an NRZ signal, shown in Fig. 1.5a, can be sliced at the midpoint of the waveform (0 mVppd) to determine whether each bit is a "0" or a "1". The baud rate of an NRZ signal is equivalent to the data bit rate (after any encoding), and the maximum fundamental frequency of the signal is half of this baud rate.

Multilevel signaling schemes are also possible. Figure 1.21 shows the signal eyes for *duobinary* and four-level *phase amplitude modulation* (PAM-4).

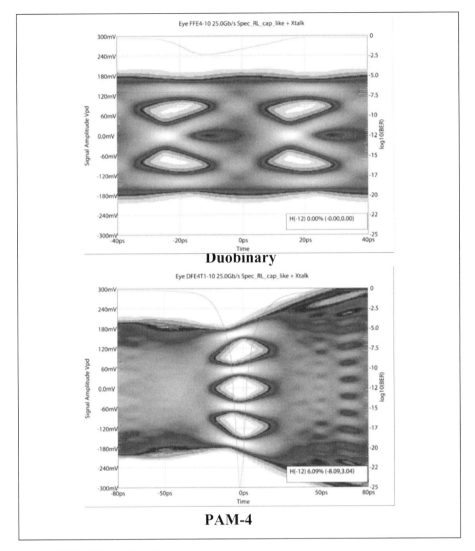

Fig. 1.21 Multilevel signaling eyes

Duobinary uses three signal levels, requiring two slicer circuits to determine the signal level. A "1" is indicated by any change in signal level, while a "0" is indicated by no change in signal level. The baud rate of a duobinary signal is the same as that of an equivalent NRZ signal. However, because the signal cannot transition from one extreme of the dynamic range to the other extreme in the same unit interval, the maximum fundamental frequency of the signal is one-quarter of this baud rate.

PAM-4 uses four signal levels, requiring three slicer circuits to determine the signal level. Each signal level represents the transmission of two bits, either "00," "01," "10," or "11." The baud rate of a PAM-4 signal is half that of an equivalent NRZ signal, and the maximum fundamental frequency of this signal is 50% of this baud rate (25% of the baud rate of an equivalent NRZ signal).

Because the signal spectrum of a duobinary and PAM-4 signal is reduced from that of an NRZ signal, the insertion loss of the channel affects the signal less. This is offset by the fact that splitting the dynamic range of the signal into multiple levels results in a reduction of the eye height at the transmitter. If the slope of the insertion loss curve is sufficiently steep, then classical analysis indicates an advantage for the multilevel signal over that of the NRZ signal. The eye height of a PAM-4 signal, for example, is reduced by approximately 9 dB at the transmitter from that of an equivalent NRZ signal. If the difference between the channel insertion loss at the NRZ baud rate vs. the PAM-4 baud rate is greater than 9 dB, then PAM-4 would result in more eye opening at the receiver.

Note, however, that the classical analysis described above assumes no equalization is used in the HSS device. If equalizers are used, then the effects of these equalizers on channel response must be taken into account. Decision feedback equalization in the receiver has the effect of flattening the overall insertion loss of the system as shown in Fig. 1.22. While multilevel signaling may have an advantage for a given channel when equalization is not used, this advantage may not exist when NRZ is employed with a DFE in the receiver.

Most HSS devices integrated in ASIC chips use NRZ signaling, and all of the interface standards discussed in this text are based on NRZ signaling. Standards development efforts up to and including the 10–11 Gbps range have found NRZ with DFE to produce better results than multilevel signaling approaches. However, as HSS devices continue to target higher and higher data rates, this may or may not be true in the future.

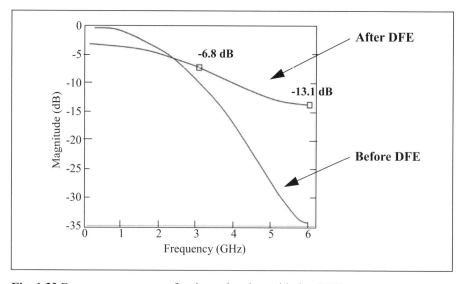

Fig. 1.22 Frequency response of a channel and considering DFE

1.6 Exercises

1. Assume the logic on the chip processes data using a 125-MHz clock. Data is to be transferred from one chip to another chip across a parallel data bus using the on-chip clock frequency. What parallel data bus width is required to achieve the following bandwidths on the interface?

   (a) 1 Gbps?   (b) 10 Gbps?   (c) 40 Gbps?

2. Assume the logic on the chip processes data using a 125-MHz clock. Data is to be transferred from one chip to another chip across a parallel data bus. A maximum of 20 pins is to be used to transfer the data. If necessary to meet this restriction, assume internal chip data is multiplexed as described in Fig. 1.2 using the minimum possible multiplexor ratio. What is the multiplexor ratio, number of pins required, and I/O data rate for each of the following interface bandwidths? You should also comment on whether the solution is practical given your knowledge of current ASIC technologies.

   (a) 1 Gbps?   (b) 10 Gbps?   (c) 40 Gbps?

3. Assume a parallel data bus is constructed as shown in Fig. 1.2. The propagation delay from the clock source to the clock inputs of the flip-flops in chip #1 which launch the data is 5 ns. The propagation delay from the clock source to the clock inputs of the flip-flops in chip #2 which capture the data is 2 ns. The propagation delay of the data in chip #1 (from the clock input of the flip-flops to the I/O pin) is 2 ns. The setup time of the inputs to chip #2 (for the I/O pin relative to the clock input of the flip-flops capturing the data) is 1 ns. The propagation delay of the channel is negligible.

   (a) What is the maximum data rate per pin at which this interface can operate?

   (b) The propagation delays and setup times in this problem may vary by 10% as voltage and temperature vary. Assuming these chips must operate under all environmental conditions, what is the maximum data rate per pin under the worst case combination of conditions?

   (c) Now assume the parallel data bus is constructed as shown in Fig. 1.3. The propagation delay of the interface clock generated by chip #1 is similar to that of the data. The propagation delay from the interface clock input pin to the clock inputs of the flip-flops capturing the data in chip #2 is 2 ns. All other timing parameters remain the same. What is the maximum data rate per pin at which this interface can operate?

   (d) The I/O cells used to implement the above interface have a maximum
       operating frequency of 200 MHz. Given this restriction, should the
       interface in (c) use a SDR clock or a DDR clock? What would the
       frequency of each of these clocks need to be?

4. Figure 1.4 illustrates data being transmitted by a DDR clock. Logic to
   transmit this data can be constructed using a combination of positive-edge
   and negative-edge clocked flip-flops, and multiplexors. Draw a schematic
   diagram for logic to implement this function. (Note: This logic cannot use
   an SDR clock.)

5. A parallel data bus between chip #1 and chip #2 is 16-bits wide. How
   many pins does each of the following architectures require?

   (a) Single-ended signals with no forwarded interface clock.

   (b) Single-ended signals with one forwarded interface clock for each
       eight bits of data.

   (c) Differential signals with one forwarded interface clock.

   (d) Differential signals with one forwarded interface clock for each eight
       bits of data.

6. Define a truth table for the *edge select logic* function in Fig. 1.7. For
   purposes of this exercise, you can assume signal transitions are noise free.

7. Explain the difference between a *simplex* and a *full duplex* Serdes core.

8. Assume that the input to a serializer stage in the transmitter has a 16-bit
   input and a 4-bit output. The datapath through the FFE logic is 4-bits, and
   the driver stage serializes these 4-bits down to one serial bit. The baud
   rate for this transmitter is 5 Gbps. What is the frequency of the SDR clock
   used by the driver stage? What is the frequency of the clock used by the
   FFE logic? What is the frequency of the sample clock for the parallel
   data?

9. Figure 1.11 illustrates a three-tap FFE with baud-spaced taps. Draw a
   similar figure for a six-tap FFE with taps spaced at half-baud intervals.

10. A CDR circuit is sampling a signal which is operating at a baud rate of
    3.125 Gbps. The eye width of this signal is 0.35 UI at the receiver, and the
    CDR is sampling the signal in the exact center of the eye when a long
    string of "0" bits begins. Plesiosynchronous clocks are employed, and the
    frequency difference between the transmitter and the receiver clock rates
    is 300 ppm. What is the maximum run length of 0's under these conditions
    before the CDR sample point drifts past the edge of the eye.

11. The high and low voltages of a differential signal are provided below for various systems. For each pair of voltages, calculate the corresponding V_{cm} and V_{diff} values.

    (a) 1.1 V, 0.3 V

    (b) 1,050 mV, 700 mV

    (c) 650 mV, 150 mV

    (d) 900 mV, 800 mV

12. Assuming the worst case insertion loss curve in Fig. 1.20, what is the approximate insertion loss at the maximum fundamental frequency of a 12-Gbps signal? Contrast this to the approximate insertion loss at the maximum fundamental frequency of a 3-Gbps signal.

13. Explain the difference between *jitter* and *jitter tolerance*.

14. Qualitatively explain why equalization can correct for various pattern-dependent forms of deterministic jitter.

15. Would you assume that the sum of *deterministic jitter* and *random jitter* is equal to *total jitter*? Explain your answer. (You may want to peek at Chap. 8.)

Chapter 2
HSS Features and Functions

In Chap. 1, a number of basic features and functions of High-Speed Serdes (HSS) cores were discussed. In this chapter, the HSS EX10 10-Gbps core is described. This core is a fictitious example which implements specific features using specific input/output pins and programmable registers. The HSS EX10 is presented for tutorial purposes, with composite feature descriptions drawn from a number of real HSS core examples.

The operation of various features of the EX10 is described in sufficient detail for the reader to gain an appreciation of nuances associated with using such a core. Since this is primarily a tutorial example, features are not described in as much detail as would be found in the databook for a real HSS core, and in some cases the operation and programming of certain functions has been simplified where additional complexity would not serve an educational purpose. From this chapter, the reader should gain an appreciation for the types of features that may or may not be found on a particular core, the applications they are intended to support, and the details associated with using these features which should be understood by the chip designer.

2.1 HSS Core Example: HSS EX10 10-Gbps Core

Similar to many real HSS core implementations, the EX10 is comprised of subfunction blocks known as *slices*. A typical core consists of one and sometimes more phase-locked loop (PLL) slices, some number of Transmitter (TX) Slices, and/or some number of Receiver (RX) Slices. Simplex cores contain either transmitters or receivers; full duplex cores contain both. Note the physical definition of a "slice" as described for a specific core may vary slightly from this logical definition.

The naming convention used for EX10 pins is of the form:

<slice><signal_name>

where the "slice" prefix is either "HSS," "TXx," or "RXx." The nomenclature "HSS" is associated with signals on the PLL slice which are common to all lanes of the EX10. Transmitter and receiver slice signals are prefixed with "TXx" and "RXx," where the "x" indicates the channel identifier for the associated slice. This text uses the convention that an EX10 core containing more than one TX and/or RX slice assigns channel identifiers starting with "A" and incrementing upward. For example, the TXxDCLK signal name refers generically to a signal on the TX slice, and a core with four TX slices would name these signals TXADCLK for channel A, TXBDCLK for channel B, etc.

The EX10 configuration described in this chapter includes one PLL slice, four TX slices, and four RX slices. Each slice has associated control and status signals. For

D. R. Stauffer et al., *High Speed Serdes Devices and Applications,*
© Springer 2008

this example, many control and status signals are mapped into internal registers of the core, while other signals are directly accessible as I/O pins of the core. (Some functions are accessible both through I/O pins and through internal registers.) A simple register read/write interface is provided to access internal registers of the core. The reader should note that the allocation of functions between I/O pins and internal registers is arbitrary; actual implementations may or may not allocate functions in the same way, and many HSS cores do not have internal registers at all.

In the first section of this chapter, the EX10 I/O pins and register functions are defined for each slice. Subsequent sections of this chapter describe functional features of the EX10 in more detail. These descriptions can be used as reference material when reading the subsequent chapters.

2.1.1 HSS EX10 Input/Output Pin Descriptions

Pins for the EX10 PLL slice are shown in Fig. 2.1 , and described in Table 2.1 . Pins for the EX10 Transmitter slice are shown in Fig. 2.2 , and described in Table 2.2. Pins for the EX10 Receiver slice are shown in Fig. 2.3 , and described in Table 2.3. Transmitter and Receiver slice pin descriptions use the prefix "TXx" or "RXx" (respectively) to indicate the generic signal name, where the "x" is replaced by "A," "B," "C," or "D" to indicate the particular instance of the slice within the core.

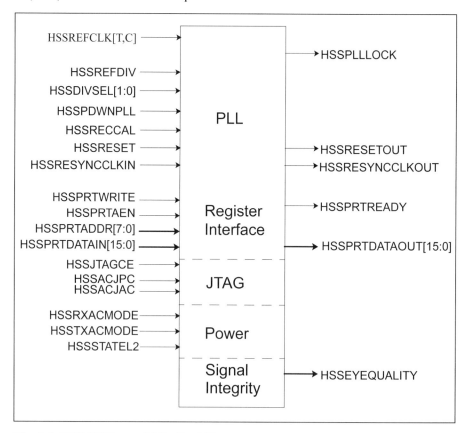

Fig. 2.1 HSS EX10 core PLL slice I/Os

Table 2.1 HSS EX10 PLL slice core pin definitions

Pin name	Type	Description
PLL signals		
HSSREFCLK[T,C]	In	Differential reference clock input to the PLL. HSSREFCLKT is the true leg of the differential signal pair; HSSREFCLKC is the complement leg
HSSREFDIV	In	Control signal to the PLL reference clock divider. See HSSDIVSEL[1:0]below for application. 0=normal operation, 1=divide reference clock by 2
HSSDIVSEL[1:0]	In	PLL VCO C1 clock vs. HSSREFCLK ratio selector HSSREFDIV = 0 HSSREFDIV = 1 00 = 16x 00 = 8x 01 = 20x 01 = 10x 10 = 32x 10 = 16x 11 = 40x 11 = 20x
HSSPDWNPLL	In	HSS PLL Power Down 0=normal operation, 1=power down the HSS PLL Slice
HSSRECCAL	In	HSS PLL Calibration Request 0=normal operation, 1=pulse high for minimum of eight HSSREFCLK cycles to force PLL recalibration.
HSSRESET	In	Asynchronous reset input signal. This signal must be asserted for a minimum of eight HSSREFCLK periods any time after initial power on. 1=reset, 0=normal operation
HSSRESYNCCLKIN	In	This input is pulsed to cause a *resync* to occur. Pulse must be synchronous to HSSRESYNCCLKOUT pin. 0=normal, 1=resync. Multiple core resynchronization requires HSSREFDIV = 0
HSSPLLLOCK	Out	PLL locked indicator. 0=unlocked, 1=locked
HSSRESETOUT	Out	This signal is asserted high during the VCO coarse calibration and during the beginning of the reset sequence. 0=normal, 1=reset in progress

Table 2.1 HSS EX10 PLL slice core pin definitions

Pin name	Type	Description
HSSRESYNCCLKOUT	Out	This clock output is used to synchronize the HSSRESYNCCLKIN signal
Register access bus signals		
HSSPRTWRITE	In	Parallel Port Write. 0=read addressed register, 1=write addressed register
HSSPRTAEN	In	Parallel Port Address Enable. This is the result of external address decode to select a given core instance for access. 0=inactive, no read/write cycle, 1=active, access addressed register
HSSPRTADDR[7:0]	In	Parallel Port Address
HSSPRTDATAIN[15:0]	In	Parallel Port Input Data Bus
HSSPRTREADY	Out	Register port ready to access transmitter and receiver registers. (PLL registers can be accessed at any time the PLL is running.) 0=not ready, 1=ready (after reset sequence completed)
HSSPRTDATAOUT[15:0]	Out	Parallel Port Output Data Bus
JTAG signals		
HSSJTAGCE	In	JTAG Test configuration enable. 0=normal operation, 1=JTAG test mode. This configures all necessary internal logic to support JTAG test, eliminating the need to configure multiple individual controls
HSSACJPC	In	JTAG mode clock signal
HSSACJAC	In	JTAG ACmode control signal. 0=dc coupled mode, 1=AC coupled mode
Power control		
HSSRXACMODE	In	Sets the Receiver termination voltage. 0=dc coupling mode (V_{TR}), 1=ac coupling mode ($0.8*V_{TR}$)
HSSTXACMODE	In	Sets the Transmitter internal bias. 0=dc coupling mode, 1=ac coupling mode

Table 2.1 HSS EX10 PLL slice core pin definitions

Pin name	Type	Description
HSSSTATEL2	In	Power down signal which powers off part of the PLL slice in compliance with implementation of a PCI Express L2 link state. Also forces power down of transmitter and receiver slices. 0=normal operation, 1=core is in L2 link state
Signal Integrity		
HSSEYEQUALITY	Out	HSS RX interrupt status signal. 0=inactive. No new status information available for any RX links in the core. 1=active. New status information is available for at least one RX in the core. When active, register 0x0F for each RX link can be read to determine updated status

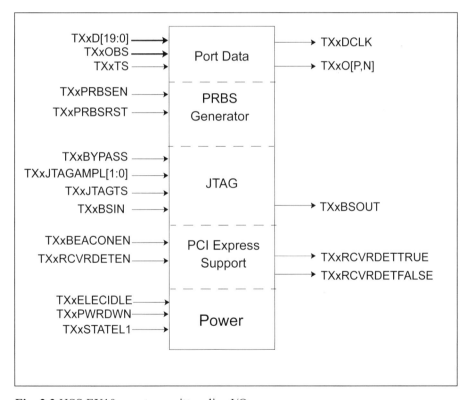

Fig. 2.2 HSS EX10 core transmitter slice I/Os

Table 2.2 Transmitter slice specific core pin definitions

Pin name	Type	Description
Port data signals		
TXxD[19:0]	In	Parallel input data. TxD(0) is the LSB, and is transmitted first on the serial output. Unused inputs should be tied to an inactive level
TXxO[P,N]	Out	Output differential pair – connects to chip I/O
TXxDCLK	Out	Word clock used to capture parallel input data TxD(19:0). Data captured on rising edge of this clock. The frequency of this clock is determined by the C1 clock frequency, the selected bus width (8, 10, 16, or 20 bits), and the selected data rate (full, half, quarter, or eighth-rate)
TXxOBS	In	"Out of Band Signalling": Drives transmitter outputs to the DC common Mode voltage as required by certain applications. 0=normal, 1=OBS mode enabled
TXxTS	In	Disables the transmitter output drivers. 1=normal operation, 0=disable (transmitter outputs are pulled up to AVTT through internal 50-Ω termination resistors)
PRBS generator		
TXxPRBSEN	In	TX Logic BIST enable signal. 0=normal, 1=enables internal loopback test
TXxPRBSRST	In	TX Logic BIST reset signal. 0=normal, 1=resets and restarts the BIST process
JTAG signals		
TXxBSIN	In	Serializer Bypass Data. When TXxBYPASS is set to 1, this data from the JTAG Boundary Scan Register cell is transmitted on serial output.
TXxBSOUT	Out	Connected to the input of the JTAG Boundary Scan Register cell.
TXxJTAGTS	In	Driver Tristate control. This pin is only active if HSSJTAGCE = 1. The state of this pin overrides the state of the TXxTS pin. 0=disable serial output driver when in JTAG test mode, 1=normal operation when in JTAG test mode

Table 2.2 Transmitter slice specific core pin definitions

Pin name	Type	Description
TXxJTAGAMPL[1:0]	In	When in JTAG Mode, these bits give the chip designer the ability to select 1 of 4 output driver amplitude levels overriding the Transmit Power Register. 00 = 30%, 01=44%, 10=72%, 11=100%
TXxBYPASS	In	Serializer Bypass Enable. 0=normal, 1=the data present on TXxBSIN is transmitted on serial output
PCI express support		
TXxBEACONEN	In	Transmit Beacon: When enabled, drives a beacon signal on the transmit serial data lines. 0 = normal operation, 1 = transmit beacon signal
TXxRCVRDETEN	In	Transmit Receiver Detect Enable: Drives a transition on the serial data and measures the charge time of the line in order to determine whether a receiver is connected. 0 = normal operation, 1 = initiate a Receiver Detect sequence
TXxRCVRDETTRUE	Out	Transmit Receiver Detect True Status: Asserted while TXxRCVRDETEN is high if the result of the Transmit Receiver Detect operation is that a receiver *is* detected. 0 = operation not in progress, not yet complete, or no receiver is detected; 1 = operation is complete and a receiver is detected
TXxRCVRDETFALSE	Out	Transmit Receiver Detect False Status: Asserted while TXxRCVRDETEN is high if the result of the Transmit Receiver Detect operation is that a receiver *is not* detected. 0 = operation not in progress, not yet complete, or receiver is detected; 1 = operation is complete and a receiver is not detected
Power control		
TXxPWRDWN	In	Transmit Power State: Power down signal which powers off the Transmitter slice. 0 = normal operation, 1 = power down

Table 2.2 Transmitter slice specific core pin definitions

Pin name	Type	Description
TXxSTATEL1	In	Transmit Power State: Power down signal which powers off the Transmitter slice in compliance with implementation of a PCI Express L1 link state. 0=normal operation, 1=transmitter is in L1 link state
TXxELECIDLE	In	Transmit Electrical Idle: Forces transmit serial data to an electrical idle signal level in compliance with implementation of a PCI Express L0s link state. 0 = normal operation, 1 = electrical idle state

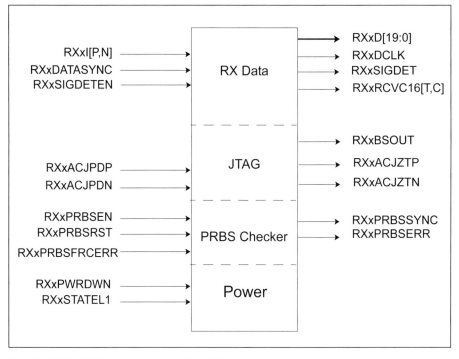

Fig. 2.3 HSS EX10 core receiver slice I/Os

Table 2.3 Receiver slice specific core pin definitions

Pin name	Type	Description
Port data signals		
RXxI[P,N]	In	Input differential pair – Connects to chip I/O
RXxD[19:0]	Out	Parallel output data. RxD(0) is the LSB, and is received first from the serial input
RXxDATASYNC	In	Data synchronization control pin. Each rising edge of this signal causes 1 bit to be discarded from recovered data, resulting in a 1-bit clock alignment adjustment. 0=normal, 1=discard 1 bit
RXxDCLK	Out	Word clock used to clock parallel output data RXxD(19:0). Data valid on rising edge of this clock. The frequency of this clock is determined by the C1 clock frequency, the selected bus width (8, 10, 16, or 20 bits), and the selected data rate (full, half, quarter, or eighth-rate)
RXxSIGDET	Out	Signal Detect indicator. 0=no signal, 1=active signal
RXxRCVC16[T,C]	Out	SONET Reference Clock Output. This differential output signal provides a divided down version of the recovered RX data clock to support SONET applications, which must synchronize TX and RX channels. This clock can be configured via RX register 0x02. This is a differential signal output: RXxRCVC16T is the true leg of the differential signal pair; RXxRCVC16C is the complement leg
JTAG signals		
RXxACJPDP	In	JTAG scan input path for positive side of differential input
RXxACJPDN	In	JTAG scan input path for negative side of differential input
RXxACJZTP	Out	JTAG scan output path for positive side of differential input
RXxACJZTN	Out	JTAG scan output path for negative side of differential input

Table 2.3 Receiver slice specific core pin definitions

Pin name	Type	Description
RXxBSOUT	Out	JTAG Receive Boundary Scan Out. When HSSJTAGCE=1, these outputs assume the logic state seen on the corresponding receiver input. When HSSJTAGCE=0, these outputs assume logic state "0." Toggle rate for these signals is limited to 100 MHz or less
PRBS checker		
RXxPRBSEN	In	RX Logic BIST enable signal. 0=normal (PRBS controlled via registers); 1=enables internal loopback test
RXxPRBSRST	In	RX Logic BIST reset signal. 0=normal; 1=resets RXxPRBSSYNC and RXxPRBSERR latches, and restarts the BIST process
RXxPRBSFRCERR	In	RX Logic BIST force error signal. This enables verification of the PRBS error detector. 0=normal; 1=forces errors in internal loopback path by changing the loopback mux selector to the loop back selection
RXxPRBSERR	Out	RX Logic BIST error flag. Once RXxPRBSSYNC is achieved, subsequent errors cause this signal to be latched at 1. Passing condition is for RXxPRBSSYNC=1 and RXxPBSERROR=0. 0=no error; 1=errors detected (latched, requires RXxPRBSRST to clear)
RXxPRBSSYNC	Out	RX Logic BIST sync flag. 0=BIST pattern checker not in sync; 1=BIST pattern checker has achieved sync since last RXxPRBSRST
Power control		
RXxPWRDWN	In	Receive Power State: Power down signal which powers off the Receiver slice. 0 = normal operation, 1 = power down
RXxSTATEL1	In	Receive Power State: Power down signal which powers off the Receiver slice in compliance with implementation of a PCI Express L1 link state. 0=normal operation, 1=transmitter is in L1 link state

Table 2.3 Receiver slice specific core pin definitions

Pin name	Type	Description
RXxSIGDETEN	In	Signal Detect Enable 0=Signal Detect power control using *Signal Detect Power Down* bit in *SIGDET Control Register,* 1=Signal Detect circuit powered on

2.1.2 HSS EX10 Register Descriptions

The HSSPRTADDR[7:0] inputs of the HSS EX10 core select the internal register to be written or read. The HSSPRTADDR[7:4] address bits select which slice is being addressed, and the HSSPRTADDR[3:0] bits select the particular register within the slice. All registers are 16 bits wide, although in many cases not all bits of the register have a defined function.

Table 2.4 describes how HSSPRTADDR[7:4] map to register space of the HSS EX10 slices. Note that values 0xC and 0xD are broadcast addresses which allow all TXx or RXx slice registers (respectively) to be written in parallel with one write cycle. Not all values of HSSPRTADDR[7:4] are used.

Table 2.4 HSS EX10 address map

Addr (7:3)	Maps to slice
0x0	TXA Slice Registers
0x1	TXB Slice Registers
0x2	RXA Slice Registers
0x3	RXB Slice Registers
0x4	TXC Slice Registers
0x5	TXD Slice Registers
0x6	RXC Slice Registers
0x7	RXD Slice Registers
0x8	PLL Slice Registers
0xC	Write all TXx Slice Registers in parallel.
0xD	Write all RXx Slice Registers in parallel.

Table 2.5 describes the registers and bit definitions for registers in the PLL slice. Table 2.6 describes the registers and bit definitions for registers in the transmitter slice. Table 2.7 describes the registers and bit definitions for registers in the receiver slice.

Table 2.5 HSS EX10 PLL slice register definitions

Addr (3:0)	Bits	R/W	Reset value	Description
0x0	16	R	0x0000	**VCO Coarse Calibration Status Register**
	0	R	0	CCALCOMP, calibration complete signal where: 0=calibration not complete, 1=calibration complete
	1	R	0	CCALERROR, calibration error occurred. 0=no errors, 1=calibration error occurred
	2	R	0	LOCK_DETECTED signal: 0=not locked, 1=locked
	15:3	R		Unused
0x1	16	R/W	0x0000	**VCO Coarse Calibration Control Register**
	0	R/W	0	Recalibrate signal, pulse high for minimum of eight reference clocks then return low to initiate an autocalibration sequence. 0 = normal (default), 1 = force PLL recalibration
	15:1	R		Unused
0x2	16	R/W	0x00FF	**Link Enable Register**
	0	R/W	1	Link enables to TXA. 0=disabled, 1=enabled
	1	R/W	1	Link enables to TXB. 0=disabled, 1=enabled
	2	R/W	1	Link enables to RXA. 0=disabled, 1=enabled
	3	R/W	1	Link enables to RXB. 0=disabled, 1=enabled
	4	R/W	1	Link enables to TXC. 0=disabled, 1=enabled
	5	R/W	1	Link enables to TXD. 0=disabled, 1=enabled
	6	R/W	1	Link enables to RXC. 0=disabled, 1=enabled
	7	R/W	1	Link enables to RXD. 0=disabled, 1=enabled
	15:8	R		Unused

Table 2.5 HSS EX10 PLL slice register definitions

Addr (3:0)	Bits	R/W	Reset value	Description
0x3	**16**	**R/W**	**0x0000**	**Link Reset Register**
	0	R/W	0	Link reset to TXA. 0=normal, 1=reset
	1	R/W	0	Link reset to TXB. 0=normal, 1=reset
	2	R/W	0	Link reset to RXA. 0=normal, 1=reset
	3	R/W	0	Link reset to RXB. 0=normal, 1=reset
	4	R/W	0	Link reset to TXC. 0=normal, 1=reset
	5	R/W	0	Link reset to TXD. 0=normal, 1=reset
	6	R/W	0	Link reset to RXC. 0=normal, 1=reset
	7	R/W	0	Link reset to RXD. 0=normal, 1=reset
	15:8	R		Unused

Table 2.6 HSS EX10 transmitter slice register definitions

Addr (3:0)	Bits	R/W	Reset value	Description
0x0	**16**	**R/W**	**0x0008**	**Transmit Configuration Mode Register**
	1:0	R/W	00	Rate Select 00=Full rate (default) 01=Half rate 10=Quarter rate 11=Eighth rate
	3:2	R/W	10	Parallel Data Bus Width 00=8 bit 01=10 bit 10=16 bit (default) 11=20 bit
	15:4	R		Unused

Table 2.6 HSS EX10 transmitter slice register definitions

Addr (3:0)	Bits	R/W	Reset value	Description
0x1	**16**	**R/W**	**0x0000**	**Transmit Test Control Register** Note: TXxPRBSEN pin = "1" overrides this register and forces PRBS7+ to be transmitted
	2:0	R/W	000	Test Pattern Selector 000 = PRBS7+ (noninverted) (default) 001 = PRBS7– (inverted) 010 = PRBS23+ (noninverted) 011 = PRBS23– (inverted) 100 = PRBS31+ (noninverted) 101 = PRBS31– (inverted) 110 = 1010101.... 111 = repeating pattern of 64 "1"s followed by 64 "0"s
	3	R/W	0	Test Pattern Generator Enable 0=disable generator and select Customer Parallel Data (default), 1=enable generator and select Test Pattern Data
	4	R/W	0	PRBS Reset. 0=normal (default), 1=reset applied to Test Pattern generator
	15:5	R		Unused
0x2	**16**	**R/W**	**0x0000**	**Transmit Coefficient Control Register**
	0	R/W	0	Apply Load This bit applies the register-loaded values of coefficients, power, polarity and FFE mode to the coefficient recalculation logic, and presents this new value to the analog circuits
	1	R/W	0	Reset Coefficient Logic 0=normal (default), 1=reset
	15:2	R		Unused

Table 2.6 HSS EX10 transmitter slice register definitions

Addr (3:0)	Bits	R/W	Reset value	Description
0x3	16	R/W	0x0020	**Transmit Driver Mode Control Register**
	1:0	R/W	00	FFE mode Select 00=FFE2 (default) 01=FFE3 10=reserved 11=Force Hi-Z
	4:2	R/W	000	Slow Slew Control Used to limit the minimum Transmitter output rise and fall time. 000 = 24ps min. (default) 001= 36ps min. 101= 50ps min. 011= 60ps min. 111=100ps min.
	15:5	R		Unused
0x4	16	R/W	0x0000	**Transmit Tap0 Coefficient Register**
	3:0	R/W	0000	FFE Tap 0 Coefficient This register's value is applied to the analog logic after "Apply Load" (Transmit Coefficient Control Register 0x02 bit 0) is pulsed. The value read from this register is the actual value being driven to the analog logic. Value is unsigned magnitude. See the Transmit Polarity Register 0x08 for sign values
	15:4	R		Unused
0x5	16	R/W	0x003F	**Transmit Tap1 Coefficient Register**
	5:0	R/W	111111	FFE Tap 1 Coefficient See the description for the Transmit Tap0 Coefficient Register (0x04)
	15:6	R		Unused
0x6	16	R/W	0x0000	**Transmit Tap2 Coefficient Register**
	4:0	R/W	00000	FFE Tap 2 Coefficient See the description for the Transmit Tap0 Coefficient Register (0x04)
	15:5	R		Unused

Table 2.6 HSS EX10 transmitter slice register definitions

Addr (3:0)	Bits	R/W	Reset value	Description
0x7	**16**	**R/W**	**0x007F**	**Transmit Power Register**
	6:0	R/W	0x7F	Transmit amplitude value (unsigned positive magnitude). Valid values are 0x20 minimum to 0x7F maximum.
	15:7	R		Unused
0x8	**16**	**R/W**	**0x0007**	**Transmit Polarity Register**
	0	R/W	1	Polarity (sign) value for FFE Tap 0 Coefficient. (0 = negative, 1 = positive) This register's value is applied to the analog logic after "Apply Load" (Transmit Coefficient Control Register 0x02 bit 0) is pulsed. The value read from this register is the actual value being driven to the analog logic
	1	R/W	1	Polarity (sign) value for FFE Tap 1 Coefficient. (0 = negative, 1 = positive) See this register's bit 0 description
	2	R/W	1	Polarity (sign) value for FFE Tap 2 Coefficient. (0 = negative, 1 = positive) See this register's bit 0 description
	15:3	R		Unused

Table 2.7 HSS EX10 receiver slice register definitions

Addr (3:0)	Bits	R/W	Reset value	Description
0x0	**16**	**R/W**	**0x0038**	**Receive Configuration Mode Register**
	1:0	R/W	00	Rate Select 00=Full rate (default) 01=Half rate 10=Quarter rate 11=Eighth rate
	3:2	R/W	10	Parallel Data Bus Width 00=8 bit 01=10 bit 10=16 bit (default) 11=20 bit
	5:4	R/W	11	DFE/non-DFE Mode Selector: 00=DFE5, 01=DFE3, 10 or 11 =non-DFE
	15:6	R		Unused

Table 2.7 HSS EX10 receiver slice register definitions

Addr (3:0)	Bits	R/W	Reset value	Description
0x1	**16**	**R/W**	**0x0000**	**Receive Test Control Register** Note: RXxPRBSEN pin = "1" overrides this register and enables checking of a PRBS7+ pattern
	2:0	R/W	000	Test Pattern Selector 000 = PRBS7+ (noninverted) (default) 001 = PRBS7– (inverted) 010 = PRBS23+ (noninverted) 011 = PRBS23– (inverted) 100 = PRBS31+ (noninverted) 101 = PRBS31– (inverted) 110 or 111 =Unused
	3	R/W	0	PRBS Check Enable 0=disabled (default), 1=enabled
	4	R/W	0	PRBS Reset. 0=normal (default), 1=reset applied to PRBS Checker
	5	R/W	0	Full Duplex wrap enable. 0=normal. Selects primary input to the RX, and disables the internal TX to RX wrap buffer (default), 1=wrap. Enables the wrap back driver in the TX to drive the internal wrap path to this RX
	6	R	0	State of RXxPRBSSYNC pin – PRBS checker status 0=PRBS checker not synchronized to incoming data, 1=PRBS checker synchronized and locked to incoming PRBS data
	7	R	0	State of RXxPRBSERR pin – PRBS checker status 0=PRBS pattern match or PRBS checker status = 0, 1=PRBS error detected after PRBS synchronized to incoming data
	8	R	0	State of RXxPRBSFRCERR pin – PRBS force error input signal status 0=PRBS normal operation, 1=PRBS error forced by opening wrap path
	15:10	R	0x00	Unused

Table 2.7 HSS EX10 receiver slice register definitions

Addr (3:0)	Bits	R/W	Reset value	Description
0x2	**16**	**R/W**	**0x0000**	**Sonet Clock Control Register** Enable and frequency selection for RXxRCVC16[T/C] clock output
	1:0	R/W	0x0	Sonet Clock rate selector 00=C4 (default), 01=C8, 10=C16, 11=C4
	2	R/W	0	Sonet Clock output enable 0=disabled, 1=enabled
	15:3	R		Unused
0x3	**16**	**R/W**	**0x0000**	**Phase Rotator Control Register**
	0	R/W	0	Spread Spectrum Clocking Enable: 0=Spread Spectrum Clocking support disabled, 1=Spread Spectrum Clocking support enabled. Should not be enabled unless SSC data is applied. RX Jitter tolerance is improved in non-SSC mode
	1	R/W	0	Reset Flywheel: 0=normal (default, the flywheel is enabled), 1=assert reset to the phase rotator flywheel (disable the flywheel)
	2	R/W	0	Freeze Flywheel: 0=normal (default), 1=freeze the phase rotator flywheel at its current update rate. This can be used to prevent periods of inactivity from altering the state of the flywheel
	15:3	R		Unused
0x4	**16**	**R**	**0xXXXX**	**Phase Rotator Position Register** These registers are continuously updated by the DFE algorithms. To accurately read the values in these registers, the DFE logic should be stopped by setting "DFE Stand By" (bit 2 of register 0x06) to "1."
	5:0	R	0xXX	Rotator Data channel Position: Snapshot sample of DATA channel phase rotator position. This is a six-bit vector indicating which of the 64 possible positions the phase rotator is in
	7:6	R		Unused
	13:8	R	0xXX	Rotator AMP channel Position: Snapshot sample of AMP channel phase rotator position. This is a six-bit vector indicating which of the 64 possible positions the phase rotator is in
	15:14	R		Unused

Table 2.7 HSS EX10 receiver slice register definitions

Addr (3:0)	Bits	R/W	Reset value	Description
0x5	**16**	**R/W**	**0x0000**	**Signal Detect Control Register**
	4:0	R/W	00000	Signal Detect Level: Unsigned value of comparator threshold used in SIGDET circuit.
	5	R/W	0	Signal Detect Power Down 0=enable (default – required for DFE mode), 1=power down the Signal Detect circuit
	15:6	R		Unused
0x6	**16**	**R/W**	**0x000X**	**DFE Control Register**
	0	R/W	0	DFE Control Logic Reset: 0=normal, 1=triggers a reset of the DFE logic
	1	R	X	Not Random Data Status: Proper training of the DFE engine requires sufficiently random data flow. In order to prevent the DFE engine from responding to periods of non-random data, a "random data detector" function is built into the logic. This bit is read to indicate current detection value of this logic. 0=Data is "random," 1=Data is not "random"
	2	R/W	0	DFE Stand By: 0=normal operation (default), 1=standby mode. All internal DFE operations are halted at the next internal break point. Clocks continue to run, but state machines are held idle
	3	R/W	0	Sample DFE request. 0=inactive (default), 1=a rising edge causes the pipeline sampling logic to capture a new snapshot, and makes the results available in registers 0x07 and 0x08.
	4	R	0	Sample DFE request completed. 0=inactive, or not ready yet (normal), 1=requested sample snapshot is now valid and available in registers 0x07 and 0x08
	15:5	R		Unused

Table 2.7 HSS EX10 receiver slice register definitions

Addr (3:0)	Bits	R/W	Reset value	Description
0x7	**16**	**R**	**0xXXXX**	**DFE Data and Edge Sample Register**
	7:0	R	0xXX	DFE Data Samples These are the latest samples captured from the pipeline registers in response to Sample DFE Request (reg 0x06 bit 4)
	15:8	R	0xXX	DFE Edge Samples These are the latest samples captured from the pipeline registers in response to Sample DFE Request (reg 0x06 bit 4)
0x8	**16**	**R**	**0x00XX**	**DFE Amplitude Sample Register**
	2:0	R	XXX	DFE Amplitude Samples These are the latest samples captured from the pipeline registers in response to Sample DFE Request (reg 0x06 bit 4)
	5:3	R	XXX	DFE Amplitude Sample Qualifiers These are the latest samples captured from the pipeline registers in response to Sample DFE Request (reg 0x06 bit 4)
	15:6	R		Unused
0x9	**16**	**R/W**	**0x0000**	**Digital Eye Control Register**
	4:0	R/W	0x00	Minimum eye height interrupt threshold This is an unsigned vector value (positive) defining the minimum acceptable eye amplitude, as measured by the DFE logic, before "Eye Amplitude Error Flag" bit is set in the Internal Status Register (reg 0x0F, bit 6) and interrupt is triggered. 0x0: 0 threshold (default, no interrupt set) 0x1 – 0xE: 1/16 of range per step 0xF: threshold set at 15/16 of full range
	9:5	R/W	0x00	Minimum eye width interrupt threshold This is an unsigned vector value (positive) defining the minimum acceptable eye width, as measured by the Dynamic Data Centering logic algorithm, before "Eye Width Error Flag" bit is set in the Internal Status Register (reg 0x0F, bit 5) and interrupt is triggered. 0x0: 0 threshold (default, no interrupt set) 0x1 – 0xF: eye width threshold in rotator steps, approximately 0.03UI per step
	14:10	R	0x00	EYE WIDTH: Latest available eye width measurement, in units of rotator steps
	15	R		Unused

Table 2.7 HSS EX10 receiver slice register definitions

Addr (3:0)	Bits	R/W	Reset value	Description
0xA	**16**	**R**	**0x00C0**	**DFE Tap 1 Register** This register is read to obtain the status of the DFE Tap 1 coefficient
	5:0	R	0x00	Dac Tap 1: magnitude of DFE Tap 1 coefficient
	7:6	R	11	Sign Tap 1: This sign represents the asserted sign of DFE Tap 1 as it is applied to the summer, and changes under normal operation in DFE mode. 10 = negative, 01 = positive, 11 = zero
	15:8	R		Unused
0xB	**16**	**R**	**0x0060**	**DFE Tap 2 Register** This register is read to obtain the status of the DFE Tap 2 coefficient
	4:0	R	0x00	Dac Tap 2: magnitude of DFE Tap 2 coefficient
	6:5	R	11	Sign Tap 2: This sign represents the asserted sign of DFE Tap 2 as it is applied to the summer, and changes under normal operation in DFE mode. 10 = negative, 01 = positive, 11 = zero
	15:7	R		Unused
0xC	**16**	**R**	**0x0030**	**DFE Tap 3 Register** This register is read to obtain the status of the DFE Tap 3 coefficient
	3:0	R	0x00	Dac Tap 3: magnitude of DFE Tap 3 coefficient
	5:4	R	11	Sign Tap 3: This sign represents the asserted sign of DFE Tap 3 as it is applied to the summer, and changes under normal operation in DFE mode. 10 = negative, 01 = positive, 11 = zero
	15:6	R		Unused
0xD	**16**	**R**	**0x0030**	**DFE Tap 4 Register** This register is read to obtain the status of the DFE Tap 4 coefficient
	3:0	R	0x00	Dac Tap 4: magnitude of DFE Tap 4 coefficient
	5:4	R	11	Sign Tap 4: This sign represents the asserted sign of DFE Tap 4 as it is applied to the summer, and changes under normal operation in DFE mode 10 = negative, 01 = positive, 11 = zero
	15:6	R		Unused

Table 2.7 HSS EX10 receiver slice register definitions

Addr (3:0)	Bits	R/W	Reset value	Description
0xE	**16**	**R**	**0x0030**	**DFE Tap 5 Register** This register is read to obtain the status of the DFE Tap 5 coefficient
	3:0	R	0x00	Dac Tap 5: magnitude of DFE Tap 5 coefficient
	5:4	R	11	Sign Tap 5: This sign represents the asserted sign of DFE Tap 5 as it is applied to the summer, and changes under normal operation in DFE mode. 10 = negative, 01 = positive, 11 = zero
	15:6	R		Unused
0xF	**16**	**R**	**0x0000**	**Internal Status Register** This register is used to report the status of certain internal operations. When status bits in this register change, HSSEYEQUALITY is asserted. A write to this register resets the HSSEYEQUALITY output.
	0	R	0	Phase Rotator Calibration Complete. 0=calibration not completed, 1=phase rotator offset calibration process is completed
	1	R	0	VGA locked First This register is set when the VGA achieves lock, and is cleared only by reset. It indicates lock was achieved at least once
	3:2	R	00	Unused
	4	R	0	DFE training complete This register is set when the DFE logic determines that its H coefficients have converged since reset. 0=initial DFE convergence not yet achieved, 1=initial DFE convergence achieved
	5	R	0	Eye Width Error Flag: 0=normal: Measured Data Eye Width at or above interrupt threshold set in the Digital Eye Control Register (reg 0x09); 1=error: Measured Data Eye Width below interrupt threshold set in the Digital Eye Control Register (reg 0x09)
	6	R	0	Eye Amplitude Error Flag: 0=normal: Measured Data Eye Height at or above interrupt threshold set in the Digital Eye Control Register (reg 0x09); 1=error: Measured Data Eye Height below interrupt threshold set in the Digital Eye Control Register (reg 0x09)
	15:7	R	0x000	Unused

2.2 HSS EX10 Transmitter Slice Functions

In this section, the functions of the transmitter slice of the HSS EX10 10-Gbps core are described. This core supports transmit bit rates as low as 8.5 Gbps and as high as 11.1 Gbps. Frequent references are made to the pin descriptions and register definitions found in Sect. 5.5.42.1. Although the HSS EX10 is only a tutorial example, the reader should compare the functions of the EX10 to real HSS cores with which the reader is familiar. Although implementations may vary, many similar functions will be found. In reading this chapter, the reader should acquire an understanding of the types of functions that may exist and some of the key features related to these functions.

A conceptual block diagram of the HSS EX10 transmitter is shown in Fig. 2.4. The parallel data input of the transmitter (TXxD[19:0]) has a 20-bit data width which may be programmed to capture 8, 10, 16, or 20 bits of user data (based on the setting of the *Parallel Data Bus Width* control bits in the *Transmit Configuration Mode Register*) on the rising edge of the word transmit clock (TXxDCLK). Other cores may support different data widths.

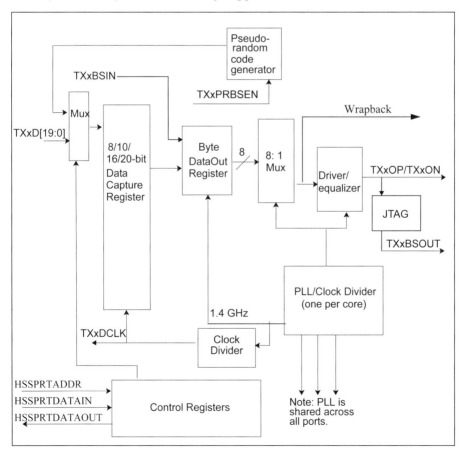

Fig. 2.4 Transmitter concept diagram

The data in Fig. 2.4 is sampled and latched into a shift register synchronous with the high-speed transmit clock. The low order byte is synchronously loaded into a data out register that is clocked at one-eighth the bit rate, while the higher ordered bytes are synchronously shifted to lower byte positions. The 8 bits in the data out register are transferred to the driver where they are further serialized and transmitted at up to 11.1 Gbps. The transferred eight-bit data byte is processed starting with the least significant bit (LSB) first, followed by the next higher significant bit, and so forth. The driver/equalizer multiplexes the 8-bit stream and creates a current-mode differential signal that is frequency equalized for the assumed media channel. The equalization is completely pro-grammable but typically implements a finite impulse response (FIR) preem-phasis filter using reduced current levels for longer run lengths. The core expects valid user data to be available on the parallel interface on each cycle.

In addition to the datapath, Fig. 2.4 also includes the PLL (actually located in the PLL slice), the PRBS generator (discussed later), and an interface for reading and writing control registers. Note that not all cores use control registers to implement slice control. Control signals can also be implemented as individual pins on the core and controlled by chip logic or by protocol cores. Which approach is used varies from one core family to the next, and is somewhat based upon the required number of control and status signals. Using internal registers in the core is a more efficient solution when an excessive number of control and status signals are needed.

2.2.1 Transmitter Parallel Data

The TXxD[19:0] bus shown in Fig. 2.4 is a 20-bit datapath input to the Transmitter Slice. Other cores may use a different names for this bus and may have different bus widths. Consistent with the naming convention described previously, the "x" in this naming convention represents the "channel id." The HSS EX10 has four TX channels with channel identifiers "A" through 'D." Each TXxD bus also has an associated TXxDCLK clock output.

The HSS EX10 core allows the user to select one of several options for the width of the TXxD bus to be used in a given application. For this example, the 20-bit TXxD bus can be programmed to use 8, 10, 16, or 20 bits of this bus. Multiples of 10 bits are useful for applications which use 8B/10B data coding, and multiples of 8 bits are useful for other applications. For the various programmed bus widths, Table 2.8 describes which TXxD bits are used and the corresponding TXxDCLK frequency as a function of baud rate.

The TXxDCLK clock latches the data on the TXxD data bus. Figure 2.5 illustrates the clock/data relationship for this interface. The variable f_{tx} represents the frequency of transmission or transmit data baud rate. The TXxDCLK frequency is a fraction of this as determined by the *Parallel Data Bus Width* in Table 2.8. (This is also affected by the *Rate Select* bits in the *Transmit Configuration Mode Register*.)

Table 2.8 Data bus width function for transmitter section

Bus width bits of "TX configuration mode register'	TXxD19... TXxD16	TXxD15... TXxD10	TXxD9... TXxD8	TXxD7... TXxD0	TXxDCLK frequency
00	X[a]	X[a]	X[a]	D7...D0	$f_{tx}/8$
01	X[a]	X[a]	D9...D8	D7...D0	$f_{tx}/10$
10	X[a]	D15...D10	D9...D8	D7...D0	$f_{tx}/16$
11	D19...D16	D15...D10	D9...D8	D7...D0	$f_{tx}/20$

[a]"X" represents do not care

Although all the transmitter slices in the core are frequency locked, each transmitter slice operates independently, and it is generally not possible to assume that the phase of each TXxDCLK is the same. The maximum phase difference between any two TXxDCLK outputs at the core boundary is specified in the core databook, and results from differences in signal buffering and wiring parasitics within the core. These phase difference limits assume the channels were *resynchronized* as part of the initialization of the interface; otherwise no particular phase relationship can be assumed. (This feature of HSS EX10 core is described later in this chapter.)

The TXxDCLK phases between two channels which are not contained within the same core must also consider chip-level reference clock skew differences generated by clock tree distribution and static phase error differences of the different PLL Slices. Core-to-core phase difference limits are specified in the core databook; these values again assume cores have been resynchronized as part of the initialization sequence.

The HSS EX10 core allows the databus width to be changed at any time. If the application requires the databus width to change dynamically, core documentation must be consulted to determine how the core behaves during this transition. In particular, the latency before the change takes effect, and the behavior of TXxDCLK during the transition must be considered. If the phase difference between TXxDCLK outputs is of concern to the application, resynchronization may be required.

Another consideration of Transmitter Slice usage is the bit order in which bits of the parallel data bus are serially transmitted. The HSS EX10 core always transmits bit 0 first. The user must be cautious that this is consistent with the interface standard being implemented. If necessary the datapath connections to TXxD inputs must be rearranged to obtain the desired bit order for transmission.

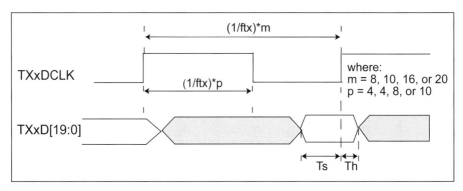

Fig. 2.5 Input data interface timing

2.2.2 Transmitter Signal Characteristics

Signal level and slew rate requirements vary for the various serial data standards. Also, power dissipation can reduced by using smaller differential signal levels, and crosstalk can be reduced by restricting slew rates. These parameters are programmable for the HSS EX10 core, allowing the user to adapt the core to the application.

2.2.2.1 Programmable Normalized Driver Power (NDP) Setting

The driver output power of the HSS EX10 transmitter is programmed using the *Transmit Power Register*. When driving an ideal 100-Ω terminated network, these output power settings set the differential voltage swing at the driver output. The register field in the *Transmit Power Register* contains 7 bits which allow the selection of 64 discrete power settings (some settings are not supported). Each transmitter slice in the EX10 may be separately programmed. Table 2.9 shows some typical register values for sample signal power levels.

Many applications require that signal power levels be adjusted based on the frequency characteristics of the channel being driven by the transmitter slice. Imperfect terminations, for instance, may cause the differential voltage swing to be different from the ideal value shown in Table 2.9. Sometimes the transmitter drives a backplane, and channel characteristics may vary based on the card slot into which the card containing the transmitter is plugged. Also, the core vendor may update the register values corresponding to specific values of differential voltage swing in Table 2.9 as the result of core qualification. For these reasons, most applications require that the signal power levels be programmable. The HSS EX10 core includes registers inside the core; in cases where power levels are controlled from input pins, it is the chip designer's responsibility to ensure programmability of these control pins.

The power supply voltage, termination voltage, and the desired driver power level interact. The power supply and termination voltages must somewhat greater than the maximum voltage driven by the transmitter. This limits the supply voltage ranges that can be used for larger driver power settings. For the HSS EX10, these limitations are specified in Table 2.10.

2.2.2.2 Output Slew Rate Control

In order to support legacy protocols and reduce crosstalk, the minimum slew rate of the HSS EX10 driver stage is programmable. Table 2.11 shows examples of slew rate settings, and also shows typical applications corresponding to these settings.

Table 2.9 Settings for typical output amplitudes

Minimum inner eye amplitude (mVppd)	Transmit power register setting for 8.5 Gbps	Transmit power register setting for 11.1 Gbps
400	45	55
600	70	80
800	95	110
900	110	127
980	127	(N/A)
Test conditions: VDD = 1.1 V, AVTT = 1.4 V, worst case temperature and process, K28.5 data pattern, main tap only			

Table 2.10 Desired output amplitude vs. required circuit supply voltages

Minimum achievable output amplitude[a] (mVppd)	Min. VDD (V)	Max. VDD (V)	Min. VTT[b] (V)	Max. VTT (V)
1,000	1.10	1.30	1.65	1.95
600			1.10	1.95
[a]VTT must always be equal to or greater than VDD [b]'1010...' data pattern, 11.10 Gbps, FCPBGA package, power setting of 127				

Table 2.11 Output slew rate control settings

Slow slew control bits of "Transmit Driver Mode Control Register" for Port x	Typical application	Approximate minimum slew Rate[a] (PS)
"000"	Full rate	24
"001"	Infiniband SDR and DDR	36
"011"	XAUI at 3.125 Gbps	60
"101"	Fibre channel at 4.25 Gbps	50
"111"	Fibre channel at 1.06 Gbps	100
[a]20– 80% transition		

2.2.3 Transmitter FFE Programming

The HSS EX10 transmitter includes a feed forward equalization (FFE) to reduce intersymbol interference (ISI) at the receiver. This preemphasis technique uses a FIR filter to compensate for the high frequency roll-off of the transmission channel. Control inputs are provided to allow adjustment of the driver FFE filter coefficients on a per-port basis. The following equation describes the relationships of the FFE Coefficients.

$$H(Z) = K\,(C_0\,z^{+1} + C_1\,z^{0} + C_2\,z^{-1}) \tag{2.1}$$

This Z-transform equation describes an FFE with three taps, including one pre- and one post tap. Other cores may have FFEs with fewer or more taps.

The driver amplitude (K) is adjustable in the range of 0 – 1,320 (nominal) mV peak-to-peak differential using the *Transmit Power Register* described previously in this chapter. The relative weights and polarities of C_0 to C_2 are configured using the three *Transmit TapX Coefficient Registers* ($X = 0, 1, 2$), and the *Transmit Polarity Register*. The actual range of relative coefficient weights is defined in Table 2.12.

The resolution of the levels is also indicated in Table 2.12, with each tap having equal bit weighting. The driver circuit design enforces a constant driver output power for any combination of coefficients and polarities, provided the sum of coefficients $C_0 - C_2$ is 63 or higher (power decreases proportionally below this sum). As coefficients are initialized or updated, the logic calculates the appropriate internal amplitude (K) to maintain the overall output power at the level defined in the *Transmit Power register.*

Table 2.12 Transmitter FFE summary

Tap coefficient #	0	1	2
Max current (mA)	7.5	30	15
Relative max (%)	25	100	50
DAC resolution (bits)	6	8	7
Tap allocation	Precursor	Main tap	Postcursor

Table 2.13 Transmit driver modes

Mode	FFE mode select bits of "Transmit Driver Mode Control Register" for Port x	FFE taps activated
FFE2	"00"	0, 1
FFE3	"01"	0, 1, 2
(Reserved)	"10"	(Indeterminate)
Hi-Z	"11"	None (see Sect. 2.2.4.1)

To save power in certain applications, the FFE can be configured into several configurations using the *Transmit Driver Mode Control Register* (see Table 2.13). By using fewer taps, power is reduced.

2.2.3.1 Loading Transmit Coefficients

Many of the TX slice parameter control registers are not effective immediately after being written. Application of register values to the analog circuits is controlled by writing the *Apply Load* bit in the *Transmit Coefficient Control Register*. This allows the FFE and power level configuration to be completely loaded in the individual registers and then simultaneously applied to the analog circuits. For the HSS EX10, the values loaded into the *Transmit TapX Coefficient Registers*, the *Transmit Power Register*, the *Transmit Polarity Register*, and the *FFE Mode Select* field of the *Transmit Driver Mode Control Register* are not applied to the analog circuits until the *Apply Load* bit is pulsed. The read-back value of these registers reflects the value being applied to the analog circuits. If these registers are written and then read before the *Apply Load* bit is pulsed, the old values are read. (The value of the *Slow Slew Control* field of the *Transmit Driver Mode Control Register* is applied to the analog circuits immediately, and is not gated by the *Apply Load* bit.)

For cores which do not contain internal registers, filter coefficients may be set using input control pins. Sequencing updated values onto these pins is the responsibility of the chip designer.

FFE coefficient values are not intended to be changed dynamically while data is being transmitted (except as part of a speed negotiation or link optimization process). Significant changes to the TX slice configuration can cause loss of data at the receiver until such time as the receiver adapts to the new waveform characteristics.

2.2.3.2 FFE Coefficient Negotiation Support

Higher speed protocol standards sometimes support negotiation of transmitter FFE settings based on characteristics of the signal at the receiver. IEEE 802.3ap Backplane Ethernet is an example of such an application. Implementations of such applications must update FFE coefficients in response to a full-duplex training protocol by reading and writing the corresponding registers.

The training protocol used for Backplane Ethernet is introduced in Sect. 5.3.5.1. Some cores provide register control signals to more easily implement the actions defined by this protocol, which include incrementing/decrementing coefficient values. Such features are not defined for the HSS EX10 core.

2.2.4 Transmitter Power Control

Power management is becoming an increasingly important part of chip designs. Various features are incorporated into HSS EX10 core to facilitate turning off all or part of Transmitter Slice when the corresponding interfaces are not in use.

2.2.4.1 Transmit Driver Disable Mode

The TXxTS pin forces transmit core drivers to an "off" state in which the driver generates zero differential voltage and does not actively sink current on its outputs (TXxOP and TXxON). This state can also be forced by the *FFE Mode Select* bits of the *Transmit Driver Mode Control Register.*

While the transmit driver is disabled, the transmitter outputs are still terminated and the outputs are pulled to the V_{TT} voltage rail through that termination. This mode only disables the driver outputs, the rest of the transmitter circuitry is still functioning.

2.2.4.2 Selective Power Down

The HSS EX10 core has the capability to selectively power down independent ports using the *Link Enable Register* in the PLL slice. This function may also be performed using the TXxPWRDWN pin.

This power down mode is different from the Transmit Driver Disable Mode discussed previously in that the entire transmitter slice is shut down. Each transmitter slice within the core shares a common PLL, and generally the per-link transmitter power budget quoted in core documentation includes a prorated portion of the power for the PLL. Powering down the transmitter slice therefore reduces the power dissipation of the link by approximately 70–85% of the per-link power budget. The remaining 15–30% is consumed by the associated PLL and clock buffering circuitry, which is not powered down. If the entire core is to be powered down, each of the transmitter and receiver ports must be "disabled," and then the PLL Slice must be disabled using the HSSPDWNPLL control signal.

The TXxDCLK output freezes at either a "0" or "1" value while the transmitter port is disabled. A "glitch" or "sliver" can occur on TXxDCLK during the transition into the power down state. Any chip logic outside of the core which uses this clock must be designed to take this into account.

On initial power on reset, all ports are enabled, thus allowing all ports to go to their reset state. Subsequent re-enabling of individual ports should be followed by a corresponding port reset to ensure proper operation. If synchronization is required between the re-enabled port and other ports, then a resynchronization sequence is also required. (HSS EX10 resynchronization is described later in this chapter.)

2.2.5 Half-Rate/Quarter-Rate/Eighth-Rate Operation

In *full rate* mode, the transmitter serializes and transmits data at a rate determined by the cycle time of the high-speed clock generated by the PLL slice. The baud rates over which the transmitter can operate in full rate mode are limited by the Voltage-Controlled Oscillator (VCO) frequency range supported by the design of the PLL slice. Wider VCO frequency ranges require more complex circuit designs, and therefore this range is generally somewhat limited. However, there are applications where operation is required at slower baud rates, or where the interface must support switching between a full-rate baud rate and legacy baud rates in order to support connections to legacy equipment.

The HSS EX10 includes modes of operation which allow the core to operate at a fraction of the full baud rate. Half-rate, quarter-rate, and eighth-rate link operation is supported. These modes cause the TXxD parallel data to be undersampled and shifted/multiplexed onto the serial data output at half, quarter, or one-eighth the rate of full-rate operation. Selection of the rate mode is performed using the *Rate Select* bits in the *Transmit Configuration Mode Register*, and may be performed on a link-by-link basis. If lower speeds of operation (with larger VCO frequency divisors) are required, external logic may be used to further divide the sample rate of the parallel data.

Users of fractional rate modes should pay careful attention to the following areas of core behavior when using these modes: TXxDCLK behavior, and FFE filter behavior.

2.2.5.1 TXxDCLK Behavior for Half/Quarter/Eighth Rate

Activation of half-rate mode causes the transmit core to double the time period of the TXxDCLK output as well as double the timing of the serializer logic on a per-port basis. This has the effect of reducing the throughput by a factor of two. Activation of quarter-rate mode causes the time period of TXxDCLK to quadruple and reduces throughput by a factor of four; eighth-rate mode causes the time period of TXxDCLK to octuple and reduces throughput by a factor of eight.

When using external logic to implement additional fractional rate modes, care should be taken to determine what mode of the HSS core is used to implement the fractional rate mode. For example, the external logic may force the HSS core into full-rate mode when implementing a 16th-rate mode through external logic. The HSS rate mode being used affects the TXxDCLK frequency and therefore correct understanding is necessary in order to properly analyze timing of the interface.

If the application requires changing the rate mode, the chip designer also needs to understand how TXxDCLK behaves during the transition, and whether glitches are possible on this clock. For the HSS EX10 core, the TXxDCLK completes its current cycle, then remains at a steady "0" while the requested change is enabled, and then resumes normal operation at the new rate. Logic connected to this clock experiences a temporary disruption in the clock, but there is no glitching or slivering of the TXxDCLK during the rate/bus width transition. Data transmitted during this mode change is invalid. If synchronization is required between the links following a rate change, then a resynchronization sequence is also required. (HSS EX10 core resynchronization is described later in this chapter.)

Other core implementations may allow TXxDCLK to glitch or sliver during the transition. In such cases, the chip designer may wish to gate the clock outside the core during this transition. Alternatively, logic connected to this clock may be reset after the transition completes to ensure proper operation.

Internal PLL circuits are unaffected by the rate mode selection. The reference clock frequency should not be changed for fractional rate operation. Power dissipation, although reduced, does not scale linearly with data rate.

2.2.5.2 FFE Behavior for Half/Quarter/Eighth Rate

For the HSS EX10 core, the FFE is always a "T-spaced" filter with respect to the signal rate (T = period of signal rate). In other words, the FFE tap spacing is adjusted based on the rate select mode such that it always runs at the baud rate. These circuits continue to work correctly if the port is set to half-rate, quarter-rate, or eighth-rate speed modes.

For other cores, the user should carefully consult core documentation to determine behavior of the FFE for fractional rate modes. If the FFE clocking is not switched, a "T-spaced" filter would become a "T/2-spaced" filter when used in half-rate mode. This would affect the calculation of filter coefficients.

2.2.6 JTAG 1149.1 and Bypass Mode Operation

The HSS EX10 core transmitter supports a bypass mode of operation in which the parallel data capture and serialization logic is bypassed, and the serial data output is forced to the logic value defined by bypass data input pins on the core. This bypass function is sometimes used to support lower speed source–synchronous interfaces for legacy applications. This bypass function is also required for compliance with JTAG 1149.1 [1] and JTAG 1149.6 [2].

2.2.6.1 Transmit Bypass Path

The TXxBSIN data input and the TXxBYPASS transmit bypass control bit are provided on the HSS EX10 core to support the transmitter bypass feature and are limited to toggle rates of 100 MHz or less. When TXxBYPASS is active, data provided to the TXxBSIN inputs forces the state of the TXxOP and TXxON outputs. A truth table for this function is shown in Table 2.14.

Table 2.14 Transmit bypass path selection

TXxBSIN	TXxBYPASSx	TXxOP-TXxON
X	0	Normal operation
0	1	–
1	1	+

2.2.6.2 JTAG Mode

JTAG 1149.1 requires that all chip outputs be controllable from a *JTAG Boundary Scan Register* (BSR) under the control of a *JTAG TAP Controller*. JTAG 1149.1 is used to perform stuck-at fault testing during the manufacture of printed circuit boards. The transmitter bypass feature is of particular use to provide JTAG compliance on interfaces that are driven by HSS transmitters.

A block diagram of the JTAG function for the HSS EX10 core is shown in Fig. 2.6. The HSSJTAGCE pin forces the HSS core into JTAG 1149.1 mode in which all transmitters are forced into bypass mode. In this mode, TXxBSIN determines the data being driven, TXxJTAGTS determines whether the transmitter is enabled or in Hi-Z mode, and TXxJTAGAMPL[1:0] determines the amplitude of the signal. For JTAG compliance, TXxBSIN and TXxJTAGTS inputs must be driven from JTAG BSR cells, and TXxJTAGAMPL pins are generally tied to VDD or GND. If the TXxJTAGAMPL pins are to be driven from another source, the logic function driving these pins must drive a known value during JTAG test mode. During JTAG test, chips on the circuit board are entirely controlled through their JTAG interfaces and do not go through any operational reset sequences. JTAG signals cannot be driven by flip-flops in the chip (which do not get reset during JTAG test) unless these flip-flops are forced to a known value by a JTAG Compliance Enable signal.

The HSSJTAGCE pin also forces the FFE into a suitable mode for JTAG operation. The FFE is set to FFE2 mode, Tap 2 coefficients are set to values determined by the TXxJTAGAMPL pins, other tap coefficients are set to 0, and the output polarity is set to positive. This has the effect of driving the serial data output to match the TXxBSIN data with no distortion due to filtering.

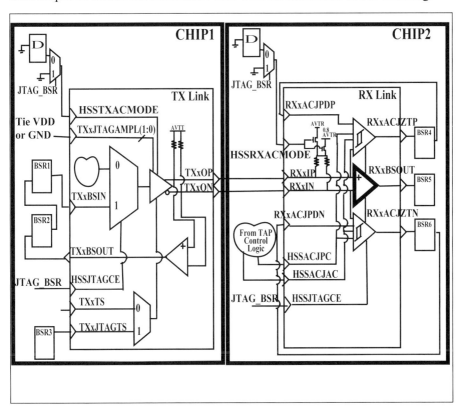

Fig. 2.6 Data path during boundary scan, DC coupled configuration

The pins used for implementation of the JTAG transmitter bypass function may vary from core to core, and may or may not be shared with the pins used for functional operation. The HSS EX10 includes the TXxJTAGAMPL and TXxJTAGTS pins for use during JTAG. These pins, serve the same function as the *Transmit Power Register* and TXxTS pins, respectively. The JTAG pins, and their equivalent functional pins, are multiplexed in the core by the HSSJTAGCE signal. Other cores may not provide separate pins, and may require multiplexors outside the core to perform similar selection.

2.2.7 PRBS/Loopback Diagnostic Features

Diagnostic features are an essential part of any Serdes design. Such features support system fault isolation requirements, and facilitate chip testing and characterization. Essential diagnostic features for any Serdes include:
- Pseudorandom Bit Sequence (PRBS) Pattern generation/checking
- Support for loopback of data at key points within the system datapath

2.2.7.1 Transmit Test Patterns

The HSS EX10 supports the eight test patterns described in Table 2.15, as selected by bits in the *Transmit Test Control Register*. These patterns are available at all supported data rates. The repeating test patterns (1010... and the 64 1's followed by 64 0's patterns) generate data transition run length extremes, with the first pattern generating data transitions at the fastest possible rate and the other pattern generating data transitions at a very slow rate. While these patterns are useful, the frequency spectrum of these patterns is monotonic and does not stress the receiver as much as a pattern with a more diverse spectral content. PRBS patterns provide more varied spectral content, and therefore are more representative of real data.

Table 2.15 Internal test pattern generator

Transmit test control reg [2:0]	Pattern generated
000	PRBS7+
001	PRBS7– (inverted)
010	PRBS23+
011	PRBS23–
100	PRBS31+
101	PRBS31–
110	1010101....
111	(Repeating pattern of 64 '1's then 64 '0's)

PRBS patterns are produced using a linear feedback shift register. The nomenclature PRBS-n is used as a short-hand reference to indicate a PRBS pattern produced by a standard polynomial, where n is the order of the polynomial. PRBS-31, PRBS-23, and PRBS-7 patterns are commonly used in many systems. An order n polynomial implementation requires an n-bit shift register along with a few XOR gates. The longest run length of 0's or 1's that can occur in an n-order PRBS pattern is also n. Having different PRBS patterns available allows the user to select the pattern that is most appropriate given the data encoding used by the application. Using a PRBS-7 pattern to test a SONET system which transmits scrambled data would be overly optimistic; using a PRBS-31 pattern to test a Fibre Channel system which carries 8B/10B data and therefore guarantees short run lengths would be overly pessimistic.

2.2.7.2 Loopback Paths and PRBS Checkers

The HSS EX10 core is a full-duplex configuration containing both transmitters and receivers, and therefore includes the capability to loop the outputs of the transmitter slices to the inputs of the receiver slices. This permits testing through the entire datapath of the chip including the analog circuits. If the transmitter is generating a PRBS or other diagnostic pattern, this pattern can be validated using the PRBS Checker in the receiver.

Other cores implementing simplex core configurations containing only transmitters may include simple receivers and PRBS checker circuits for the sole purpose of providing a check of the PRBS pattern being sent.

2.2.8 Out of Band Signalling Mode (OBS)

The Serial Attached SCSI (SAS) and Serial ATA standards specify an out of band signaling (OBS) mode that forces the driver output to a specific DC state in order to signal certain conditions to the receiver. The TXxOBS pin forces this state on the HSS EX10 core. When this pin is asserted, both the serial output signal legs are driven to the same DC voltage that is nominally the value of the common mode voltage during normal operation. The transmitter data inputs are ignored in this mode. When TXxOBS returns to the low state, the transmitter resumes normal operation.

2.2.9 Features to Support PCI Express

The HSS EX10 core includes support for features which are either required or optional for implementations of the PCI Express standard. These PCI Express related features in the HSS transmitter are described below.

2.2.9.1 Electrical Idle

When in an electrical idle state, transmitter outputs are driven to the common mode voltage, as described further in Sect. 5.5.3.2. The TXxELECIDLE pin controls entry and exit into this state on the HSS EX10 core. The common mode voltage in the electrical idle state is only at the correct level if the link is AC Coupled (as required by the PCI Express standard). Note that assertion of TXxELECIDLE stops the TXxDCLK for the link.

2.2.9.2 Receiver Detection

The transmitter is required to support detection of whether a receiver is connected at the other end of the link. As described in Sect. 5.5.3.2, this is performed by driving an abrupt change in the DC Common Mode voltage of the link, and monitoring the amount of time it takes for the voltage on the wire to settle to the new value.

The TXxRCVRDETEN input initiates the receiver detection operation on the HSS EX10 core. One or the other of the TXxRCVRDETTRUE and TXxRCVRDETFALSE output pins transitions high to indicate completion of the receiver detect process; the former indicating a receiver is present, and the latter indicating no receiver is connected to the link.

2.2.9.3 Beacon Signaling

Beacon signaling is optional in the PCI Express specification as described in Sect. 5.5.3.2. The transmitter sends a low-frequency high/low waveform called a beacon signal on the link to indicate a desire to exit the L2 power state and return to a full-on state. The beacon signal has a pulse width of at least 2ns and no more than 16 μs. HSS EX10 core supports this feature and sends a compliant beacon signal when the TXxBEACONEN pin is asserted.

2.2.9.4 PCI Express Power States

PCI Express defines a number of link power states which are described in Sect. 5.5.4. The HSS EX10 core provides a TXxSTATEL1 pin which removes power from most of the transmitter. Link logic may assert this pin when in the L1 power state in order to reduce power dissipation. (Other pins exist on the receiver and PLL slices to support PCI Express power states and are described elsewhere in this chapter.)

2.3 HSS EX10 Receiver Slice Functions

A conceptual block diagram of a receiver slice for the HSS EX10 core is shown in Fig. 2.7. This receiver is representative of the functionality of the receiver slice designs for higher baud rates.

The receiver slice performs clock and data recovery (CDR) on the incoming serial data stream. The quality of this operation is a dominant factor for the bit error rate (BER) performance of the system. For enhanced performance, several features are combined in this receiver architecture.

The differential data is received by an automatic gain control amplifier to compensate for lossy media. Data is oversampled by a digital circuit that detects the edge positions in the data stream. This digital circuit selects the optimum data sample, and generates early and late signals to indicate the status of the recovered clock alignment which are used to control the output phase positions in a feedback loop. This feedback loop includes a filter to reduce high-frequency jitter phenomenon, and a flywheel to improve jitter tolerance and handling of long run length patterns. The effect of this feedback loop is to maintain a static edge position in the oversampled data array by continuous adjustment of the sampling phase locations.

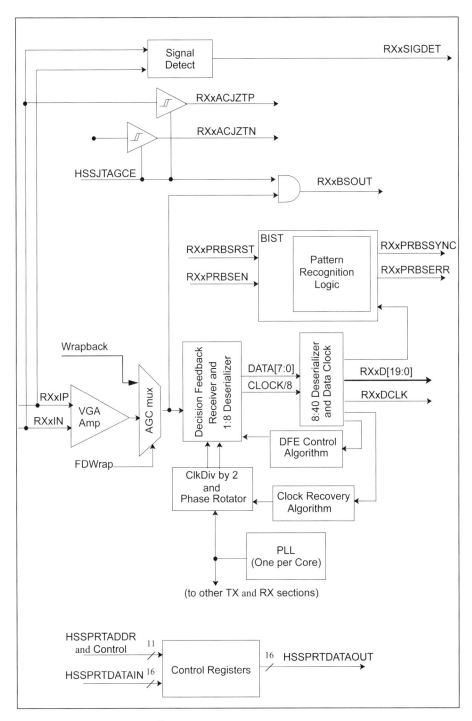

Fig. 2.7 Receiver concept diagram

The HSS EX10 core incorporates a Decision Feedback Equalizer (DFE). The DFE used for this core is a 5-tap DFE; filter coefficients update dynamically based on signal characteristics of the serial data. Cores operating at lower baud rates may or may not include DFE circuits.

The parallel data output of the receiver (RXxD[19:0]) has a 20-bit data width which may be programmed to output 8, 10, 16, or 20-bits of user data (based on the setting of the *Parallel Data Bus Width* bits of the *Receive Configuration Mode Register*) on each clock cycle. Serial data running at up to 11.1 GHz is processed and deserialized to the appropriate data width, and then clocked out on the RXxD bus synchronous with RXxDCLK. The first serial bit received is steered to the LSB of this bus, the next bit is steered to the next higher significant bit, and so forth.

In addition to the datapath, Fig. 2.7 also includes the PLL located in the PLL slice), the PRBS checker, JTAG 1149.6 receivers, Signal Detect logic, and an interface for reading and writing control registers.

2.3.1 Receiver Data Interface

The RXxD[19:0] bus shown in Fig. 2.7 is a 20-bit datapath output of the RX Slice. Other cores may use a different names for this bus and may have different bus widths. Consistent with the naming convention described previously, the "x" in this naming convention represents the "channel id." The HSS EX10 has four receiver channels with channel identifiers "A" through "D." Each RXxD bus also has an associated RXxDCLK clock output.

The HSS EX10 core allows the user to select one of several options for the width of the RXxD bus to be used in a given application. The 20-bit RXxD bus can be programmed to use 8, 10, 16, or 20 bits. For the various programmed bus widths, Table 2.16 describes which RXxD bits are used and the corresponding RXxDCLK frequency as a function of baud rate.

The RXxDCLK clock launches the data on the RXxD data bus. Figure 2.8 illustrates the clock/data relationship for this interface. The variable f_{rx} represents the frequency of reception or receive data baud rate. The RXxDCLK frequency is a fraction of this as determined by the *Parallel Data Bus Width* in Table 2.16. (This is also affected by the *Rate Select* bits in the *Receive Configuration Mode Register*.)

The RX slices in the core independently derive frequencies of operation from the incoming serial data stream. The RXxDCLKs for these channels may or may not be frequency locked to each other depending on the application. If serial data for a set of channels is launched by a common frequency reference, then the RXxDCLKs are frequency locked. However, the CDR circuits of the individual channels continue to make independent decisions as to when to update sampling phase, and therefore it is generally not possible to assume that the phase of each RXxDCLK is the same. The maximum phase difference between any two RXxDCLK outputs at the core boundary therefore cannot be predicted, even if the RXxDCLK outputs were in phase after reset. Use of RXxDATASYNC also affects the phase of RXxDCLK.

The HSS EX10 core allows the data width to be changed at any time. If the application requires the data width to change dynamically, core documentation must be consulted to determine how the core behaves during this transition. In particular, the latency before the change takes effect, and the behavior of RXxDCLK during the transition must be considered.

Table 2.16 Data bus width function for receiver section

Bus Width Bits of 'RX Configuration Mode Register'	RXxD19... RXxD16	RXxD15... RXxD10	RXxD9... RXxD8	RXxD7... RXxD0	RXxDCLK Frequency
00	0000	000000	00	D7...D0	$f_{rx}/8$
01	0000	000000	D9...D8	D7...D0	$f_{rx}/10$
10	0000	D15...D10	D9...D8	D7...D0	$f_{rx}/16$
11	D19...D16	D15...D10	D9...D8	D7...D0	$f_{rx}/20$

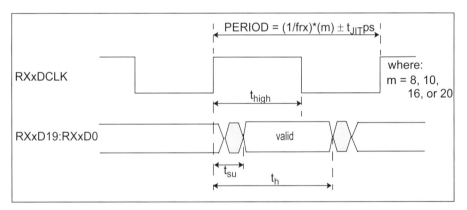

Fig. 2.8 Output data interface timing

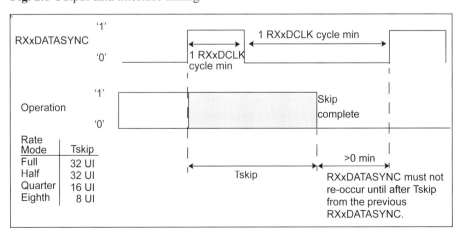

Fig. 2.9 RXxDATASYNC timing diagram

Another consideration of Receiver Slice usage is the bit order in which bits of the parallel data bus are serially received. For the HSS EX10 core, the first bit received is always steered to bit 0 of the RXxD bus. The user must be cautious that this is consistent with the interface standard being implemented. If necessary the datapath connections to the RXxD inputs must be rearranged to obtain the desired bit order.

Receive data alignment may be adjusted through the use of the RXxDATASYNC control input, and may be independently adjusted on each receiver slice. When the RXxDATASYNC signal transitions from a "0" to a "1," the deserializer within the receive logic ignores the most recent serial input and holds the contents of the deserializer for one-bit clock cycle. The RXxDATASYNC operation has the effect of changing the deserializer alignment by one-bit position.

The RXxDATASYNC operation also has the effect of stretching the corresponding RXxDCLK cycle. This behavior is dependent on the rate mode selected. In full-and half-rate modes, the RXxDCLK is stretched 2 UI after every RXxDATASYNC operation. In quarter-and eight-rate modes, the RXxDCLK is stretched 1 UI after every RXxDATASYNC operation. Because the cycle is always stretched, the minimum pulse width of the RXxDCLK signal is never reduced. Figure 2.9 illustrates the timing requirements for the RXxDATASYNC input.

2.3.2 DFE and Non-DFE Receiver Modes

In full-rate and half-rate modes the HSS EX10 receiver operates in one of three equalization modes; non-DFE, DFE3, or DFE5. In quarter-rate and eighth-rate modes only non-DFE mode is supported. In non-DFE mode, the received input signal is processed through the variable gain amplifier (VGA) then captured in the front end logic and deserialized. In DFE3 and DFE5 modes, the deserialized samples are processed using DFE algorithms that automatically adjust the receiver threshold to better compensate for severe pattern-dependent distortions in the channel. These modes are selected using by the *DFE/non-DFE Mode Select* bits in the *Receive Configuration Mode Register* as shown in Table 2.17.

Table 2.17 DFE/non-DFE mode select

Receiver mode	DFE/non-DFE mode selector bits of "Receive Configuration Mode Register" for Port x	
DFE5	0	0
DFE3	0	1
non-DFE	1	0
non-DFE (default)	1	1

DFE5 and DFE3 modes use 5 and 3 filter taps, respectively for feedback. A DFE-n mode makes decisions regarding whether a bit is "1" or "0" based on the history of the last n bits received. DFE5 provides maximum performance in terms of BER, while DFE3 may be used in cases where signal degradation is not as severe. The non-DFE mode uses a fixed threshold receiver for cases where signal degradation is small enough not to require further equalization. The VGA amplifier is used in all modes of operation.

2.3.3 Serial Data Termination and AC/DC Coupling

The RX slice includes a termination resistance between the two legs of the differential input as well as biasing to the termination supply voltage. The common mode voltage that is seen on the differential signal tracks the termination supply voltage, and for this reason the termination supply may be applied through separate power pins on the HSS core. On some cores, additional inputs may be provided to select the voltage range for this supply.

The HSS EX10 uses the nomenclature AVTT to designate the termination supply at the transmitter and AVTR to designate the termination supply at the receiver. These supply voltages are usually set to the same voltage to prevent circulating DC currents and to insure proper operation and reliability of the receiver. Note that there may be power supply sequencing requirements between the AVTT/AVTR supplies and the main power supply.

Some applications and system designs use AC coupling on serial links to eliminate some of these restrictions. An AC coupled system includes decoupling capacitors inserted inline with the serial data signals such that there is no DC path between the transmitter and the receiver device. The decoupling capacitor must have sufficient capacitance to pass the lowest spectral frequency expected in the data. When AC coupling is employed, the common mode bias point of the transmitter and the receiver do not need to be the same.

In a DC coupled link, the current path through the transmitter, the channel, and the termination network in the receiver determines the resulting common mode voltage. The HSS EX10 receiver is designed using the assumption that the transmitter provides a termination which contributes to the voltage bias at the receiver. The current path through the transmitter does not exist when AC coupling is employed, and therefore the transmitter termination no longer contributes to the DC voltage bias for the circuit. The HSS EX10 core has an HSSRXACMODE control input on the PLL slice which enables biasing the receiver's termination network when the core is used in AC coupled applications. This pin enables an additional current path which biases the receiver inputs to a common mode voltage equal to around 80% of the AVTR voltage, which is optimal for operation of this receiver.

2.3.4 Signal Detect

Each individual receiver circuit includes signal detection circuitry. The function of signal detection is to continuously monitor the attached media channel and to provide feedback concerning link status. A "1" on the

RXxSIGDET output indicates that signal transitions are occurring on the link, and the amplitude of the differential signal being received exceeds the detection threshold; otherwise the RXxSIGDET output is "0".

The signal detect output is determined by looking at the average value of the received AC signal. Transitions in the state of the RXxSIGDET status have some inherent latency, which can be attributed to the low-pass filtering techniques used to distinguish signal from noise.

Many standards define the minimum signal level above which the signal must be detected, however, different standards use different values for this threshold. For the HSS EX10 core, the signal detect circuit threshold can be adjusted to be compliant with various industry standards by programming the *Signal Detect Level* bits in the *Signal Detect Control Register.* Examples of signal detect threshold settings for this core are shown in Table 2.18.

Chip designers sometimes use RXxSIGDET to gate data or to disable processing of data by downstream logic. However, signal detect thresholds are sensitive to common mode voltage levels on the receiver inputs, and the RXxSIGDET function often requires circuit level tuning based on hardware characterization. For these reasons, erratic operation may be observed for early users of a core, and it is generally good practice to provide a programmable "chicken switch" to disable any such gating.

Table 2.18 Signal detect register settings

Industry standard compatibility	Signal detect threshold, AC coupled		SDLVLxD bits of "SIGDET Control Register" for Port x				
	"Bad" signal	"Good" signal	Bit 4	Bit 3	Bit 2	Bit 1	Bit 0
Infiniband (2.5 Gbps)	<85 mVppd	>175 mVppd	0	0	1	1	0
SAS (1.5 and 3.0 Gbps)	<120 mVppd	>240 mVppd	0	1	1	0	0
Fibre channel (1.0625, 2.125, 4.25 Gbps)	<90 mVppd	>250 mVppd	0	1	0	0	1

2.3.5 Receiver Power Control

Power management is becoming an increasingly important part of chip designs. The HSS EX10 includes various features to facilitate turning off all or part of Receiver Slices when the corresponding interfaces are not in use.

2.3.5.1 Selective Power Down

The HSS EX10 core has the capability to selectively power down independent ports using the *Link Enable Register* in the PLL slice. This function may also be performed using the RXxPWRDWN pin. Each receiver slice within the core shares a common PLL, and generally the per-link receiver power budget quoted in core documentation includes a prorated portion of the power for the PLL. Powering down the receiver slice therefore reduces the power dissipation of the link by approximately 70–85% of the per-link power budget. The remaining 15–30% is consumed by the associated PLL and clock buffering circuitry, which is not powered down. If the entire core is to be powered down, each of the transmitter and receiver ports must be "disabled," and then the PLL Slice must be disabled using the HSSPDWNPLL control signal.

The RXxDCLK output freezes at either a "0" or "1" value while the receiver port is disabled. A "glitch" or "sliver" can occur on RXxDCLK during the transition into the power down state. Any chip logic outside of the core which uses these clocks must be designed to take this into account.

On initial power on reset, all ports are enabled, thus allowing all ports to go to their reset state. Subsequent re-enabling of individual ports should be followed by a corresponding port reset to ensure proper operation.

2.3.5.2 Powering off Signal Detect

If the Signal Detect function in the HSS EX10 core is not needed in non-DFE mode, it may be turned off using the *Signal Detect Power Down* bit of the *Signal Detect Control Register*.

DFE filters sometimes use the signal detect function, and in such cases the DFE will not operate correctly if the signal detect circuit is turned off, especially in high crosstalk environments. Note, however, that there are exceptions where turning off signal detect may produce improved performance: high loss, high-frequency channels operating in a low crosstalk environment being one such example.

The RXxSIGDETEN input pin overrides the *Signal Detect Power Down* control register bit and performs the same function. If this pin is asserted, signal detect circuitry is forced to a power on state. This is useful in applications such as PCI Express where RXxSIGDET is used to detect a beacon signal while the remainder of the link is powered down. The control register bit cannot be written or read while the link is powered down.

2.3.6 JTAG 1149.1/1149.6 and Bypass Mode Operation

The HSS EX10 core receiver supports a bypass mode of operation in which the data deserialization logic is bypassed, and the serial data input value is driven to a bypass data output of the core. This bypass function is sometimes used to support lower speed source–synchronous interfaces for legacy applications. This bypass function is also required for compliance with JTAG 1149.1 and JTAG 1149.6 [1, 2].

2.3.6.1 Receive Bypass Path

The RXxBSOUT bypass data output supports the receiver bypass feature and is limited to toggle rates of 100 MHz or less. The logic value of the differential data on the serial data input is driven to the RXxBSOUT pin.

2.3.6.2 JTAG 1149.1 Mode

JTAG 1149.1 requires that all chip inputs must be observable in a *JTAG BSR* under the control of a *JTAG TAP Controller*. JTAG 1149.1 is used to perform stuck-at fault testing during the manufacture of printed circuit boards. The receiver bypass feature is of particular use to provide JTAG compliance on interfaces that are connected to HSS receivers.

A block diagram of the JTAG function for the HSS EX10 core is shown in Fig. 2.6. The HSSJTAGCE pin forces the HSS core into a mode which support JTAG 1149.1 (or 1149.6), including forcing all receivers into bypass mode. In this mode, the RXxBSOUT outputs reflect the value being received by the differential receiver. For JTAG compliance, RXxBSOUT outputs must be captured in JTAG BSR cells.

2.3.6.3 JTAG 1149.6 Mode

The JTAG 1149.1 DC stuck fault testing cannot be used to test links which include decoupling capacitors in the channel. JTAG 1149.6 transmits pulse waveforms which propagate through the decoupling capacitor and are received by a JTAG 1149.6 receiver circuit. JTAG 1149.6 is also be used to perform independent testing of the true/complement legs of the serial signal.

The JTAG1149.6 receiver circuit is shown in Fig. 2.6, along with the associated RXxACJZTP and RXxACJZTN outputs of the core. These outputs are latched in the JTAG 1149.6 BSR, and fed back to the RXxACJPDP and RXxACJPDN inputs of the core.

Figure 2.10 illustrates the propagation of JTAG 1149.6 pulse signals through the decoupling capacitors of the link.

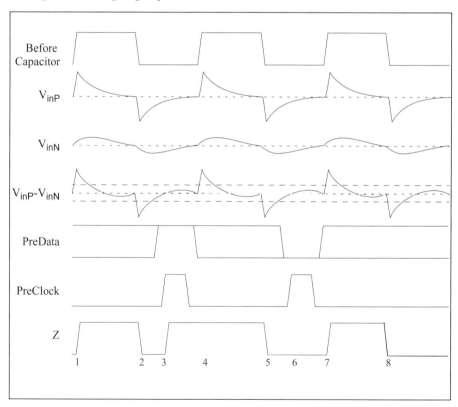

Fig. 2.10 JTAG test receiver waveforms

The signals in Fig. 2.10 are defined as follows:

> Before Capacitor = HSS TX output before the AC coupling capacitor
> Vinp = RXxIP or RXxIN (differential HSS RX input)
> Vinn = Delayed internal Vinp signal at Test Receiver (TR) comparator
> Vinp–Vinn = Signal detected by the TR's edge detection circuit
> PreData = RXxACJPDP or RXxACJPDN pin
> PreClock = HSSACJPC pin
> Z = RXxACJZTP or RXxACJZTN pin (output of TR)

The horizontal points in Fig. 2.10 are:

1. Z transitions high in response to the rising edge of the input signal
2. Z transitions low in response to the falling edge of the input signal
3. Z transitions high due to the hysteretic memory being loaded via PreData/PreClock
4. Nothing happens at this rising edge because output is already high
5. Z transitions low in response to falling edge of the input signal
6. Nothing happens at in response to PreData/PreClock since output is already at desired value
7. Z transitions high in response to the rising edge of the input signal
8. Z transitions low in response to the falling edge of the input signal

The JTAG 1149.6 receiver has hysteresis built in as shown in Fig. 2.11. RXxACJZTP/N outputs are generated per the truth table in Table 2.19.

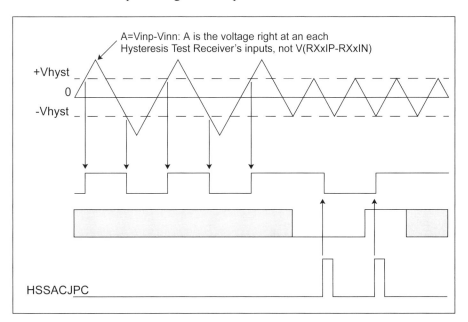

Fig. 2.11 Hysteresis diagram of the ACJTAG test receiver

Table 2.19 Truth table for AC JTAG test receiver when HSSJTAGCE=1

RXxIP/N	HSSACJPC	RXxACJPDP/ RXxACJPDN	RXxACJZTP/ RXxACJZTN
Transition from 0 to 1 (HSSACJAC=1) or Steady State 1 (HSSACJAC=0)	0	X	1
Transition from 1 to 0 (HSSACJAC=1) or Steady State 0 (HSSACJAC=0)	0	X	0
No Transitiona[a]	1	0	0
No Transition[a]	1	1	1

[a]HSSACJP=1 and RXxIP/N transition at the same time should not have occurred because they must be a mutually exclusive event in ac JTAG test

2.3.7 Half-Rate/Quarter-Rate/Eight-Rate Operation

Similar to the discussion related to the transmitter, the HSS EX10 includes modes of operation which allow the core to operate at a fraction of the baud rate for full-rate mode. Half-rate, quarter-rate, and eighth-rate link operation is supported. These modes cause the RXxD serial data to be undersampled as it is shifted into internal shift registers used for deserialization. The net effect is that serial data is sampled at half, quarter, or one-eighth the rate of full-rate operation. Selection of the rate mode is performed using the *Rate Select* bits in the *Receive Configuration Mode Register*, and may be performed independently for each link. If lower speeds of operation are required, external logic may be used to further divide the sample rate of the parallel data.

2.3.7.1 RXxDCLK Behavior for Half/Quarter/Eighth Rate

Activation of half-rate mode causes the receive core to double the time period of the RXxDCLK output as well as double the timing of the deserializer logic on a per-port basis. This has the effect of reducing the throughput by a factor of two. Activation of quarter-rate mode corresponding causes the time period of RXxDCLK to quadruple and reduces throughput by a factor of four; eighth-rate mode causes the time period of RXxDCLK to octuple and reduces throughput by a factor of eight.

When using external logic to implement additional fractional rate modes, care should be taken to determine what mode of the HSS core is used to implement the fractional rate mode. For example, the external logic may force the HSS core into full-rate mode when implementing a 16th-rate mode through external logic. The HSS rate mode being used affects the RXxDCLK frequency and therefore correct understanding is necessary in order to properly analyze timing of the interface.

If the application changes the rate mode, the chip designer needs to understand how RXxDCLK behaves during the transition, and whether glitches are possible on this clock. For the HSS EX10 core, the RXxDCLK completes its current cycle, then remains at a steady "0" while the requested change is enabled, and then resumes normal operation at the new rate. Logic connected to this clock experiences a temporary disruption in the clock, but there is no glitching or slivering of the RXxDCLK during the rate/bus width transition. Data received during the mode change is invalid.

Other core implementations may allow RXxDCLK to glitch or sliver during the transition. In such cases, the chip designer may wish to gate the clock outside the core during this transition. Alternatively, logic connected to this clock may be reset after the transition completes to ensure proper operation.

Internal PLL circuits are unaffected by the rate mode selection. The reference clock frequency should not be changed for fractional rate operation. Power dissipation, although reduced, does not scale linearly with data rate.

2.3.8 PRBS/Loopback Diagnostic Features

Diagnostic features are an essential part of any Serdes design. Such features support system fault isolation requirements, and facilitate chip testing and characterization. Essential diagnostic features for any Serdes include:

- Pseudorandom Bit Sequence (PRBS) Pattern generation/checking
- Support for loopback of data at key points within the system datapath

2.3.8.1 Test Patterns

The receiver contains a flexible test pattern data checker. The data checker is used to self-check the receiver port using an internal loopback test utilizing data generated by its companion transmitter port. The test pattern checker may also be used to self-check an entire TX-RX link with an external wrap back test using an external channel between a compatible transmitter and receiver. For the HSS EX10 core, the receiver controls to implement these tests are in the *Receive Test Control Register*. Six test patterns can be checked as was described for the transmitter in Table 2.15: PRBS7 normal and inverted, PRBS23 normal and inverted, and PRBS31 normal and inverted.

2.3.8.2 Loopback Paths and PRBS Checkers

The HSS EX10 core is a full-duplex configuration containing both transmitters and receivers, and therefore includes the capability to loop the outputs of the transmitter slices to the inputs of the receiver slices. This permits testing through the entire datapath of the chip including the analog circuits. If the transmitter is generating a PRBS or other diagnostic pattern, this pattern can be validated using the PRBS Checker in the receiver. Table 2.20 illustrates programming for the various loopback modes of the HSS EX10 core.

Table 2.20 Register settings for BIST tests

Test mode	TX test control register			RX test control register			
	Test pattern [2:0]	PRBS Gen enable [3]	PRBS reset [4]	Test pattern [2:0]	PRBS check enable [3]	PRBS reset [4]	FD wrap Sel [5]
Receiver internal loopback BIST	"000" through "101"	"1"	0->1->0	(must match TX pattern)	"1"	0->1->0	"1"
External loopback	"000" through "101"	"1"	0->1->0	(must match TX pattern)	"1"	0->1->0	"0"
Normal operation	'XXX'	"0"	"0"	'XXX'	"0"	"0"	"0"

Other cores implementing simplex core configurations containing only receivers may include simple transmitters and PRBS generator circuits for the sole purpose of providing a check of the PRBS pattern being sent.

2.3.9 Phase Rotator Control/Observation

The phase rotator is a critical element embedded within the receiver architecture and by design serves a critical role in the clock recovery function. Although it is intended to operate automatically and without external intervention, there are cases when some external control is desirable.

The rotator contains a flywheel mechanism which, when enabled, provides a means of stepping the rotator in the absence of incoming data based on prior history of the timing characteristics of the data being received. This is intended to improve jitter tolerance and allow for extended run length data patterns. This flywheel can be disabled or reset by the *Freeze Flywheel* and *Reset Flywheel* bits of the *Phase Rotator Control Register*. A snapshot of the current phase rotator position may be read from the *Phase Rotator Position Register*.

2.3.10 Support for Spread Spectrum Clocking

The receiver includes support for spread spectrum clocking (SSC). In a system using SSC, the reference clock frequency of the transmitter is frequency modulated at a low frequency around the nominal transmission rate. The receiver's CDR circuit must have enough range to track this frequency shift and sample serial data correctly. SSC is typically used in systems to reduce electromagnetic interference (EMI) by spreading the radiated energy of the reference clocks and the transmitted data over a range of frequencies. SSC is required by some standards (including SATA and SAS).

By default, the SSC support is turned off to maximize the jitter tolerance margin and ensure that the CDR recovers from a loss of lock without a reset.

For the HSS EX10 core, SSC is supported up to ±6,000ppm around the local reference clock modulated with a triangular waveform of no more than 33kHz. This is sufficient to support common SSC tolerance ranges such as ±3,000ppm or 350/− 5,650ppm. When SSC is being used, the *Spread Spectrum Clocking Enable* bit in the *Phase Rotator Control Register* must be set.

2.3.11 Eye Quality

Channel simulation methods were discussed in Chap. 1, as well as the concept of an "eye" at the output of the receiver. This "eye" does not actually exist as an analog signal within the design, but rather was a virtual mathematical "eye" in the digital logic domain at the output of the DFE. As such, there is no method for measuring the signal characteristics of this eye with test equipment in real hardware.

The channel simulations previously discussed are intended to characterize a system and allow the system designer to meet signal integrity goals. An inherent part of this process is to choose FFE coefficients which are optimal for a given channel design. Note, however, that this optimization is based on the characterized channels being representative of all systems. If significant variations exist due to manufacturing tolerances, the chosen coefficients may not always be optimal. Either excess margin must be built into the system or yield fallout must be tolerated in order to overcome this. Either of these impacts system manufacturing cost.

A better approach would be to optimize FFE coefficients dynamically based on signal measurements within the system. This would require an ability to "measure" the virtual "eye" at the output of the receiver's DFE. Such measurement requires logic features built into the receiver.

The HSS EX10 core does provide eye quality measurement features. The *DFE Data and Edge Sample Register*, *DFE Amplitude Sample Register*, and the *Eye Width* field of the *Digital Eye Control Register* provide parametric information regarding the eye opening of the received signal. Values are captured in these registers when directed by the *Sample DFE Request* bit in the *DFE Control Register*. Software pulses this bit, and then polls until the *Sample DFE Request Complete* bit is set in this register. After this bit is set, valid values may be read from the other registers.

In addition, the HSS EX10 provides an alarm mechanism to flag when the eye opening is less than programmed amplitude or eye width limits. Limit values are programmed in the *Digital Eye Control Register*. When parametric measurements of the received signal violate the *Minimum Eye Width Interrupt Threshold* value, the *Eye Width Error Flag* bit is set in the *Internal Status Register*. Likewise, violations of the *Minimum Eye Amplitude Interrupt Threshold* value cause the *Eye Amplitude Error Flag* to be set. Any change to status bit values in the *Internal Status Register* for any receive slice in the core causes the HSSEYEQUALITY output to be asserted.

2.3.12 SONET Clock Output

Some applications require that data be transmitted using a frequency reference derived from the recovered clock frequency of the received data (called *loop timing*). Although the RXxDCLK output reflects the recovered clock frequency, this output of the HSS core is not differential and therefore is not readily usable as a reference clock for the transmitter.

The RXxRCVC16T/C pins of the HSS EX10 core provide a differential clock output which is a suitable reference clock running at the recovered clock frequency. The *Sonet Clock Control Register* enables and selects the frequency of this clock. If not used, the clock should be disabled to reduce power.

Applications using this clock connect the RXxRCVRC16[T,C] pins of one of the receiver slices to a PLL (for jitter clean-up), and connect the output of the PLL to the HSSREFCLK[T,C] inputs of another HSS EX10 core driving the transmit data. (This clock output should never drive the HSSREFCLK[T,C] inputs of the same core!)

2.3.13 Features to Support PCI Express

The HSS EX10 core includes support for features which are either required or optional for implementations of the PCI Express standard. These PCI Express related features in the HSS receiver are described below.

2.3.13.1 Beacon Signaling

Beacon signaling is optional in the PCI Express specification as described in Sect. 5.5.3.2. When the link is in a PCI Express L2 power state, the transmitter can request exit from this power state by sending a beacon signal.

While the receiver would typically be powered down while the link is in an L2 power state, power down of the signal detect circuit of the receiver is separately controlled. The HSS EX10 core includes an RXxSIGDETEN input which controls power to the signal detect circuit. Link logic continues to enable the signal detect circuit while in power down states, and can therefore detect beacon signaling on the link using the RXxSIGDET status output of the core.

2.3.13.2 PCI Express Power States

PCI Express defines a number of link power states which are described in Sect. 5.5.4. The HSS EX10 core provides a RXxSTATEL1 pin which removes power from most of the receiver. Link logic may assert this pin when in the L1 power state in order to reduce power dissipation. (Other pins exist on the transmitter and PLL slices to support PCI Express power states and are described elsewhere in this chapter.)

2.4 Phase-Locked Loop (PLL) Slice

The PLL slice is usually common to all transmitter and receiver links in the core. The block diagram of the PLL slice used in the HSS EX10 core is shown in Fig. 2.12. The primary purpose of the PLL slice is to provide high-speed clocks to the transmitter and receiver slices that are frequency locked to a

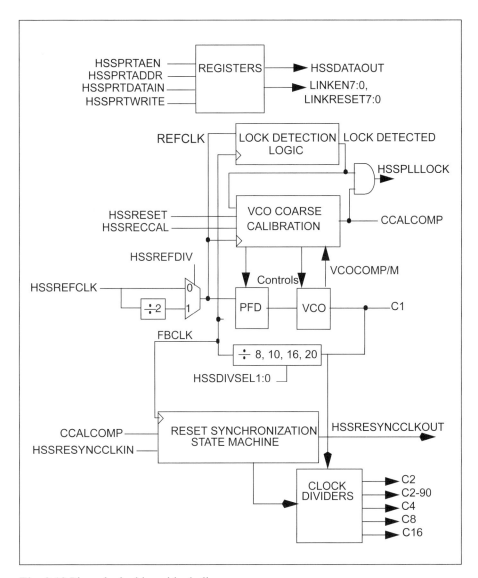

Fig. 2.12 Phase locked loop block diagram

reference clock input. However, this slice is also the natural place to put other miscellaneous logic which is common to all transmitter and receiver links. In the HSS EX10, reset sequencers and miscellaneous registers for link reset and enable are included in the PLL slice.

2.4.1 Reference Clock

The HSSREFCLK[T,C] pins are the reference clock input to the PLL slice of the HSS EX10 core. The HSSREFCLKT pin is the true leg of the differential signal pair; the HSSREFCLKC pin is the complement leg. This clock must be driven by a low-jitter differential clock source for best jitter performance.

2.4.2 Clock Dividers

There are several clock dividers in the PLL slice. On the HSS EX10 core, the HSSREFDIV and HSSDIVSEL[1:0] pins are control signals for clock dividers which can be provisioned to support a range of frequency options for HSSREFCLK[T,C].

HSSDIVSEL[1:0] controls the frequency multiplication factor between the VCO C1 clock output and the feedback clock (FBCLK). For the HSS EX10 core, 16, 20, 32, or 40 multiplier values are supported.

In addition, the HSSREFDIV input selects whether or not HSSREFCLK is divided by two prior to the VCO input. If HSSREFDIV is "0" then the frequency of HSSREFCLK is equivalent to FBCLK; otherwise HSSREFCLK is twice the frequency of FBCLK. The latter case generally results in better system jitter performance, however, the drawback is that a higher frequency HSSREFCLK must be distributed on the chip.

The HSS EX10 PLL Slice also contains additional dividers which divide the VCO C1 clock output for distribution to the TX/RX slices.

2.4.3 Power On Reset

Power-on-reset is initiated by asserting the HSSRESET pin for a minimum number of reference clock cycles as shown in Fig. 2.13. The HSSRESETOUT output of the HSS EX10 is driven active once the core enters a reset state, and remains active until after the VCO calibration completes and clock outputs of the core are at stable frequencies. This output may be used as a reset by logic surrounding the HSS core.

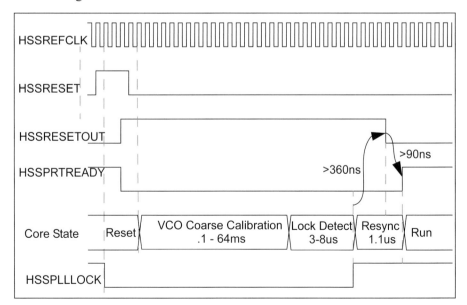

Fig. 2.13 Power on reset (POR) sequence

HSSPLLLOCK is asserted once the internal VCO clocks are determined to be frequency locked to the HSSREFCLK input.

The HSSPRTREADY output pin is deasserted while the core is in a reset state; and is reasserted when the reset sequence of the PLL slice has completed and the registers of the transmitter and receiver slices are ready to be written and/or read. Note that additional initialization of the logic in the transmitter and receiver slices occurs after this, as is described in Sect. 2.5.

2.4.4 VCO Coarse Calibration

VCO within the PLL requires calibration after a power-on-reset. As is the case with most VCO designs that must operate over a wide range of frequencies, there are a number of tuning bit inputs to the VCO which must be set to optimal values to select frequency band and reference voltage level. Setting these bits can be left as an exercise to the user for some cores; the HSS EX10 core contains state machine logic which determines the optimal tuning settings without user intervention. Calibration results are reported in the *VCO Coarse Calibration Status Register*.

Different VCO calibration algorithms may be employed by HSS cores, and it is beyond the scope of this text to describe any given algorithm. Some cores provide registers to allow the user to override values of various tune bits and/or read tune bit settings and VCO status. Details of such functions are specific to the PLL slice design.

VCO calibration occurs as part of a reset sequence of the HSS EX10 core. After the reset completes, the application may force recalibration using either the HSSRECCAL input pin or the *Recalibrate* bit of the *VCO Coarse Calibration Control Register*.

2.4.5 PLL Lock Detection

The HSSPLLLOCK output is asserted to indicate the PLL is frequency locked to the selected reference clock. The HSS EX10 core determines the lock condition by comparing the frequency of the PLL feedback clock (*FBCLK*) in Fig. 3.7 to that of the reference clock (*REFCLK*).

The PLL Lock Detect circuit for the HSS EX10 core is described in detail in Sect. 3.1.4.2. This circuit contains one counter clocked by *FBCLK*, and one counter clocked by *REFCLK*. Each of these counters is allowed to run for some specified period of time. If the number of times one of these counter is clocked is more or less than the other counter (with some tolerance), then the clock frequencies are not the same, indicating the PLL is not locked. If the two counters have counted roughly the same number of clock cycles, then the PLL is locked, and HSSPLLLOCK is asserted.

The length of time over which the clock frequencies are compared determines the accuracy of the lock condition being reported. If there is a frequency delta between the two clocks, but the delta is sufficiently small, then HSSPLLLOCK may be reported despite a minor frequency difference.

2.4.6 Reset Sequencer

A series of reset signals is distributed in the core to synchronize various circuit functions in various clock domains. There are various clock dividers located in both the PLL slice and in the individual transmitter and receiver slices. The HSS EX10 PLL slice generates reset signals which are sequentially asserted in order to ensure the circuit exits reset in an orderly manner. Various divided clock outputs of the PLL slice are gated off during portions of the reset sequence. This sequence is initiated by assertion of the HSSRESET pin, the HSSRESYNCCLKIN pin, the HSSRECCAL pin, or the *Recalibrate* bit of the *VCO Coarse Calibration Control Register*.

The reset sequencer is initiated after the VCO calibration completes. Following completion of the reset sequence, the HSSPRTREADY output is asserted. Prior to assertion of this signal, the user should not attempt to access core registers or expect serial data transmission or reception.

Note that the EX10 core has additional initialization actions which must occur in the individual transmitter and receiver slices *after* the PLL slice completes the reset sequence. Other status conditions may need to be checked to determine whether individual transmitter and receiver slices have completed their initialization sequence and are ready to use.

2.4.7 HSS Resynchronization

2.4.7.1 Transmitter Resynchronization

The requirement to synchronize the clock dividers of the HSS transmitter to minimize lane-to-lane skew was discussed previously in this chapter, and is noted again in subsequent chapters as the application requirements of various protocols are discussed. The HSS EX10 core exits the reset sequence with the transmitter clock dividers aligned. (This may not be true for other cores.) However, subsequent changes to the TX configuration (such as updating the data width or rate mode selection) affect the operation of the clock divider in the transmit logic. After updates to the transmitter configuration, the TXxDCLK outputs of the various HSS TX slices do not have any guaranteed phase relationship. Any TXxDCLK phase difference introduces skew onto the transmitter outputs which may exceed the specifications for the application.

To meet such skew specifications the transmit slice logic of the HSS cores must be resynchronized after being configured in order to realign the phases of the TXxDCLK. The difficulty of implementing such a resynchronization function, especially when the alignment must be performed across multiple HSS cores, should be noted. The clock divider logic divides the internal high-speed C2 clock down to the frequency at which the parallel data bus is clocked. The high-speed clock in the HSS EX10 core runs at baud rate divided by 2, which may be as high as 5.55 GHz for this core. In order to successfully resynchronize the clock dividers in all of the HSS transmit channels, all of the clock dividers must be reset in the same cycle of this 5.55-GHz clock.

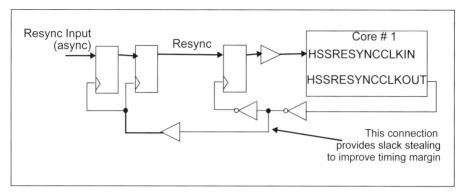

Fig. 2.14 Resynchronization logic for a single core configuration

The complexity of this problem increases when multiple HSS cores are involved. Each core has its own PLL, and while all of the PLLs are assumed to be locked to the same reference clock, there is some inherent phase variation between the high-speed clocks produced by each PLL. This reduces the window in which the reset must occur in order to reset multiple cores in the same clock cycle.

The typical solution for this problem at lower speeds is to synchronize the reset to a 5.55-GHz clock, and drive this reset synchronously to all clock dividers. However, this cannot be implemented in a typical ASIC because the clock is simply too high a frequency. The propagation delay of the reset signal from a synchronizer flip-flop to the various HSS cores would exceed the cycle time of the 5.55-GHz clock.

The HSS EX10 core solves the problem by using a reset which is synchronized to a lower frequency clock, and then distributing this reset and gating it within the HSS core to ensure all clock dividers in the HSS core are reset in the same cycle of the high-speed clock.

Two signals on the HSS EX10 core are associated with the resynchronization feature:

HSSRESYNCCLKOUT: This output of the HSS EX10 core is the clock used to synchronize the resync reset signal. Although this clock is of low frequency than the internal high-speed clock, the frequency is higher than that of the parallel data clocks. For the HSS EX10 core, the frequency of this clock is equivalent to the HSSREFCLK input. (This does *not* mean the HSSREFCLK can be used instead of using this output. The HSSRESYNCCLKOUT clock and HSSREFCLK are mesochronous.)

HSSRESYNCCLKIN: This input of the HSS EX10 core is a reset signal which is asserted to perform the resynchronization function. The signal must be synchronized to HSSRESYNCCLKOUT.

In order to resynchronize the transmitters of a single HSS core, the chip designer must implement logic as shown in Fig. 2.14. The *Resync Input* control to the circuit is first synchronized to the falling edge of

HSSRESYNCCLKOUT, and then is retimed to the rising edge before driving HSSRESYNCCLKIN. The flip-flop which drives HSSRESYNCCLKIN should be physically located near the HSS core pin on the chip.

This logic is extended in Fig. 2.15 to resynchronize multiple HSS cores. The *Resync Input* control is first synchronized to the falling edge of the HSSRESYNCCLKOUT from one of the cores, and then retimed to the rising edge of each individual HSSRESYNCCLKOUT. Flops are added between this retiming stage and the HSS core so that the retiming stage can be physically located near the centralized synchronizer, and the last flop can be physically located near the HSS core pin. Based on the physical distance between the HSS cores, additional flops may need to be added.

The configuration of this logic (first synchronizing the *Resync Input* input to the falling clock edge, and then retiming to rising edges) is key to performing the synchronization across multiple cores. While the various HSSRESYNCCLKOUT clocks are not in phase with each other, the specified variation is less than half of the clock cycle. Therefore, synchronizing to the falling edge of one of the clocks ensures that all of the retiming flops see the signal in the same cycle in their respective clock domains.

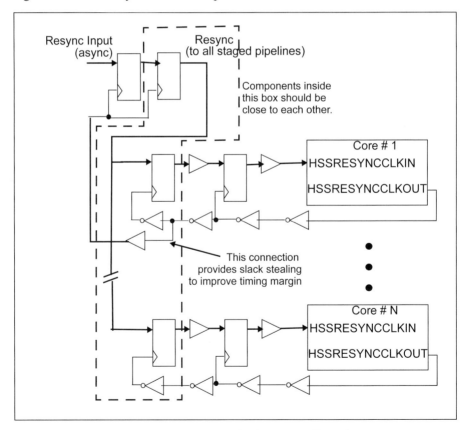

Fig. 2.15 Resynchronization logic for a multiple core configuration

2.4.7.2 Receiver Resynchronization

The resynchronization feature of the HSS EX10 core also synchronizes clock dividers in the receiver logic. However, RXxDCLK phases diverge on the receiver during normal operation because the phase rotator logic in each receiver slice makes independent update decisions as it tracks the recovered frequency of the receive data. Therefore receiver resynchronization has no practical purpose in most applications.

2.4.8 PCI Express Power States

PCI Express defines a number of link power states which are described in Sect. 5.5.4. The HSS EX10 core includes an HSSSTATEL2 pin which removes power from most of the core including the PLL. Link logic may assert this pin when in the L2 power state in order to reduce power dissipation. (This pin overrides the L1 power state controls on the TX/RX slices.)

2.5 Reset and Reconfiguration Sequences

Once the HSS links have been initialized, calibrated, and trained, then data is transmitted and received through the serial link. However, it is also instructive to understand the sequences of operations necessary to initialize and reconfigure the links, including:

- The procedure for initializing, calibrating, and training the link
- Procedures for changing operating parameters of the link (i.e., data rate, FFE coefficients, etc.)
- Procedures for entering and exiting power down modes.

This section describes these operational sequences. Although the programming details are described for the fictional HSS EX10 core, the concepts apply broadly to HSS cores from many vendors.

2.5.1 Reset and Configuration

Power-on initialization of the HSS core requires a complex series of events to occur within the core. The power-on reset signal (HSSRESET on the HSS EX10 core) resets registers within the core and initiates this series of events. First, the PLL is calibrated and locked to the reference clock while the transmit and receive logic is held in a reset state. Then, the clock dividers in the transmit and receive logic are initialized and the logic is enabled for transmitting and receiving data. Once this happens, the CDR circuit in the receive logic must train on the data eye before data can be received correctly. If the receiver contains a DFE, this circuit is also trained and filter coefficients are set.

The reset values of the HSS registers may or may not reflect the desired rate mode, bus data width, transmit amplitude, signal detect threshold, FFE coefficient values, etc., for the desired application. External intervention (by hardware state machines or by software programming) may therefore be required during the initialization process to set appropriate operating modes.

Such programming cannot occur until after the completion of prerequisite parts of the initialization process. For example, FFE coefficient values and other transmitter operating mode parameters cannot be programmed before the PLL is locked because the transmitter is held in a reset state.

A flowchart of the reset process for the HSS EX10 is shown in Fig. 2.16. Some steps may not apply to a given HSS core, and signal names or status indications may vary. However, in a general sense these steps (or a subset of them) apply in most cases. Detailed descriptions of these steps follow:

Apply power. Turn on the power supplies to the core, paying attention to any power sequencing requirements. If absolute maximum voltages are specified, voltages should never exceed these values, even for transient periods when the power is first applied.

Allow time for core input conditions to stabilize. Core inputs, including the following, must stabilize before continuing:

- Power supply voltages must be stable within recommended operating ranges.
- The HSSREFCLK input must be stable and operating within specified frequency limits. Any PLL generating this clock should be locked.
- Core input pins must be stable and at valid logic levels. Some pins may require specific values to permit the core to initialize properly.

Note that some core outputs may not be defined at this point in the sequence. For example, the RXxDCLK and TXxDCLK outputs may not be running or may be running at frequencies other than their normal specifications.

Reset the HSS core. Once the inputs and power supplies are stable, as previously described, the HSSRESET pin should be asserted and deasserted as described in Sect. 2.4.3. The PLL slice of the core then begins the reset sequence shown in Fig. 2.16, which includes coarse calibration of the VCO. All input pins on the core which affect PLL operation must be stable throughout the reset sequence.

While the above description is specific to the HSS EX10 core, any HSS core with a PLL must execute some form of calibration procedure. This procedure is initiated by a core reset or other control signal, and a status indication is asserted upon completion.

Wait for PLL reset completion. This status indication may take different forms on different cores. On the HSS EX10 core, the rising transition of the HSSPRTREADY pin indicates completion of the PLL reset sequence.

Depending on the PLL slice architecture and the VCO calibration algorithm, it is possible for VCO calibration to fail. If VCO calibration fails, the PLL never locks to the reference clock, and HSSPRTREADY is never reasserted. Chip logic and/or software should implement a timeout in order to initiate diagnostic and recovery actions should this circumstance occur.

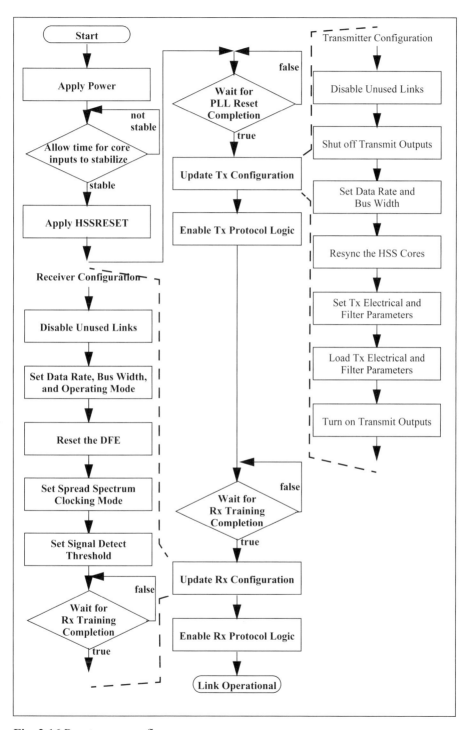

Fig. 2.16 Reset sequence flow

Registers should not be accessed while VCO calibration is in progress. Internal clocks are not stable, and registers may be held reset during this process. The notable exception to this rule applies to registers in the PLL slice which may be accessed for diagnostic and recovery purposes in the event of VCO calibration failure.

Update the transmitter configuration (if necessary). After HSSPRTREADY is asserted to indicate completion of the PLL slice reset, the transmitter begins to serialize and transmit data on the TXxD inputs based on operational parameters set by input pin values and/or default values in internal registers for the transmitter logic. At this point, values of these parameters may be reprogrammed as needed for the application. This procedure will be covered in more detail shortly.

Supply parallel data to the transmitter data inputs. Once transmitter parameters are set for the desired operating conditions, protocol logic drives transmit data on the TXxD parallel data inputs of the transmitter.

Wait for the receiver to finish calibration. While the transmitter can commence operation immediately after the reset sequence completes, the receiver must train on the incoming serial data before valid data can be received. Both CDR and DFE circuits require a training period before valid data is received. The time required for the CDR to train is usually relatively short, however, receivers with DFEs generally required a longer training period since the training sequence must execute a convergence algorithm to determine DFE coefficients. For the HSS EX10 core, DFE training may take up to 410 μs as measured from the assertion of HSSPRTREADY, and assuming there is serial data to receive on the link when the core exits reset. The HSS EX10 core provides a status indicator (bit 4 in the *Internal Status Register*) indicating completion of DFE training.

Update the receiver configuration (if necessary). After the receiver completes training, the receiver deserializes and drives data onto the RXxD outputs based on operational parameters set by input pin values and/or default values in internal registers. Values of these parameters may be reprogrammed as needed for the application at this time. This procedure will be covered in more detail shortly.

Link operational. The receiver now starts to receive valid data. The application protocol may begin its initialization, performing functions as described in Chap. 4.

2.5.2 Changing the Transmitter Configuration

The reset procedure described above included a step where the configuration of the transmitter is changed. For HSS cores that do not have internal registers, input pins would presumably be tied to their operational values as determined by the application, and no updates would be required. However, HSS cores which have internal registers initially use a default configuration (determined by register reset values) following the reset sequence. If the application requires a configuration other than this default, then reprogramming these registers is necessary.

Some applications may also require the transmitter configuration to be reprogrammed at times other than immediately after a reset. For example, some protocols negotiate link speed and therefore require reprogramming the *Rate Select* controls in the *Transmit Configuration Mode Register* and the *Receive Configuration Mode Register*. Some IEEE 802.3 Backplane Ethernet variants negotiate FFE coefficients, as is described in Sect. 5.3.5.1.

The following is a general description of the procedure used to change the transmitter configuration for the HSS EX10 core. This flow is described in Fig. 2.16 as an expansion of the *Update Tx Configuration* step. Depending on the application and the transmitter parameters being updated, some or all of these steps may be required:

Disable unused links. Some of the transmitter and/or receiver channels may be unused for some chip designs and applications. HSS cores provide a means of disabling unused links so that they do not consume power or generate noise. For the HSS EX10 core, this is performed using a *Link Enable Register* which is part of the PLL slice register map. After reset, all links are enabled, and must be reprogrammed to turn them off if they are not used.

Shut off the transmitter output. The TXxTS can be deasserted to shut off transmitter outputs while the transmitter is being reconfigured.

Set data rate and bus width. Registers are written and/or core inputs are changed to select the data rate and the TXxD parallel data bus width. These settings affect the function of the transmitter clock dividers, and change the TXxDCLK frequency. As discussed in Sect. 2.2.5.1, this clock may "glitch" or "sliver" during this transition depending on the core design. If such glitches can occur, this behavior must be taken into account when designing protocol logic to connect to the HSS transmitter.

Resynchronize the links. TXxDCLK outputs will not remain in phase with each other when the data rate and bus width are reprogrammed. Multilane applications with skew requirements between transmitter lanes require the transmitters to be resynchronized. Sect. 2.4.7 describes resynchronization of the HSS EX10 core.

Update transmitter electrical and filter parameters. The HSS EX10 core is representative of the types of electrical and filter parameters which may be provisionable on a typical transmitter. This core provides register control of the following transmitter parameters, which can be provisioned at this step in the process:

- Slew rate (*Transmit Driver Mode Control Register – Slow Slew Control*)
- Differential voltage amplitude (*Transmit Power Register*)
- FFE Mode (*Transmit Driver Mode Control Register – FFE Mode Select*)
- FFE Coefficient Values (*Transmit TapX Coefficient Registers*)

Load transmitter electrical and filter parameters. For the HSS EX10 core, the *Apply Load* bit in the *Transmitter Coefficient Control Register* must be written for most of the above parameters to be applied to the analog circuits.

Note that provisioning of transmitter filter parameters must generally occur after any updates to the data rate and bus width. FFE designs often scale operation based on baud rate, and are dependent on the data rate being set properly prior to updating the mode.

Turn on the transmitter output. If TXxTS was deasserted previously, then it is reasserted at this time. .

The above procedure can be adapted with steps omitted as necessary based on which parameters are to be updated. For instance, changing the data rate can be achieved by reducing the above sequence to only the *set data rate and bus width* step. (For multilane applications, the *resynchronize the links* step is also required.)

2.5.3 Changing the Receiver Configuration

As was the case for the transmitter, the receiver initially uses a default configuration (determined by register reset values) following the reset sequence. If the application requires a configuration other than this default, then reprogramming these registers is necessary. Additionally, some applications may also require the receiver configuration to be reprogrammed at times other than immediately after a reset.

The following is a general description of the procedure that is used to change the receiver configuration for the HSS EX10 core. This flow is described in Fig. 2.16 as an expansion of the *Update Rx Configuration* step. Depending on the application and the receiver parameters being updated, some or all of these steps may be required.

Disable unused links. As was the case for the transmitter, some transmitter and/or receiver channels may be unused for some chip designs or applications. Unused receiver channels should be disabled. On the HSS EX10 core, the *Link Enable Register* is used to disable unused links.

Note that the *Link Enable Register* on the HSS EX10 core does *not* disable the signal detection circuitry. The RXxSIGDETEN pin turns off this circuitry, if desired.

Set data rate, bus width, and operating mode. Registers are written and/or core inputs are changed to select the data rate, the RXxD parallel data bus width, and the operating mode (non-DFE, DFE3, or DFE5 on the HSS EX10 core).

The data rate and bus width settings affect the function of receiver clock dividers and change the RXxDCLK frequency. As discussed in Sect. 2.3.7.1, this clock may "glitch" or "sliver" during this transition depending on the core design. If such glitches can occur, this behavior must be taken into account when designing protocol logic to connect to the HSS receiver.

Reset the DFE. Changes to the data rate and DFE operating mode generally cause corruption of data and require reinitialization of the DFE circuit. The *DFE Reset* bit in the *DFE Control Register* of the HSS EX10 forces the DFE to reinitialize and retrain.

Set spread spectrum clocking mode. If SSC is employed by the application, this mode is enabled and any parameters are provisioned in this step.

Set signal detect threshold. Different applications specify different threshold levels to delineate between signal detected and loss of signal conditions. HSS cores often allow provisioning of the detection threshold for the signal detection circuit. This threshold should be provisioned, if applicable, in this step.

Wait for completion of CDR/DFE training. After changing the data rate, DFE mode, or resetting the DFE, the CDR and (if applicable) the DFE circuits must retrain. This requires some delay before the receive data is valid. The HSS EX10 core provides a status indicator (bit 4 in the *Internal Status Register*) indicating completion of DFE training.

As was the case for the transmitter, the above procedure can be adapted with steps omitted as necessary based on which parameters are to be updated.

2.6 References and Additional Reading

A comprehensive list of interface standards documents for various network protocols can be found in Sect. 5.6. Refer to that list for more information on standards mentioned in this chapter.

In addition, the following standards documents are referenced in this chapter:

1. "IEEE Std 1149.1-2001 IEEE Standard Test Access Port and Boundary-Scan Architecture", Institute for Electrical and Electronic Engineers, 2001.

2. "IEEE Std 1149.6-2003 IEEE Standard for Boundary-Scan Testing of Advanced Digital Networks", Institute for Electrical and Electronic Engineers, 2003.

Interested IBM employees and IBM ASIC customers may also wish to consult the following IBM HSS databooks. The HSS EX10 core described in this chapter was loosely based on these cores.

3. "High Speed Serdes (HSS) – 8.5 to 11.1 Gbps for Cu-08 Core Databook", SA15-5852-04, IBM.

4. "High Speed Serdes (HSS) – PCI Express Gen 2 for Cu-08 Core Databook", SA15-5846-02, IBM.

2.7 Exercises

1. An HSS EX10 transmitter slice is operating at 8.5 Gbps. What is the frequency of the TXxDCLK output corresponding to the following configuration values for the *Transmit Configuration Mode Register*? What pins of the TXxD bus are used for each of these configurations?

   (a) 0x00   (b) 0x04   (c) 0x01.   (d) 0x0E

   (e) 0x0D   (f) 0x03   (g) 0x0F.   (h) 0x05

2. An HSS EX10 receiver slice is operating at 10.3125 Gbps. What is the frequency of the RXxDCLK output corresponding to the following configuration values for the *Receive Configuration Mode Register*? What pins of the RXxD bus are used for each of these configurations?

   (a) 0x00 (b)0x04 (c)0x01 (d)0x0E (e)0x0D (f)0x03 (g)0x0F (h)0x05

3. Specify a series of register write cycles (specifying register address and data) that resets the FFE coefficient logic and then programs the registers of the HSS EX10 to set and apply the following parameters:
   - FFE Coefficients (decimal): $C_0 = -3$, $C_1 = +18$, $C_2 = +14$
   - Transmit Amplitude: 0x70
   - Slew 50 ps min.

4. The *Link Enable Register* and the *Link Reset Register* are both implemented in the HSS EX10 PLL slice even though these registers control the transmitter and receiver slices. Why?

5. What value should be written to the *Link Enable Register* to disable the following combinations of channels?

   (a) Disable TXA, TXB, and RXA

   (b) Disable TXD and RXD

   (c) Disable all channels except TXA and RXD

6. Specify a series of register write cycles (specifying register address and data) that switches all transmitter channels from Full-Rate mode to Quarter-Rate mode.

7. Specify the reference clock frequencies, PLL slice HSSDIVSEL and HSSREFDIV pin values, and *Transmit Configuration Mode Register* setting necessary to achieve the following baud rates. Note that there may be multiple correct answers:

   (a) 8.500 Gbps   (b) 2.125 Gbps   (c) 1.250 Gbps

   (d) 3.125 Gbps   (e) 10.3125 Gbps

8. Assume HSS EX10 PLL slice pins are tied as follows: HSSDIVSEL = 00, HSSREFDIV = 1. Specify the reference clock frequencies and *Receive Configuration Mode Register* setting necessary to achieve the following baud rates.

   (a) 8.500 Gbps   (b) 2.125 Gbps   (c) 1.250 Gbps

   (d) 3.125 Gbps   (e) 10.3125 Gbps

9. Two HSS EX10 cores are used in a SONET application requiring a baud rate of 2.488 Gbps. The transmitter must transmit data at exactly the same baud rate as the received data:

   (a) Draw the clock connections between the two HSS EX10 cores that are necessary to implement this system.

   (b) For the HSS EX10 being used to receive the SONET data, specify the reference clock frequency, PLL slice HSSDIVSEL and HSSREFDIV pin values, the *Receive Configuration Mode Register* setting, and the *SONET Clock Mode Register* setting.

   (c) For the HSS EX10 being used to transmit the SONET data, specify the PLL slice HSSDIVSEL and HSSREFDIV pin values, and the *Transmit Configuration Mode Register* setting.

10. Why is it not possible to connect JTAG Boundary Scan Cells directly onto the serial data?

11. Name all of the HSS EX10 transmitter slice pins associated JTAG 1149.1 or 1149.6 implementation, describe the function of these pins, and describe how these pins are connected.

12. Name all of the HSS EX10 receiver slice pins associated JTAG 1149.1 or 1149.6 implementation, describe the function of these pins, and describe how these pins are connected.

13. What additional test coverage is provided by JTAG 1149.6 over that of JTAG 1149.1? Which pins of the HSS EX10 are specifically associated with JTAG 1149.6 and are not used for JTAG 1149.1?

14. Assume the HSS EX10 receiver is programmed such that data is received on RXxD[19:0]. This receiver is used in an application where the bit sequence "11000000110011111100" is used as a training pattern. Design a state machine which has an input *training_active*, and an output which controls the RXxDATASYNC pin on the HSS EX10 receiver slice. When *training_active* is asserted, the state machine pulses RXxDATASYNC as needed until the training pattern is aligned on the correct bit boundary.

15. Table 2.18 specifies two thresholds for the RXxSIGDET signal detect function on the HSS EX10 receiver slice:

    (a) Explain the difference between the "good" signal and "bad" signal thresholds.

    (b) What can you say about the state of RXxSIGDET when the amplitude of the received signal is between the "good" signal and "bad" signal thresholds.

16. The HSS EX10 TXA and RXA slices are to be placed in a wrap mode and tested using a PRBS23+ sequence. Specify a series of register write cycles (specifying register address and data) that executes this sequence.

17. An HSS EX10 transmitter and receiver are to be externally connected in a wrap configuration and tested using a PRBS31-sequence. Specify a series of register write cycles (specifying register address and data) that executes this sequence.

18. The HSS EX10 TXA and RXA slices are to be tested using pins on the core to control the PRBS test. Draw a timing diagram illustrating the sequence of events on the relevant pins.

19. Specify a series of register read and write cycles (specifying register address and data) that captures and reads the phase rotator position for the HSS EX10 receiver slice.

20. Specify a series of register read and write cycles (specifying register address and data) that captures and reads the DFE data, edge, and sample values for the HSS EX10 receiver slice.

21. Specify a series of register read and write cycles (specifying register address and data) that sets alarm thresholds for the digital eye amplitude and width. The eye amplitude alarm should be triggered if the eye amplitude is less than 25% of the full signal range, and the eye width alarm should be triggered if the eye width is less than 0.33 UI.

22. Specify a series of register read and write cycles (specifying register address and data) that sets alarm thresholds for the digital eye amplitude and width. The eye amplitude alarm should be triggered if the eye amplitude is less than 50% of the full signal range, and the eye width alarm should be triggered if the eye width is less than 0.39 UI.

23. The relationship between HSSREFCLK, FBCLK, and various C1, C4, etc. clocks is illustrated in Fig. 2.12. For each reference clock frequency and divider setting below, determine the frequencies of the FBCLK, C1, C2, C4, C8, and C16 clocks:

    (a) f(HSSREFCLK) = 500 MHz, HSSDIVSEL = 11, HSSREFDIV = 0

    (b) f(HSSREFCLK) = 900 MHz, HSSDIVSEL = 11, HSSREFDIV = 1

    (c) f(HSSREFCLK) = 550 MHz, HSSDIVSEL = 10, HSSREFDIV = 0

    (d) f(HSSREFCLK) = 1 GHz, HSSDIVSEL = 00, HSSREFDIV = 0

    (e) f(HSSREFCLK) = 1 GHz, HSSDIVSEL = 01, HSSREFDIV = 0

24. Assume the VCO in the HSS EX10 PLL slice requires an operating range of 8.0–11.0 GHz for the C1 clock. How would you configure the HSS EX10 to achieve a baud rate of 5 Gbps?

25. Draw a logic diagram for the resynchronization logic necessary to resynchronize four HSS EX10 cores which are subdivided into two groups of two cores each. Each group is to be resynchronized independently.

26. Draw a logic diagram for the resynchronization logic necessary to resynchronize four HSS EX10 cores which are subdivided into two groups of two cores each. A *resync_mode* control input is used to select which of two resynchronization configurations is to be used. If *resync_mode* = 0, then each group of two cores is resynchronized independently. If *resync_mode* = 1, then all four cores are resynchronized together.

27. Specify a series of register read and write cycles (specifying register address and data) that sets the following configuration for all of the HSS EX10 transmitter slices:
    - Half rate mode
    - 16-bit data bus width
    - FFE coefficients (decimal): $C_0 = -3$, $C_1 = +33$, $C_2 = +14$
    - Transmit amplitude: 0x68
    - Slew 50 ps min.

28. Specify a series of register read and write cycles (specifying register address and data) that sets the configuration listed in Exercise 27 for the TXA and TXB slices of an HSS EX10 core, and sets the following configuration for the remaining transmitter slices:
    - Quarter-rate mode
    - 10-bit data bus width
    - FFE coefficients (decimal): $C_0 = +7$, $C_1 = +35$, $C_2 = -9$
    - Transmit amplitude: 0x5C
    - Slew 50 ps min.

29. Specify a series of register read and write cycles (specifying register address and data) that uses the *FFE Mode* in the *Transmit Driver Mode Control Register* of the HSS EX10 to disable all transmitter outputs, and then update the configuration of all transmitter slices as described in Exercise 27. Transmitter outputs should not be enabled until all configuration updates have been made.

30. Explain why the PLL reset sequence must complete before the transmitter and receiver configurations can be updated.

31. Specify a series of register read and write cycles (specifying register address and data) that sets the following configuration for all of the HSS EX10 receiver slices. Your sequence should include a DFE reset.
    - Full-rate mode
    - 20-bit data bus width
    - DFE3 mode
    - Spread spectrum clocking enabled
    - Signal detect enabled and set for threshold value 0x12

32. Specify a series of register read and write cycles (specifying register address and data) that sets the configuration listed in Exercise 31 for the RXA and RXB slices of an HSS EX10 core, and sets the following configuration for the remaining receiver slices. Your sequence should include a DFE reset:
    - Full-rate mode
    - 16-bit data bus width
    - DFE5 mode
    - Spread spectrum clocking disabled
    - Signal detect disabled

33. Specify a series of register read and write cycles (specifying register address and data) that disables all transmitter outputs (using appropriate registers), switches the data rate of all transmitter and receiver slices of an HSS EX10 core to quarter rate mode, and enables the transmitter outputs again. This sequence should not change any other configuration parameters, and should reset the DFE if needed.

34. Provide a state diagram for a state machine which executes the sequence shown in Fig. 2.16, ending after *PLL Reset Completion* block. State transitions should be determined by appropriate HSS EX10 core status signals, where applicable, and signals from timer circuits otherwise.

Chapter 3
HSS Architecture and Design

The tutorial example of the HSS EX10 core was introduced in Chap. 2, and was described from a user's point of view. The description of this core is continued in this chapter to cover architecture, major subsections, circuit basics, and core construction. This description serves as an example of an approach to Serdes architecture and design since a comprehensive treatment of the subject would require an entire textbook.

As depicted in Fig. 3.1, the Serdes core contains three major sections (or *slices* using the nomenclature introduced in Chap. 2). The HSS EX10 is configured in a full duplex arrangement with both transmitter and receiver functions present. Simplex cores contain either transmit or receive functions.

The phase-locked loop (PLL) slice provides the low jitter clocks for both transmit and receive functions.

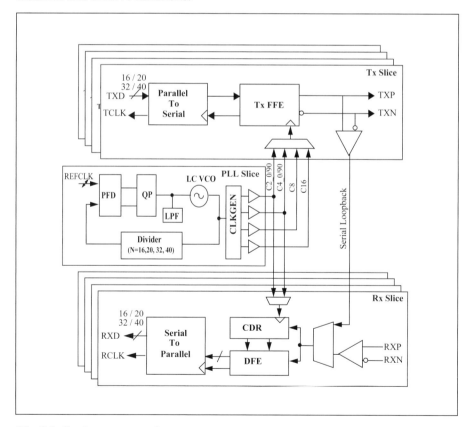

Fig. 3.1 Serdes core overview

D. R. Stauffer et al., *High Speed Serdes Devices and Applications,*
© Springer 2008

The Transmitter (Tx) Slice performs the parallel to serial conversion as well as supporting various feed forward equalizer (FFE) functions. Serialized data is driven from the transmitter to the receiver for built-in self-test (BIST).

The final slice is the Receiver (Rx) slice where receive side equalization (either a continuous-time equalizer or a decision feedback equalizer (DFE)) improves the bit error rate (BER). Once the receive signal is equalized, the serial stream is then driven through the serial to parallel converter (also called *deserialization*).

Each of these slices is covered in detail in the subsequent sections.

3.1 Phase-Locked Loop (PLL) Slice

The PLL slice contains a number of critical "centralized" functions that are shared by the various transmit and receive slices. The PLL slice of the HSS EX10 core services four pairs of transmitter and receiver slices. In practice, a PLL slice may typically service up to eight pairs of Tx/Rx slices for full duplex configurations, or up to 16 simplex transmitter or receiver slices. By sharing the PLL slice across numerous channels, the area and power dissipation can be amortized over the various Tx/Rx pairs, leading to substantially improved efficiency. Of course, the down side to sharing the PLL is that the frequency of operation for all channels is restricted to a single baud rate and binary subrates. The PLL slice of the HSS EX10 drives a single frequency C1 clock to each Tx/Rx slice, and each slice is provisioned to operate at either full rate, half rate, quarter rate, or eighth rate as described in Sects. 2.2.5 and 2.3.7. In most applications this restriction causes no impacts. In applications where per-channel baud rate programmability is desired, the chip designer ends up trading off the area/power of additional PLL slices for this programmability.

A high-level block diagram of the PLL slice is shown in Fig. 3.2. There are six major functions that reside in the PLL slice, including the PLL Macro,

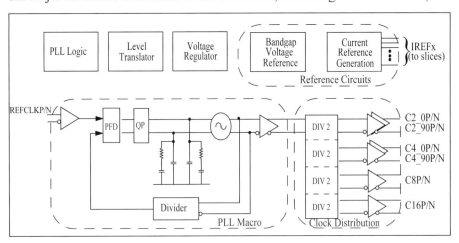

Fig. 3.2 PLL slice block diagram

Clock Generation and Distribution, Reference Circuits, Voltage Regulator, PLL Control Logic, and Digital Control Signal Level Translators. A detailed description of each block is covered in the following sections.

3.1.1 PLL Macro

The *PLL Macro* performs frequency synthesis and is the most important macro in the PLL slice. At a high level, the input to the PLL is an on-chip reference clock (at 1/16th to 1/40th of the baud rate) from which the PLL synthesizes the full rate C1 differential clock. The nomenclature used extensively in this chapter is to encode the subrate frequency in the clock name. For instance, a SerDes running at 10 Gbps would have a C1 clock at 10 GHz. Subrate clock examples are:

C2 = 10 GHz / 2 = 5 GHz,
C4 = 10 GHz / 4 = 2.5 GHz,

and so forth. Using the C1 clock, the clock distribution macro provides the various subrate clocks using a cascade of differential divide-by-2 circuits (ripple divide approach).

The architecture of the PLL Macro shown in Fig. 3.3 utilizes a classic *Charge Pump PLL* approach described in numerous textbooks [1–5]. One of the most critical design parameters of this PLL is that it must deliver a high quality clock with minimal random jitter (RJ). At speeds above approximately 4 GHz, meeting stringent RJ specifications is only feasible using an LC-tank- based VCO. Many monolithic integrated PLL designs use a ring of voltage- controlled delay elements to form the oscillator. These ring-based VCOs are compact and can be low power, however, their phase noise performance is not sufficient for I/O class links due to the resulting RJ. The VCO in the HSS EX10 PLL slice is controlled via a combination of coarse and fine control values. The fine control is driven into the VCO as an analog control voltage. In addition, to lower the gain of the VCO, a band selection scheme is implemented using a 4-bit binary weighted input vector which controls the coarse tuning of the VCO by selecting one of sixteen possible varactor combinations, resulting in 16 bands.

The remainder of the PLL follows a classic Charge Pump PLL construction. Per Fig. 3.3, the REFCLK signal enters into the PLL through the differential to single ended converter. The loop is then formed with the phase-frequency detector (PFD), the Charge Pump (QP), a passive second-order loop filter, the LC VCO, and an appropriate feedback divider. To maintain a constant PLL bandwidth, the charge pump current is varied such that the product of the charge pump current and the feedback divider value is a constant. The various output frequency ranges, charge pump currents, and feedback divide ratios are optimized to minimize the RJ.

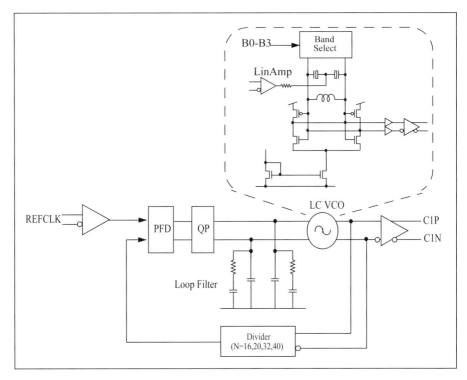

Fig. 3.3 PLL macro block diagram

3.1.2 Clock Distribution Macro

The clock distribution macro contains two main functions. The first of these is the ripple divider which uses the C1 differential clock to produce the symmetric C2_0 (in-phase) and C2_90 (quadrature or I/Q phase) clocks. Fig. 3.4 illustrates a simple symmetric current mode logic (CML) master/slave flip-flop that performs this function. The design of the I/Q clock generator separates a single master/slave flip-flop into two individual latches. The outputs of each of these latches provide the I/Q clocks. Careful attention to the loading at each output is necessary to assure the outputs remain locked in quadrature; any loading imbalance would skew the separation.

Another major source of I/Q error arises due to duty cycle distortion (DCD) on the C1 clock. The two latches trigger on the rising and falling edges of this clock, and any DCD inducing timing variation of these edges relative to their ideal placement directly impacts the separation.

Note that all of the signals are differential and therefore both polarities of each clock are produced. Thus, this clock generation function produces the four phases of clock needed for the receiver slice with phase offsets of 0, 90, 180, and 270°.

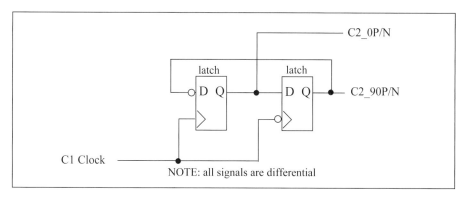

Fig. 3.4 Symmetric I/Q clock generation

Using similar divide-by-2 structures, the C4_0 and C4_90 clocks shown in Fig. 3.2 are created. The remaining C8 and C16 clock outputs are created using this topology; however, in the case of the HSS EX10 core, only one phase is required for these clocks.

The outputs of the clock divider circuits are buffered to drive the clock transmission lines to the transmitter and receiver slices. Since the transmitter slice uses these clocks as a timebase, the duty cycle of these clocks is important. Any DCD will produce jitter on the serial data output of the transmitter slice. Additional circuitry may be added in some cases to actively control the clock duty cycle. In addition, depending on the core configuration and number of ports being serviced by the given PLL slice, multiple buffers are commonly used to drive the associated load.

Depending on the operating mode provisioned for the Tx/Rx slices, some of the clock dividers in Fig. 3.2 may be powered down to save power. For example, if the Tx/Rx slices are operating in full-rate mode (using the C2 clock variants), the C4, C8, and C16 dividers can be powered down. Note that power control logic needs to consider the operating modes of all channels on a given core in order to determine which clocks may be powered down.

3.1.3 Reference Circuits

The PLL Slice reference circuits generate various current and voltage references required for proper operation. Fig. 3.5 depicts a simplified version of this reference circuitry, the heart of which is a bandgap voltage reference circuit. This circuit provides an ultra-stable output voltage based on the silicon bandgap, providing a constant voltage independent of power supply, temperature, and process. This voltage is ~1.22 V for the HSS EX10.

The bandgap voltage provides the reference input to the PLL Voltage regulator, the output of which is the primary power supply for the PLL. (A 1.2V VRR12 supply is used in the HSS EX10 PLL.) The voltage regulator circuit uses a linear regulator scheme consisting of an op-amp, an NFET pass element, and a passive feedback divider. For the HSS EX10, the total current demand for the PLL circuit is in the range of 25–40 mA. To provide adequate performance, the NFET device is chosen to be an low-Vt (LVT) variant and typically is extremely large with device widths typically measured in millimeters. The feedback divider reduces the common mode voltage of the op-amp input, which improves gain and overall op-amp performance.

A key performance metric for a voltage regulator is the power supply rejection ratio (PSRR), which is a measure of noise rejection on the power supply input. The voltage regulator in the HSS EX10 provides a minimum of 15 dB PSRR over the AVdd power supply input, which helps minimize supply induced noise on the sensitive PLL control inputs. As is the case for all linear voltage regulators, the design must maintain significant gain and phase margin in the negative feedback paths, and therefore various compensation capacitors are needed which are not shown in Fig. 3.5.

The final section in Fig. 3.5 is the Current Reference generator. The output of this circuit consists of one or more current sink outputs, typically in the 50–100 μ A range. As shown in the figure, current is pulled into the NFETs. In each of the various Tx/Rx slices, a PFET current mirror is used to create local replicas of the main IREF signal. This topology is typically chosen to minimize the number of IREF mirrors (one PFET mirror feeds each NFET mirror) from the bandgap reference to the point of use inside the Tx/Rx slices.

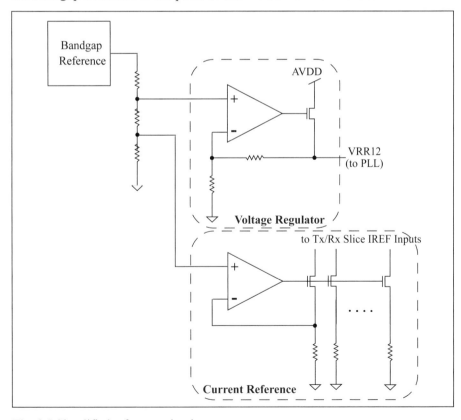

Fig. 3.5 Simplified reference circuitry

3.1.4 PLL Logic Overview

The PLL logic macro is a synthesized logic macro that accomplishes a number of critical PLL logic functions. It also provides the necessary controls to various analog circuits in the PLL as well as transfer of configuration and status data. Functions include power-on-reset, reset synchronization, VCO coarse calibration, and PLL lock detection. The VCO coarse calibration and PLL lock detection functions are key, and are described in more detail below.

3.1.4.1 VCO Coarse Calibration

As described previously, an LC-based VCO is required to meet stringent RJ specifications. A coarse/fine control scheme is used to minimize VCO gain. Calibration of the VCO coarse control is a critical function implemented in the PLL logic. The LC-based VCO has a four-bit (16 band) digital control vector which must select the appropriate band upon power-up. This selection is performed in two steps, starting by checking each band for a valid PLL Lock (as described in Sect. 3.1.4.2). The algorithm then chooses the best band for which PLL Lock can be achieved. In order to determine the best band, the center frequencies of the bands must be measured and the difference between the band center and the operating reference clock frequency is calculated. The smallest delta frequency difference for the reference clock frequency is considered to be the best band selection.

As an example, the graph in Fig. 3.6 shows various bands as a function of frequency. The x-axis represents the VCO band control voltage range. The far left indicates the minimum control voltage and hence minimum VCO frequency for each band. Likewise, on the far right is the maximum control voltage and maximum frequency. The y-axis depicts the VCO frequency. The delta frequency is measured by first forcing the VCO control voltage to its maximum value, and the corresponding maximum frequency of the band is measured. Next, the VCO control voltage is forced to its minimum value, and the minimum frequency of the band is measured. These two frequencies are averaged and the result represents the center frequency of the band. The delta between the center frequency of the band and the reference frequency is calculated. The coarse calibration algorithm simply picks the band that has the smallest difference between the desired frequency and band center.

The dotted line for *Fdesired* in Fig. 3.6 indicates an example of a desired PLL operating frequency. This frequency can be serviced by three different bands, indicated by the squares at the intersections of the dotted line and the lines for bands 3, 4, and 5. For this example, the coarse calibration algorithm would begin at band 0. Since dotted line for Fdesired does not intersect the line for band 0, the PLL cannot lock for this band. Similarly, bands 1 and 2 cannot achieve lock. When the algorithm reaches band 3, PLL lock is achieved with a relatively large positive control voltage. The algorithm computes the difference between the locked frequency and center of the band (the solid vertical line), and then continues examining bands 4 and 5. After exhausting all possibilities, the algorithm selects band 4 given that the intersection of the Fdesired line and the band line is closest to the center frequency.

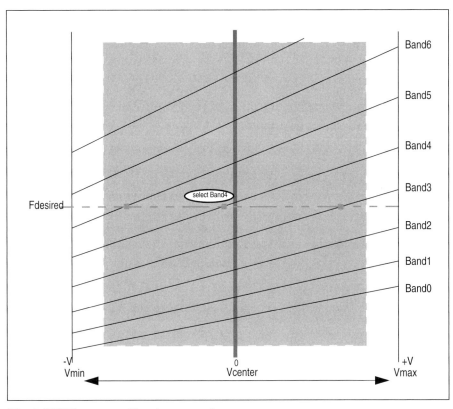

Fig. 3.6 VCO coarse calibration example

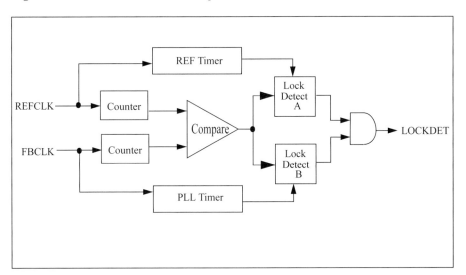

Fig. 3.7 Block diagram of PLL lock detect

3.1.4.2 PLL Lock Detection

Figure 3.7 depicts the PLL Lock Detection circuitry implemented in the PLL Logic of the HSS EX10 core. As shown in the figure this lock condition is determined using the 2-bit counters, one for the reference clock and one for the feedback clock within the PLL. The two counters are initialized at different values with a separation count of two. If these counters do not equal each other within a specified timeout period, then the PLL lock signal is asserted high to indicate lock. If the clocks are not frequency locked, then at some point the compare function will detect equal values in the counters and the lock detection is deasserted.

The reference clock and feedback clock also clock the respective timer circuits shown in the figure. These timers generate a periodic timeout signal. If the comparator does not detect equal counter values within the timeout period, the corresponding PLL lock indicator is set. If both PLL lock indicators are set, then the LOCKDET output is asserted. This lock indication is generated using timers in both clock domains to ensure both clocks are oscillating. The LOCKDET output shown in Fig. 3.7 drives the HSSPLLLOCK output of the HSS EX10 core.

The length of the timeout period determines the accuracy of the lock condition being reported. If there is a frequency delta between the two clocks, but the delta is sufficiently small, then LOCKDET may be reported despite a minor frequency difference. For this to happen, the frequency delta would need to be small enough so that it does not accumulate two clock cycles of difference within the timeout period. The HSS EX10 core uses 10-bit timer counters, so this delta would need to be less than 2 divided by 1024, or approximately 1,953 ppm.

3.2 Transmitter Slice

Figure 3.8 illustrates a simplified block diagram of the HSS EX10 transmitter slice. The transmitter is implemented using a mixed signal approach, combining custom high-speed CML and static CMOS circuitry with a synthesized standard cell digital logic macro. To achieve the high rates of speed involved (at 10 Gbps a C2 clock frequency of 5 GHz is required), most circuits are implemented using a CML topology.

The PC2, PC4, PC8, and PC16 clock inputs in the lower right corner of Fig. 3.8 are driven by the C2, C4, C8, and C16 clock outputs of the PLL slice clock distribution macro. The MUX selects the appropriate clocks as determined by the *Rate Select* bits of the *Transmit Configuration Mode Register* described in Table 2.6. The selected clock enters the clock divider block. This clock is driven onto the C2 clock output of the clock divider and clocks the three stage shift register which forms the basis of the FFE. This clock is also divided to provide C4, C8, and LC8 clocks to other transmitter circuits.

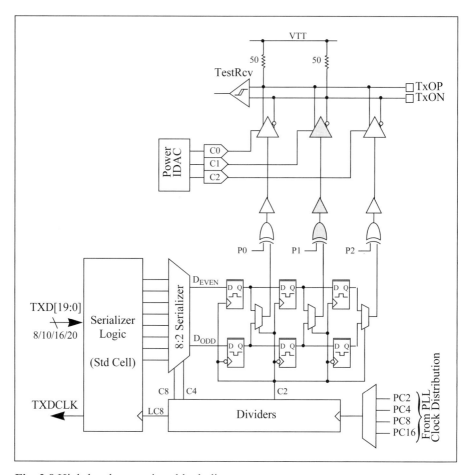

Fig. 3.8 High-level transmitter block diagram

The driver stage shown in Fig. 3.8 uses a C2 half-rate clock as was described by the second driver architecture in Fig. 1.16. The FFE shift register consists of three stages and is 2-bits wide, and the multiplexors shown in Fig. 3.8 between shift register stages implement the last stage of serialization. The half-rate clock is used to clock these multiplexors, providing the timing reference for the outbound bits. On the primary FFE tap, D_{EVEN} is selected when the clock is high and D_{ODD} is selected when the clock is low.

There are three taps in the FFE shown in Fig. 3.8: The cursor tap is highlighted in gray and there is one precursor tap and one postcursor tap. Each tap is serviced by a segment consisting of an XOR gate (which provides individual polarity control), a predriver stage, and a driver output stage. The three segments are summed together at the output node driving the TXxOP/N pins. The magnitude of the tap weights is controlled by a combination of the Power IDAC (controlled by the *Transmit Power Register*) and individual Coefficient DACs (controlled by the *Transmit Tap0–2 Coefficient Registers*).

An FFE coefficient update macro (residing in the standard cell logic section) adjusts coefficient values at the circuit to enforce a constant driver output power for any combination of coefficients and polarities as was described in Sect. 2.2.3. As coefficients are initialized or updated, this logic calculates the appropriate internal amplitude (K) to maintain the overall output power (and differential output peak-to-peak swing) at the constant level defined in the *Transmit Power Register.*

Driver termination is nominally set at 50Ω and achieved by using on-chip uncalibrated precision resistors. This corresponds to a 100-Ω differential termination. The termination is biased by the AVtt power supply.

3.2.1 Feed Forward Equalizer (FFE) Operation

The tap ranges and resolutions for the HSS EX10 FFE circuit were described in Table 2.12, and this table is repeated in Table 3.1. In a non-equalized case, the main tap provides 100% of the driver output amplitude. As shown in the table, this limit corresponds to the maximum current that this tap can provide. Limits for other taps were chosen after empirical studies of equalization solutions for a variety of channel design examples.

Table 3.1 Transmitter FFE summary

Tap coefficient #	0	1	2
Max current (mA)	7.5	30	15
Relative max (%)	25	100	50
DAC resolution (bits)	6	8	7
Tap allocation	Precursor	Main tap	Postcursor

Figure 3.9 illustrates a simplified generic FFE example. A simplified half-rate FFE architecture is shown in Fig. 3.9a, and is representative of the circuit in Fig. 3.8. This example explores a two-tap FFE case (i.e., *FFE Mode Select* set to "FFE2" in the *Transmit Driver Mode Control Register* as described in Table 2.6). The cursor and postcursor taps are set at +0.8 and − 0.2, respectively. Figure 3.9b illustrates the various signal waveforms and the resulting FFE output. The *Serial Data* or $x(t)$ signal in Fig. 3.9b is the serial data input to the FFE circuit. The $x(t–T)$ and $–x(t–T)$ signals are noninverted and inverted polarities of the original signal delayed by one-bit time. Finally, the *Tx OUT* or $y(t)$ signal is the Tx Output waveform, and can be expressed by the following mathematical summation:

$$\text{Tx Out} = y(t) = 0.8\, x(t) - 0.2\, x(t - T), \tag{3.1}$$

where the signals $y(t)$, $x(t)$, and $x(t–T)$ were defined above, and T is one-bit time (or 1 UI).

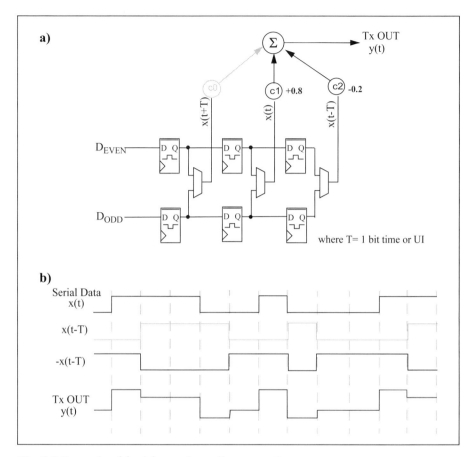

Fig. 3.9 Example of feed forward equalizer operation

This example clearly illustrates how postcursor equalization affects the output waveform. Each time the input waveform transitions up or down, the instantaneous output increases in amplitude to a peak value of unity (the sum of $0.8 + 0.2$), producing a larger amplitude for one-bit time. The resulting waveform has increased high frequency content, which compensates for the high frequency loss of the channel.

To illustrate the performance advantage of using FFE in a system, before and after measurements were made on a 20-in. FR4 lossy backplane with approximately 10 dB of loss at 3.2 GHz (6.4 Gbps example). Figure 3.10 depicts two eye diagrams from real hardware measurements. Eye diagrams were introduced in Sect. 1.2.3, and are a common method of illustrating transmit and receive signals in serial links.

In Fig. 3.10a, there is a substantial amount of eye closure due to high frequency loss of the channel. The eye is barely open, and as a result the channel is operating with very little margin.

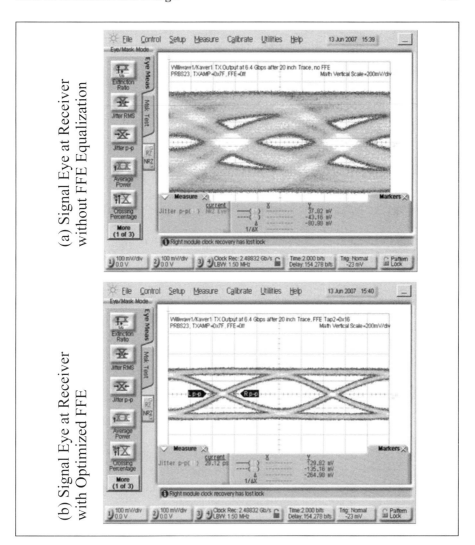

Fig. 3.10 Feed forward equalizer measured results

Figure 3.10b shows the behavior of the system when the FFE is optimally configured. The operating margin of this link is substantially improved as seen by the eye diagram. Note that the outer edge of the eye amplitude has been reduced in response to the subtractive nature of the FFE. In fact, while extensive FFE equalization can be applied to very lossy channels, eventually the receiver sensitivity limits system performance as the FFE optimized eye shrinks vertically. Therefore, there exists a finite amount of FFE equalization that can be employed. This limit is related to the receiver sensitivity and maximum transmitter launch voltage.

3.2.2 Serializer Operation

Serialization of the even and odd data streams from Fig. 3.8 is illustrated in Fig. 3.11. A simplified schematic of the 2:1 Serializer is shown in Fig. 3.11a. This simple tree structure has been successfully used at speeds over 40 Gbps in CMOS technologies and over 132 Gbps in SiGe technologies. The key design constraint that enables using this architecture for high baud rates is as follows: the *Tsq* propagation delay of the multiplexor (from the select input to the multiplexor output) must be less than the *Tcq* propagation delay of the latch (from the clock input to the latch output).

This is illustrated in Fig. 3.11b. The two bits of parallel data, D_{EVEN} and D_{ODD}, are assumed to be time-aligned into the serializer and are synchronized to the half-rate C2 clock. The first two latches capture the parallel D_{EVEN} and D_{ODD} signals, creating De and Do outputs on the rising edge of the C2clk signal. The Do′ signal is generated by resampling the Do signal on the falling edge of the C2 clock. These two signals are skewed by 1 UI and provide the inputs to the 2:1 MUX. The C2 clock controls the select input of this MUX such that when the De input is selected when the clock is low, and Do′ is selected when the clock is high. If *Tsq* < *Tcq*, then the multiplexor always selects a stable input signal, resulting in clean, glitch-free operation. This ping-pong action is illustrated in Fig. 3.11b.

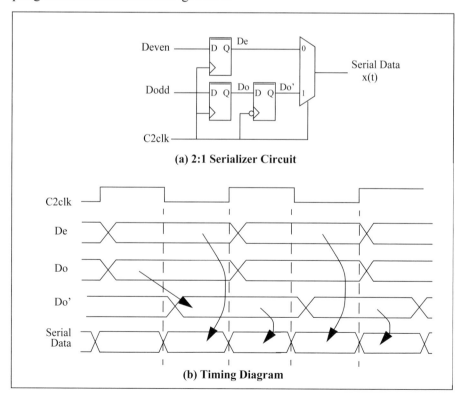

(a) 2:1 Serializer Circuit

(b) Timing Diagram

Fig. 3.11 Detailed 2:1 serializer stage operation

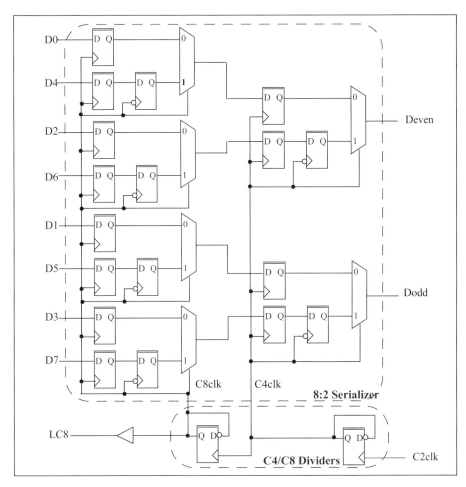

Fig. 3.12 Detailed 8:2 serializer and clock divider stage operation

Working backward from the output serialization stage, the input to the 8:2 serializer stage consists of eight time-aligned parallel bits, and the output consists of the D_{EVEN} and D_{ODD} half-rate serial streams feeding the FFE shift register. The 8:2 serializer is implemented using a cascade of four 2:1 serializers feeding two 2:1 stages. This circuit is illustrated in Fig. 3.12, along with associated C4 and C8 clock dividers. The circuit topology and timing constraints for the multiplexor stages in this circuit are similar to that of the 2:1 serializer in Fig. 3.11a. (Not shown in the figure are various clock buffers required to drive the heavy clock loading on the C8 and C4 clocks.)

The LC8 clock in Fig. 3.12 is a buffered (single-ended CMOS) signal that is used by the synthesized logic macro to clock the transmit data (D0–D7). The timing specification across this interface is critical. For the HSS EX10 core running at 10.0 Gbps, the C8 clock operates at 1.25 GHz, and the design of this interface must assure proper setup and hold timing is met.

Continuing to move backward in Fig. 3.8, the Serializer Logic macro provides the initial stages of serialization. The HSS EX10 core supports various transmit data bus widths as selected by the *Parallel Data Bus Width* bits of the *Transmit Configuration Mode Register* described in Table 2.6. The Serializer logic performs the appropriate multiplexing to output 8-bit parallel data regardless of the selected data width. There are many logic configurations which achieve this function; the most commonly used circuit is a shift register which loads parallel data of arbitrary width, and shifts data by 8 bits at a time as data is clocked to the 8:2 multiplexor stage.

In addition to serialization, the Serializer logic macro also provides the support for Pseudorandom Bit Sequence (PRBS) pattern generation. Capabilities of this circuit were described in detail in Sect. 2.2.7.

3.3 Receiver Slice

Fig. 3.13 illustrates a simplified version of the HSS EX10 receiver slice. Note that the DFE feedback paths are not shown to simplify the diagram. As with the transmitter slice, the receiver slice is implemented using a mixed signal approach, combining custom high-speed CML and static CMOS circuitry with a synthesized standard cell digital logic macro. Starting on the left side of the figure, serial data is received on the RxIP/N inputs. Both AC and DC coupled termination options are supported; the *CMVbias* circuit provides bias options corresponding to each coupling option. The split termination network provides a half amplitude signal to the input of the *variable gain amplifier* (VGA), while the *Input Offset Compensation* block adjusts this network to cancel input referred common offset voltage at the receiver input.

In a DFE system, the information about how to optimally set the DFE taps is contained in the amplitude information of the signal. As such, the entire front end of the receiver must operate in its linear range and not clip the signal. When channel losses in the path from the transmitter to the receiver are small, the input signal amplitude at the receiver may be large, resulting in clipping unless there is a reduction in signal swing. The VGA block performs this function and the split termination provides a 1/2 amplitude (− 6dB) attenuation to help maintain linearity.

The VGA is used to control the amplitude of the input and drives a programmable shunt peaking amplifier. This Peaking Amp in turn drives three sets of summer circuits in the DFE and CDR loops. The top two summer circuits drive the DFE data and amplitude paths, while the lower summer drives the DFE edge macro which provides the input into the CDR macro.

The DFE macro can be logically broken into two distinct functions. The first of these functions (the DFE block in Fig. 3.13) moves the decision threshold of the sampling latch around at the baud rate. The history of the previously sampled bits is used to control this decision threshold and there are 2^n possible levels, where n is the number of DFE taps. For example a five-tap DFE dynamically modifies the decision threshold in one of 32 possible combinations based on the history of previous bits sampled.

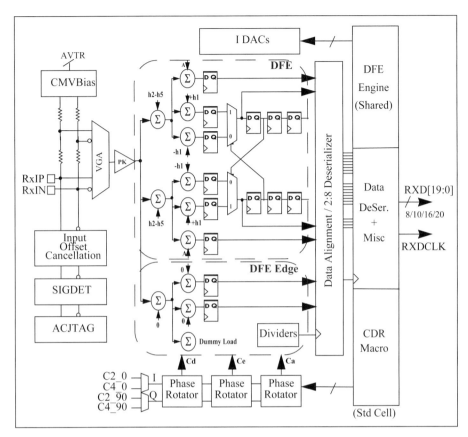

Fig. 3.13 High-level receiver block diagram

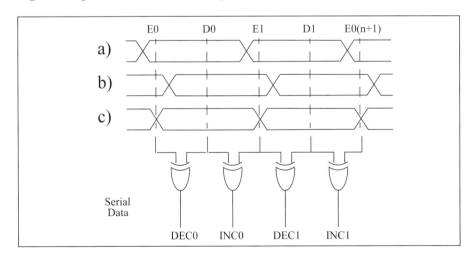

Fig. 3.14 Half-rate bang–bang phase detector operation

The second DFE function is to determine the H1 to H5 tap coefficient values, which is accomplished via an algorithm in the DFE Engine (the DFE Engine block in Fig. 3.13).

Once the data has been equalized and captured, the data/edge/amplitude samples are aligned and deserialized in the *Data Alignment/2:8 Deserializer* block. This data is further processed in digital form by the three subfunctions in the synthesized logic macro: the *CDR Macro*, the *Data Deserializer*, and the *DFE Engine* functions. The data/edge/amplitude sample latches are clocked by three independent phase rotators. These phase rotators use a set of I/Q (0/90°) clocks and provide a digitally controlled phase interpolation function, producing 64-step granularity across a 2 UI range. This corresponds to a net time resolution of the phase rotator of 32 steps per UI.

Other blocks shown in Fig. 3.13 include the *SigDet* block which implements the signal detection features (see Sect. 2.3.4) and the *AC JTAG* block which provides low-speed board level testing (see Sect. 2.3.6).

3.3.1 Clock and Data Recovery (CDR) Operation

The CDR macro uses a twofold oversampling scheme, capturing four samples across two bits (two edge and two data samples). In Fig. 3.13, the data samples are captured in the *DFE* block and the edge samples are captured in the *DFE Edge* block. Note that the *DFE* block and the *DFE Edge* block use identical summers to maintain timing alignment between the two functions; this is critical for proper data centering. As illustrated in Fig. 3.14, a simplified *bang–bang phase detector* approach can be visualized. In practice, the HSS EX10 core uses a 16-way bang–bang phase detector, with 16 data and 16 edge samples implemented in the CDR logic macro. Digital filtering in the CDR logic results in a loop bandwidth of approximately 1/1,000th of the baud rate (depending upon incoming transition density).

There are four sets of latches in the bang–bang phase detector: the D0 and D1 data latches (also used extensively in the *DFE* block) and the E0 and E1 edge latches (specifically in the *DFE Edge* block). Samples are latched in these latches using clocks which are sequenced as follows: E0 is clocked first, followed by D0, E1, and finally D1. This sequencing is shown in Fig. 3.14. As shown in the figure, XOR gates are used to detect differences in phase. When a difference exists between adjacent values, the XOR output is asserted.

The simplified bang–bang phase detector scheme is examined for three cases of potential phase alignment in Fig. 3.14. In Fig. 3.14a, both the (E0, D0), and the (E1, D1) latches are sampling the same portion of the waveform. Since the digital values are the same, the DEC0 and DEC1 signals are not asserted. Likewise the (D0, E1), and the (D1, E0) latches are sampling differing logic levels, and therefore the INC0 and INC1 signals are asserted. The INC signal causes the phase of all four clocks to shift to the left. The net effect of this is to move the E0 and E1 clocks closer to the data transitions.

Figure 3.14b shows the opposite situation where the data edges are between the E0/D0 and the E1/D1 samples. In this case, the DEC0 and DEC1 signals are asserted, and the clock phases shift to the right. The effect again is that the E0 and E1 clocks are moved toward the data transitions.

Finally, Fig. 3.14c shows the case where the signal is in lock. In the absence of noise, the lock condition for the phase detector is characterized by thrashing back and forth one step around the ideal sample phase.

The digital INC/DEC signals are then used to control the sampling phase of the clocks driving the Data/Edge latches. There are several popular schemes to digitally control the sampling phase, including phase rotators, PLLs, delay locked loops, and digital oversampling schemes. The HSS EX10 core utilizes a phase rotator- based CDR.

A phase rotator is essentially a digital mixer circuit which accepts two differential clocks with 90° separation, often referred to as In-phase and Quadrature-phase (I & Q). Shown in Fig. 3.15, the half-rate (C2) input clocks *CK_I* and *CK_Q* are mixed in various proportion to control the output phase of the rotator. The linearity (phase out vs. digital phase control in) or step size of a phase rotator is a strong function of the shape of the input clock waveforms. Optimal performance is achieved when the input signals are very close to sinusoidal waveforms. To that end, the *SLEWBUF* section controls the slew rate of the input signals and produces a sine wave on the *SCK_I* and *SCK_Q* outputs. Two 15-value IDACs (typically thermometer encoded) control the amount of *I* vs. *Q* contribution to the output, and the *POL* signals select the quadrant in which the output operates. The net result is that the digital control word INT + POL can produce any one of 64 phases at the ZP/ZN outputs.

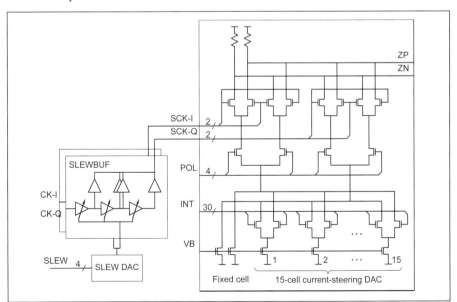

Fig. 3.15 Phase rotator conceptual diagram

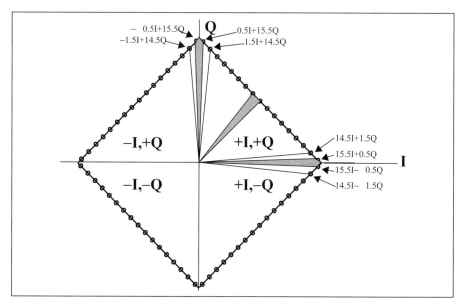

Fig. 3.16 Four quadrant phase rotator diagram

Fig. 3.16 illustrates a phase rotator quadrant diagram, showing the 64 possible combinations of phase. In order to track a frequency offset (in ppm) between the incoming signal stream and the local C2 clocks, the digital control word is periodically adjusted in a given direction. The greater the difference between the signal baud rate and the local C2 clock frequency, the more often the phase must be adjusted.

3.3.2 Decision Feedback Equalizer (DFE) Architectures

The heart of the DFE-based core is the high-speed analog function that varies the sample latch thresholds at the baud rate. Figure 3.17 illustrates how this DFE threshold movement improves the BER.

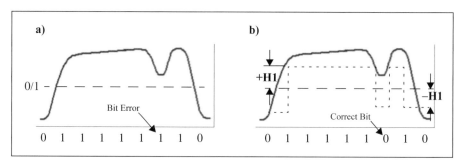

Fig. 3.17 Decision feedback equalizer threshold adjustment

Fig. 3.17a shows a differential signal which has a significant amount of intersymbol interference (ISI). The binary sequence transmitted is "01111010." Assuming a conventional latch with a 1/0 decision threshold at zero, the received sequence for this signal with ISI would be "01111110." Since the input signal never crosses the 1/0 threshold, the sample latch mistakenly captures a "1" instead of the runt "0" transition.

Now consider a single tap (H1) DFE. Per convention, the H1 tap operates on the last bit received, H2 on the bit received before this, and so forth. The latch sample threshold is adjusted as shown in Fig. 3.17b based on the value of the previous received bit. If the previous bit is a "1," the threshold is moved up by the value of H1; conversely if the previous bit is "0" then the threshold is adjusted down by the value of H1. If H1 is carefully chosen, the receiver can properly discern the single "0" transition. The key is determining the correct value for H1.

A "direct" feedback architecture implementation of this one-tap DFE is depicted in Fig. 3.18. This architecture uses a full-rate clock. The direct feedback approach is conceptually simple: Based on the digital value of the sampled data, the sampled digital value is multiplied by the H1 coefficient and fed back to the input of the sample latch. If the sampled data is "1," then the threshold of the latch for the next bit is increased by H1; if the sampled data is "0," then the threshold for the next bit is decreased by H1.

Timing requirements for the direct implementation are challenging. The loop indicated by the arrow in Fig. 3.18 must meet the following equation:

$$Tcq + Tsum + Tsu < 1 \text{ UI}, \qquad (3.2)$$

where Tcq is the latch propagation delay (clock input to the output), $Tsum$ is the propagation delay of the summer circuit, and Tsu is setup time of the latch. At 11.1 Gbps the bit time is only ~90 ps. Given an analog value is being added to the input waveform, the H1 feedback must settle to within 2–5% of the final analog value within this time constraint. This constraint is difficult to meet at higher baud rates. Although the direct feedback architecture is commonly used for slower baud rates, other architectures are used at higher baud rates.

The "Speculative" (or "Unrolled") feedback architecture implementation of a one-tap DFE is depicted in Fig. 3.19. This technique is a little more costly

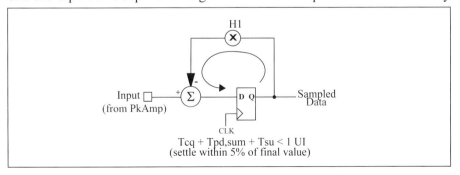

Fig. 3.18 Full-rate "direct" decision feedback equalizer

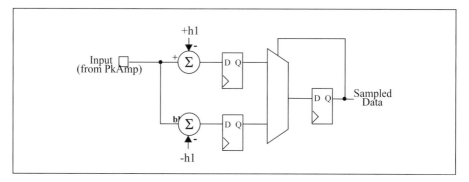

Fig. 3.19 Full-rate "speculative" decision feedback equalizer

in terms of hardware area and power, but the difficult timing of the direct approach is avoided. The speculative architecture duplicates the summers and latches. In the direct approach, the digital value of the previous received bit is fed back into the first summer. The speculative approach instantiates two latches and two summers. One summer uses a static value of +H1, while the other uses a static value of –H1. A digital multiplexor selects the output of one or the other sample latch based on whether the previous received bit was a "0" or a "1." As a result, the analog settling time and the propagation delay of the summer circuit is removed from the critical timing path.

The HSS EX10 core uses a hybrid approach combining both speculative and direct DFE tap feedback. Figure 3.20 illustrates the high level concept of the hybrid approach. As was the case in Fig. 3.13, the feedback tap paths from the

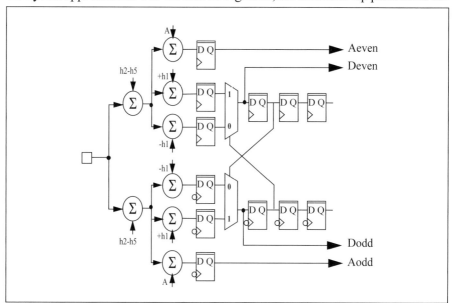

Fig. 3.20 Half-rate hybrid "speculative-direct" DFE

even/odd data shift register are not shown to simplify the diagram. In this figure, the H1 DFE tap is calculated speculatively, removing the baud spaced timing constraint. The H2–H5 taps use direct feedback. Since the H2 tap feedback is permitted to take up to two bit times, a half-rate clock is used. This substantially reduces the power dissipation and simplifies the design at the cost of doubling the number of sampling latches. Note this figure is simplified and does not show all of the crossconnections and feedback connections. The primary outputs of this DFE are two data bits: D_{even} and D_{odd}, and two amplitude A_{even} and A_{odd} signals. These signals are driven to the *Data Alignment/2:8 Deserializer* macro shown in Fig. 3.13.

Figure 3.20 also shows the A_{even} and A_{odd} latch paths in parallel to the *H1* speculation latches. The A_{even} and A_{odd} latches are used to determine the magnitude of the *H* coefficients. The DFE tap weights (*H1–H5*) are computed through amplitude and data sample correlation using a sign-error driven algorithm. Tap weights are optimized to cancel postcursors of the signal as described generally in Sect. 8.4.1.3. More detailed descriptions of equalization circuits may be found in [6–10].

3.3.3 Data Alignment and Deserialization

Once the data has been captured in the DFE sample latches, the amplitude latches, and the edge latches, the signals must be aligned and deserialized. Figure 3.21 illustrates the 1:2 DEMUX macro that is the fundamental building block of the deserialization process. As with the 2:1 MUX previously described, this 1:2 DEMUX can be cascaded to realize larger deserialization functions. The conceptual circuit topology is shown in Fig. 3.21a. Serial data is applied to two latches, each triggering on opposite edges of the clock. In order to time-align both serial streams, a third latch is added to recapture the data on a common clock edge. Figure 3.21b shows the timing diagram of the function. The net result of the deserialization is that two time-aligned half-rate data streams are created from the single full-rate data stream. The *Data Alignment/2:8 Deserializer* macro shown in Fig. 3.13 produces eight data bits, eight edge bits, four amplitude bits, and a C8 clock synchronous with the inbound serial stream.

A subsequent stage of the *Data Alignment/2:8 Deserializer* macro performs additional deserialization. As was the case for the transmitter, the HSS EX10 core supports various receive data bus widths as selected by the *Parallel Data Bus Width* bits of the *Receive Configuration Mode Register* described in Table 2.7. Deserialization expands the 8-bit data bus to the appropriate data width.

In addition to deserialization, the *Data Alignment/2:8 Deserializer* macro also provides the support for Pseudorandom Bit Sequence (PRBS) pattern checker. Capabilities of this circuit were described in detail in Sect. 2.3.8.

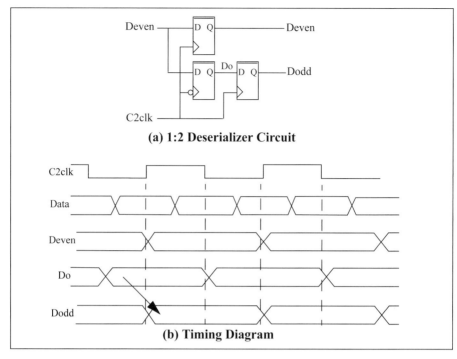

(a) 1:2 Deserializer Circuit

(b) Timing Diagram

Fig. 3.21 Detailed 1:2 deserializer operation

3.4 References and Additional Reading

The following reading is recommended for more in depth information regarding PLL circuit architectures and design:

1. "Phase-Locked Loops: Design, Simulation and Applications", R. E. Best, McGraw-Hill, New York, 2003.

2. "Design of Integrated Circuits for Optical Communication", B. Razavi, McGraw-Hill, New York, 2003.

3. "Design of Analog CMOS Integrated Circuits", B. Razavi, McGraw-Hill, New York, 2001.

4. "Charge-Pump Phase Locked Loops", F. M. Gardner, IEEE Trans. Commun., vol. COM-28, 1849–1858, 1980.

5. "A Comparison of MOS Varactors in Fully Integrated CMOS LC VCO's at 5 and 7 GHz", H. Ainspan, et. al., European Solid State Circuits Conference (ESSCIRC), Sept. 2000.

The following reading is recommended for more in depth information regarding equalizer and CDR design:

6. "Equalization and Clock Recovery for a 2.5-10Gb/s 2-PAM/4-PAM Backplane Transceiver Cell", J.L. Zerbe, et. al., IEEE J. Solid-State Circuits, Dec. 2003.

7. "An 8Gb/s Source-Synchronous I/O Link with Adaptive Receiver Equalization, Offset Cancellation and Clock Deskew", J. Jaussi, et. al., IEEE Solid State Circuits Conference (ISSCC), Feb. 2004.

8. "A 6.25Gb/s Binary Adaptive, DFE with First Post Cursor Tap Cancellation for Serial Backplane Communications", R. Payne, et. al., ISSCC Digest of Technical Papers, Feb. 2005.

9. "A 6.4Gb/s CMOS SerDes Core with Feedforward and Decision-Feedback Equalization", M.A. Sorna, et. al., ISSCC Digest of Technical Papers, Feb. 2005.

10. "A 10Gb/s 5-tap DFE / 4-tap FFE Transceiver in 90 nm CMOS", M. Meghelli, et. al., ISSCC Digest of Technical Papers, Feb. 2006.

3.5 Exercises

1. If the C1 clock in Fig. 3.2 is operating at 11 GHz, what are the frequencies of the C2_0, C2_90, C4_0, C4_90, C8, and C16 clocks?

2. For the various combinations of transmitter and receiver configurations described below, indicate which of output clocks in Fig. 3.2 can be powered down to save power:

   (a) All transmitter and receiver slices operating in full-rate mode

   (b) All transmitter and receiver slices operating in half-rate mode

   (c) Some transmitter and receiver slices operating in half-rate mode, and some in quarter-rate mode

3. Draw a timing diagram illustrating the C1 clock input and the four output phases of the C2 clock for the circuit shown in Fig. 3.4.

4. Cascade two of the circuits in Fig. 3.4 to produce C2 and C4 clocks.

5. Refer to the graph in Fig. 3.6:

   (a) Draw an alternative line for *Fdesired* which is half way between the line shown and the bottom of the graph. Which bands can lock to this frequency? Which band should be chosen by the Coarse Calibration Algorithm?

   (b) Draw an alternative line for *Fdesired* which is half way between the line shown and the top of the graph. Which bands can lock to this frequency? Which band should be chosen by the Coarse Calibration Algorithm?

6. In Fig. 3.6, assume a value for *Fdesired* which is just below the top of the graph. Can the PLL lock to this frequency? If so, what band is used?

7. The PLL lock detect circuit for the HSS EX10 is shown in Fig. 3.7. If the timeout period were determined using 16-bit counters, this would change the frequency tolerance at which the lock condition would be detected. Calculate this frequency tolerance in parts per million (ppm).

8. Repeat Exercise 7 assuming the timeout period is determined by 5-bit counters.

9. Referring to Fig. 3.9a, the multiplexor in each FFE tap segment selects between either the D_{EVEN} or D_{ODD} bit stream. However, note this must be done in such a manner that the $x(t+T)$ data bit is the next bit following the $x(t)$ bit, and the $x(t–T)$ data bit is the bit prior to the $x(t)$ bit. Indicate the correct C2 clock polarities into the flip-flops and mux select input to achieve this.

10. Draw a timing diagram illustrating the operation of Fig. 3.12.

11. Assume that the circuit in Fig. 3.9a is provisioned with the FFE coefficients: $x(t+T) = 0$, $x(t) = 0.7$, and $x(t–T) = +0.3$. Draw a timing diagram for this case similar to that in Fig. 3.9b.

12. Repeat Exercise 11 assuming the circuit in Fig. 3.9 is provisioned with the FFE coefficients: $x(t+T) = –0.2$, $x(t) = 0.8$, and $x(t–T) = 0$.

13. Refer to Fig. 3.14. As noted in the description in the text for this figure, steady-state operation of a half–rate bang-bang phase detector circuit is characterized by thrashing between timing case (a) and (b) in the figure. Draw the sample points for eight consecutive data bits illustrating this thrashing. Also show the corresponding values of the DEC0, INC0, DEC1, and INC1 control outputs.

14. In Fig. 3.17a, the bit stream "01111010" is incorrectly sampled as "01111110". Explain why the incorrectly sampled "0" bit does not transition below the 0/1 sample threshold in this figure. (You may wish to peek at Chap. 8 to answer this question.)

15. In Fig. 3.17b the bit stream is sampled correctly for the value of H1 shown:

    (a) If $H1$ is half of the amplitude shown in the figure, what are the sampled values of the bits?

    (b) If $H1$ is twice the amplitude shown in the figure, what are the sampled values of the bits?

16. For the bit stream shown in Fig. 3.17, and the circuit shown in Fig. 3.19, draw a timing diagram showing the sampled bits on the inputs and output of the multiplexor as well as the *sampled data* output of the circuit.

Chapter 4
Protocol Logic and Specifications

HSS devices are frequently used as part of the implementation of a standardized interface for a network protocol. The specification for the network protocol generally includes several layers of functionality. As described in this chapter, the HSS device implements only a portion of the *physical layer* of the network protocol. Additional logic is required to implement the remainder of the physical layer, as well as portions of the *data link layer*. For purposes of this text, this logic is referred to as *protocol logic*; the function of this logic depends on the applicable network protocol standard.

This chapter covers a broad range of topics generally related to protocol specifications and the implementation of protocol logic. Specific network protocol standards are discussed in Chap. 5. This chapter begins by describing construction of network protocol standards, including protocol layers, methods of specifying serial signals, and basic concepts regarding clocking methods and data organization. Next, typical functions implemented in protocol logic are discussed, including determining the bit/byte order of transmission; encoding and/or scrambling data for transmission (and decoding/descrambling at the receiver); error detection and/or correction; elastic FIFOs to retime data to local clock domains; and bit alignment and deskew functions in the receiver. These topics are covered at a level which is sufficient for the reader to understand basic approaches for logic design of these functions; exercises at the end of the chapter enhance this understanding.

4.1 Protocol Specifications

This section covers a number of topics related to network protocol standards, including protocol layers, methods of specifying serial data signals, and basic concepts regarding clocking methods and data organization.

4.1.1 Protocol Layers

The TCP/IP model, or Internet Reference Model, was developed by the Internet Engineering Task Force (IETF). This model partitions software and hardware functions necessary to communicate on a network into several layers, as shown in Fig. 4.1. Most serial link protocols loosely follow this model.

The *application layer* processes data in an application-specific format and encapsulates this into the common format used by the *transport layer*. The *transport layer* is responsible for end-to-end message transfer independent of the underlying network, including error control, fragmentation, and flow control. The *network layer* is responsible for packet routing; intermediate nodes in the network perform network layer processing as needed to route the packet from its source to its target. While higher layers are agnostic as to the underlying protocol, the *data link layer* is responsible for formatting the data into a specific format for transmission, and decoding the format of the received data. Finally, the *physical layer* transmits/receives data to/from the media.

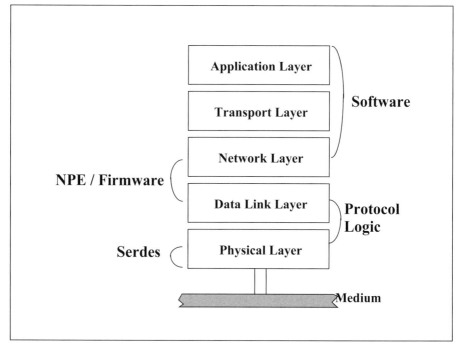

Fig. 4.1 TCP/IP reference model

Figure 4.1 also illustrates the common implementation of the various protocol layers. The physical layer is implemented, or in some cases is partially implemented, by the HSS device. Additional *protocol logic* in the chip implements the remainder of the physical layer and some or all of the data link layer, and is the subject of this chapter. Portions of the data link and network layers may be implemented by *Network Processing Elements* (NPEs), firmware, and *Content Addressable Memories* (CAMs) in network router or enterprise systems, or by software in low-end systems. Higher layers of the protocol are almost always implemented in software.

4.1.2 Serial Data Specifications

Serial data interface standards specify that the serial data must meet certain eye amplitude and eye width requirements at normative compliance points of the interface in order to claim compliance. Details of the above statement are elaborated upon by the subsections below.

4.1.2.1 Compliance Points

Serial data specifications are imposed at points in the system which are defined by the interface standard as *compliance points*. Various standards choose different points in the system to define as compliance points. In general, the choice of compliance points tends to align with the needs of the industry serviced by the standard.

When evaluating a standard (or characterizing Serdes hardware compliance with a standard), it is important to note the compliance points defined by the standard. Compliance points may be at the chip pins or may be at intermediate points within the channel such as circuit board connectors. The reference channel may include an optical segment, in which case additional compliance points would be defined for the optical signal.

The choice of compliance points affects how interface specifications translate to HSS specifications. For example, if the Serdes transmitter generates 10 ps of jitter, and the specification requires no more than 10 ps of jitter measured at a compliance point defined as a connector pin, then it is unlikely that the specification can be met. The circuit board traces, vias, and connectors add additional jitter, resulting in measured jitter at the connector exceeding 10 ps. In such cases a budget amount of jitter must be assigned to each component (the transmitter device and the circuit board) such that the total is no greater than 10 ps at the compliance point.

It is also necessary to consider the measurement conditions specified in the standard. Often, measurements are specified for ideal (or nearly ideal) conditions. The Serdes hardware must produce signals within the specified range under the conditions specified. Signals may vary from these ranges in real systems where conditions such as termination impedances and impedance discontinuities are less than ideal.

4.1.2.2 Normative and Informative Specifications

When reading serial data interface standards, it is important to note *normative* elements of the standard vs. *informative* elements of the standard. The serial link can be viewed as a collection of the following three elements: the Serdes Transmitter device (including the package), the channel, and the Serdes Receiver device (including the package). Any interface standard must impose normative requirements on at least two of the three elements of the serial link in order to ensure interoperability between components. (It is possible to provide normative requirements for all three elements, but this requires substantially more analysis to ensure the requirements are self-consistent.)

There is a *normative* specification for the element when requirements are imposed which must be met in order to claim compliance. For example, the standard may provide a normative specification of signal characteristics at the Serdes Transmitter, and a normative specification of signal characteristics that must be tolerated by the Serdes Receiver. Defining the channel is then left as an exercise to the system designer. In this case, the channel specification is an *informative* specification. The interface standard may provide informative specifications of channel characteristics as guidance, or may leave it as an exercise to the user to derive requirements from the transmitter and receiver specifications. (The channel must be capable of propagating the worst case signal produced by the Serdes transmitter to the Serdes receiver without distorting the signal beyond the worst case signal allowed at the receiver.)

Informative specifications do not need to be verified in order to claim compliance with the standard.

Note that an increasingly popular alternative at higher baud rates is to define normative specifications for the Serdes transmitter and the channel, and provide informative specifications for the signal at the Serdes receiver.

4.1.2.3 Serial Data Amplitude and Eye Width

One method of defining serial data characteristics at the compliance point is through the use of an *eye diagram*, an example of which is shown in Fig. 4.2. The example has parameterized the key geometric features of this diagram. The data signal is measured over a period of time and the signal trace is mapped onto the eye diagram. If the data signal strays into the gray zones, the data signal is not in compliance.

The eye amplitude in this diagram is limited by the maximum signal amplitude defined by the Y2 parameter, and the minimum signal amplitude defined by the Y1 parameter. The Y2 parameter should never be violated. The Y1 parameter applies to a portion of the bit time as defined by parameters on the *X*-axis of the graph; the signal is less than Y1 while it is switching. The eye width in this diagram is limited by the maximum jitter defined by the X1 parameter. The X2 parameter has the affect of placing a maximum limit on the rise and fall time of the signal. In some cases the X2 parameter may be 0.5 UI, at which point the shape of central gray zone degenerates to a diamond shape.

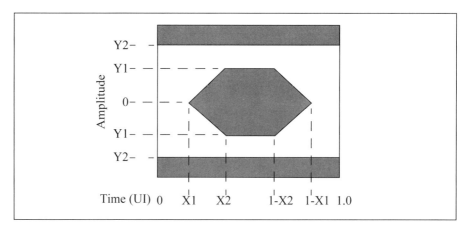

Fig. 4.2 Example of a transmitter eye mask

The drawback of using eye diagrams to specify signal characteristics is that many standards dictate use of an FFE to provide preemphasis or deemphasis of the signal. The FFE induces a deliberate distortion of the signal at the transmitter in order to partially cancel the distortion effects of the channel and provide a better signal at the receiver. The distorted signal is likely to violate the eye diagram on both the X- and Y-axis. The standard may handle this by specifying the eye mask with the FFE disabled, or by adjusting the eye mask based on filter settings.

Eye diagrams (or equivalent specifications of signal amplitude and jitter) may be used to specify the signal at the transmitter output (or some other compliance point near the transmitter). They may also be used to specify the signal at the receiver input (or some other compliance point near the receiver). In such cases, they may be used either as a normative compliance specification for a channel or for a Serdes receiver. In one case, the eye diagram at the receiver represents the worst case signal output of a compliant channel when driven by a compliant Serdes transmitter. In the other case, the eye diagram represents the worst case signal input that a compliant Serdes receiver must tolerate.

Higher speed interface standards may expect the eye at the receiver input to be closed, in which case an eye diagram cannot be specified at this point. Such standards define a reference receiver which often includes a DFE. The DFE circuit is a negative contributor to the jitter budget, in effect resulting in a more open eye at the output of the DFE circuit than at the input. Of course, the output of the DFE circuit is beyond the point where analog-to-digital conversion of the signal has been performed. The notion of an analog "eye" at this point is really a virtual concept; this eye is not a measurable analog signal. Compliance is determined through mathematical calculation of the signal eye at this point, and comparison to the eye mask specified by the standard.

4.1.2.4 Receiver Signal Detect Function

It is sometimes desirable for the receiver to detect the condition where the transmitter is not sending a signal (either because it is disabled, powered down, or unplugged). Under these conditions the loss of signal condition may be used to avoid trying to receive and process any noise on the serial data input to the Serdes receiver. Serdes receivers therefore generally provide a signal detect feature which detects when the amplitude of the received signal falls below a threshold level for a sustained period of time.

Depending on the interface standard, one or more of the following parameters may be specified using various terminology:

Minimum signal amplitude. Signal amplitude above which the received signal must be detected and correctly received.

Maximum loss of signal amplitude. Signal amplitude below which the receiver must detect a "Loss of Signal" condition.

Loss of signal response time. Length of time allowed to flag "Loss of Signal" or "Signal Detected" conditions.

These parameters place specifications on the design of the Serdes signal detection circuit. The first two of these parameters determine a range for the signal detection threshold. The threshold of the signal detection circuit cannot be higher than the specification for minimum signal amplitude, and cannot be lower than the specification for the maximum loss of signal amplitude. The response time specification determines the length of time over which the signal detection circuit monitors the signal before changing state to indicate that the signal has been lost or has returned.

Various interface standards specify differing thresholds and response times. For this reason, it is common for the Serdes receiver to implement a programmable signal detection threshold and response time.

4.1.2.5 Jitter, Wander, and Skew

Jitter is the deviation from the ideal timing of an event at the mean amplitude of the signal population. Low frequency deviations are tracked by the clock recovery circuit, and do not directly affect the timing allocations within a bit interval. Jitter that is not tracked by the clock recovery circuit directly affects the timing allocations in a bit interval.

Common usage of the term "jitter" refers only to the portion of the jitter that is not tracked by the clock recovery circuit. However, as illustrated in Fig. 4.3, the jitter frequency spectrum can be visualized as extending from DC to well above the baud rate of the clock recovery circuit. In this example, the clock recovery circuit cannot track jitter at a sinusoidal frequency above the baud rate divided by 1667. Jitter in this range can be viewed as phase variations in a signal (clock or data) after filtering the phase with a single-pole high-pass filter with the $-3\,\mathrm{dB}$ point at the jitter corner frequency.

Some standards have used the terminology *skew* and *wander* to refer to jitter at frequencies below the jitter corner frequency [1]. These parameters do not have any affect on data reception on a given serial data link. However, when multiple serial data links are used to implement an *n*-bit wide port, differences in the skew and wander among the various links affect the design of downstream protocol logic.

Skew is the constant portion of the difference in the arrival time between the data of any two signals. This can be visualized as jitter on one signal relative to the reference signal at DC (0 Hz). Skew results from differences in routing trace lengths of clocks to the Serdes cores, differences in routing trace lengths for clock distribution within the Serdes cores, and differences in routing of data in the package and circuit boards. At higher baud rates, skew of several UI is possible. If the protocol requires the outputs of multiple Serdes receivers to be aligned, then the deskew function in the protocol logic must have sufficient range to adjust the signal. Fig. 4.4 illustrates skew using a timing diagram. The constant portion of the difference in arrival times of Lane Y relative to Lane X corresponds to the skew.

Wander is the peak to peak variation in the phase of a signal (clock or data) after filtering the phase with a single-pole low-pass filter with the $-3\,\mathrm{db}$ point at the wander corner frequency in Fig. 4.3. Wander does not include skew. Wander results from manufacturing process variation, voltage, and temperature differences between the circuitry for the two links being compared.

Temperature variations can result in phase drift over time which contributes to wander. Voltage variations can result in phase drift at frequencies corresponding to the switching frequencies of the power supply. If the protocol requires the outputs of multiple Serdes receivers to be aligned, then the deskew function in the protocol logic must have sufficient range to adjust the signal to compensate for both the skew and maximum amplitude of the wander. Furthermore, FIFOs used to cross clock domains must have sufficient depth to handle the maximum amplitude of the wander without overflow or underflow conditions resulting. Wander is illustrated in the timing diagram in Fig. 4.4. The variable portion of the difference in arrival times of Lane Y relative to Lane X corresponds to the wander.

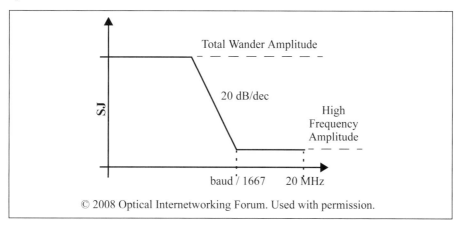

Fig. 4.3 Typical jitter spectrum

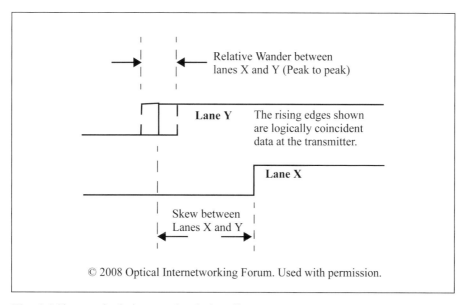

Fig. 4.4 Skew and relative wander timing diagram

4.1.3 Basic Concepts

It is relevant to define a few basic concepts of network protocols regarding clocking methods and data organization before discussing the details of implementing protocol logic functions.

4.1.3.1 Synchronous vs. Plesiosynchronous Clocks

Telecom SONET/SDH networks generally use *synchronous clocking*. Reference clocks are distributed across the entire network so that the transmitter and receiver device on the link are clocked by the same reference clock and are operating at exactly the same frequency as shown in Fig. 4.5. This is true

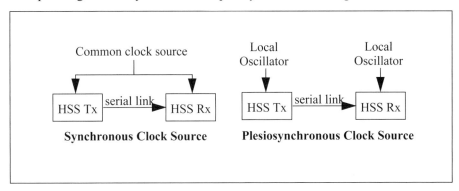

Fig. 4.5 Clocking architectures

even when the transmitter and receiver are not in the same city (or state, or continent). Reference clock distribution does sometimes fail, and locally generated reference clocks are used as backup clock sources. As would be the case for any two clock signals generated by independent oscillators, the locally generated reference clock cannot have exactly the same clock frequency. This frequency tolerance between the two clock sources is generally specified in parts per million (ppm). For telecom systems locally generated reference clocks for key network elements must use high-quality (and expensive) oscillators that have low ppm frequency tolerances. Protocol mechanisms allow bytes to occasionally be dropped or added in order to adjust for any differences in clock frequencies.

Telecom networks also have a significant number of intermediate nodes within a line segment. It would be expensive to distribute the reference clock to each and every repeater node, and therefore these nodes often use the clock recovered from the receive data as a clock source for transmission (called *loop timing*). Such systems use the RXxDCLK output of the HSS receiver to perform any processing of the data so that it is not necessary to add or drop bytes. The system then retransmits data to the next node on the optical ring using a reference clock that is frequency locked to this RXxDCLK frequency. Any node on the SONET/SDH ring may use this method of clocking, although there must be at least one node in the ring that retimes the data to a reference clock. (If there are no nodes using a reference clock then the ring would become a feedback loop and would be inoperable.)

In addition to use in telecom networks, synchronous clocking is also often used for serial links where the transmitter and receiver are in close proximity to each other. If both the transmitter and receiver chip are on the same circuit board, both HSS devices may very well share a common oscillator. Since the clock distribution network for this oscillator is localized and does not require backup sources (unlike telecom applications), these applications may use inexpensive oscillators with larger frequency tolerances than would be used in telecom networks.

Ethernet, Fibre Channel, and other datacom protocols always operate using locally generated reference clocks. These reference clocks are generated using relatively inexpensive oscillators. In such networks, the reference clock for the receiver device on the serial link is always operating at a slightly different frequency than the transmitter device. This is called *plesiosynchronous clocking*, and is also illustrated in Fig. 4.5. Protocols using plesiosynchronous clocking must allow frequent adding and dropping of bytes in order to adjust for differences in clock frequencies. The extent to which bytes may need to be added or dropped depends on the specified frequency tolerance allowed for the clock sources.

4.1.3.2 Packet vs. Continuous Transmission

HSS cores transmit and receive data continuously; it is up to higher layer protocols to determine whether the data is useful or is simply filler which can be discarded. Telecom protocols such as Synchronous Optical NETwork (SONET) and Synchronous Digital Hierarchy (SDH) require continuous transmission of a repetitive data frame containing control bytes and data bytes in defined positions within the frame. Telecom protocols are designed in this manner in order to facilitate time domain multiplexing of data into and out of the frame, and thereby supply a guaranteed bandwidth to each client connection on the link. SONET and SDH transmit 8,192 frames per second; the higher the data baud rate, the more the data in the frame, and the more client connections being multiplexed. The 8-kHz frame rate permits a voice connection to achieve a 4 kHz frequency response using one byte multiplexed into each SONET/SDH frame. The SONET/SDH frame format is explained in more detail in Sect. 5.1.2.

Ethernet, Fibre Channel, and other datacom protocols collect data into packets for transmission rather than relying on continuous transmission of the data. Each packet of data generally includes the following components:

Packet header. May include start of packet delimiter, packet type and routing information, sequence information, and other fields as defined by the protocol.

Data. Hopefully self-explanatory.

Packet trailer.: May include an end of packet delimiter and code words for checking and/or correcting for errors in the packet. Cyclic Redundancy Check (CRC) error checking is commonly used in many protocols.

The packet format for Ethernet is described in Sect. 5.3.2.1, and the packet format for Fibre Channel is described in Sect. 5.4.3.1.

Packet-based protocols send *idle symbols* between packets. These symbols do nothing except fill time, and the receiver may add or drop idle symbols as needed to adjust for the difference in frequency between the received data and the local reference clock. The protocol generally specifies a maximum packet length, a minimum number of idles between packets, and a required frequency tolerance for the local reference clock. These specifications determine the amount of buffering required to ensure a packet can be received without inducing errors due to the frequency mismatch, and ensure that sufficient adjustment can be made between packets to compensate.

An extension of the packet-based approach is to transmit *skip symbols* at regular intervals within the packet which the receiver may additionally add or drop as needed. This allows the frequency tolerance of the local reference clock to be further relaxed (and thereby permitting the use of less expensive oscillators) without having to restrict packet length.

Packet protocols such as Ethernet are *network protocols* where each packet of data has a source which originated the data and a destination to which the data is to be routed. Although the network may consist of Full Duplex links where each node can send and receive data from its neighbor, there is no concept of the packet transmission requiring any particular response that should be tracked at the lower layers of the protocol. (It is possible that the application using the network, such as a web browser, might be expecting a response. However, the network interface is not cognizant of this.) Telecom protocols also fit this description of a network protocol, with routing determined by network management software outside of the protocol, and each payload byte within the frame potentially having different add/drop points in the network.

Other protocols, such as PCI Express, are *transaction protocols*. Some nodes are *master* devices which may originate transactions, and other nodes are *slave* devices which respond to transactions. When a master originates a transaction, it expects a response from the target slave device, and tracks this as part of the protocol implementation.

4.2 Protocol Logic Functions

Topics in this section describe functions typically implemented by protocol logic, including: determining the bit/byte order of transmission; encoding and/or scrambling of data; error detection and/or correction; elastic FIFOs to retime data to local clock domains; and bit alignment and deskew functions.

4.2.1 Bit/Byte Order and Striping/Interleaving

All application protocol standards include an explicit definition of the order in which bits and bytes are transmitted on the serial link. The order in which bits within a byte (or encoded symbol) are transmitted is the *bit order* for the interface. The order in which bytes (or encoded symbols) within a multibyte (or multisymbol) word are transmitted is the *byte order* or *symbol order*.

Some protocols use a single serial data link to transmit data. Protocols requiring higher bandwidth use multiple serial data links to transmit data in a coordinated fashion. Such protocols must allow for *relative skew* (differences in the arrival times of data) between serial data arriving at the receiver inputs. Such protocols require that serial data links be *deskewed* by the receiver to compensate for this relative skew.

Protocols using multiple serial links must allocate the data to be transmitted to specific serial data links. The order and manner in which the data is allocated to the various links is called *striping* or *interleaving*. One method of interleaving is to send one byte on one serial data link, the next byte at the same time on a second serial data link, the third byte at the same time on a third serial data link, and so forth such that n bytes are sent simultaneously on n serial data links. This scheme is called *byte striping*. (If the protocol uses encoded symbols, then each symbol is sent on a separate serial data link. This is still commonly called byte-striping even though the symbols are not bytes.)

An example of byte striping is shown in Fig. 4.6. Each byte of a 32-bit data word is allocated to a separate HSS transmitter. Assuming that bits 31 through 24 are connected in order to TXxD[7:0] of an HSS EX10 core, and that other channels are connected similarly, the least significant bit of each byte is transmitted first. This means that bits 24, 16, 8, and 0 of the 32-bit data word are transmitted simultaneously by the various HSS links, followed by bits 25, 17, 9, and 1, and so forth, with bits 31, 23, 15, and 7 transmitted last.

An alternative is to transmit each n-bits of the data across n serial data links, followed by the next n-bits of the data, and so forth. This scheme is called *bit interleaving*. Bit interleaving has the advantage of requiring less buffering (and less latency) at the receiver to perform deskew. However, some protocols use data encoding to constrain the spectral characteristics of the serial data, and these characteristics cannot be guaranteed when the resulting data is bit interleaved across different serial data links.

An example of bit interleaving is also shown in Fig. 4.6. This example transmits the least significant 4-bit nibble on the four serial data links (bits 3 through 0), followed by the next 4-bit nibble (bits 7 through 4), and so forth until the last 4-bit nibble (bits 31 through 28) has been transmitted. This is shown pictorially by mapping each 4-bit nibble into a 4-bit wide queue for transmission, each column of which is then mapped to the corresponding HSS transmitter. (This queue is not implemented in hardware; it just helps visualize the bit mapping.) For the HSS EX10 core, implementation of this scheme requires that bits 31, 27, 23, 19, 15, 11, 7, and 3 of the 32-bit data word are connected to bits TXxD[7:0] of the HSS #1 transmitter, bits 30, 26, 22, 18, 15, 10, 6, and 2 of the data word are connected to bits TXxD[7:0] of the HSS #2 transmitter, and so forth. Note that bits 3, 2, 1, and 0 are mapped to the TXxD[0] inputs of their respective transmitters in order to be transmitted first.

Any transmission order and interleaving scheme may be implemented simply by mapping the data bits to the appropriate HSS transmitter inputs (and doing the reverse of this mapping at the receiver).

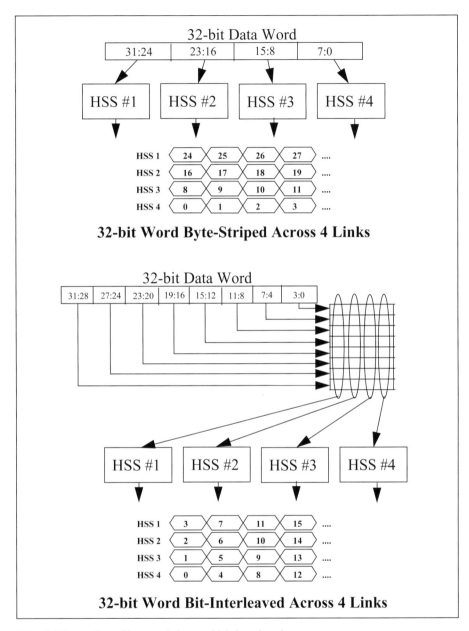

Fig. 4.6 Examples of byte striping and bit interleaving

4.2.2 Data Encoding and Scrambling

The raw data being sent on any interface is arbitrary and for most applications is *not* random. For example, a text file contains mostly printable ASCII characters. Byte codes which do not correspond to legal ASCII values do not occur. The byte codes for spaces and for "e" occur with higher frequency, while the byte code for "q" and "x" occur only occasionally. Large blocks of spaces may occur with substantially higher probability than would be predicted by the rules of a Gaussian probability distribution.

Transmission of raw data generally cannot be performed by Serdes which use clock recovery circuits to extract clock information from the serial data transitions. Any long string of 0 or 1 bits would lack any data transitions, and the data sample point for the clock recovery circuit would drift from the center of the eye, causing bit errors. At higher baud rates sudden shifts in spectral characteristics of the data can cause data to be sampled incorrectly, even when the data does contain transitions.

4.2.2.1 Block Codes

Block codes are one method of encoding data such that the maximum run length of 0 or 1 bits on the serial data link is guaranteed regardless of the data being transmitted. One of the most common block codes used by many protocol standards is the *8B/10B* code, originally patented by IBM. Every 8 bits of data is mapped into a corresponding 10-bit code word. There are two possible 10-bit code words that correspond to each 8-bit data word; one of these 10-bit code words is the *positive disparity* value, the other is the *negative disparity* value. The code scheme has rules for alternating between the positive disparity code word and the negative disparity code word. Following these rules guarantees that the number of 0 bits and the number of 1 bits in the encoded transmission is equal over time such that no DC bias builds up in the line voltage. The 8B/10B code also provides for a number of *control symbols* which may be used by the protocol. These control symbols cannot be confused with data since their 10-bit code words are unique.

Figure 4.7 illustrates the mapping of data bytes and control symbols into 8B/10B code. The input to the encoder is a byte value and a flag indicating

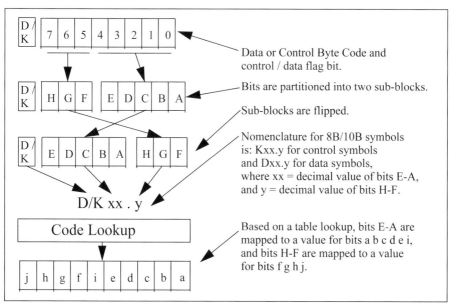

Fig. 4.7 Construction of 8B/10B code

Table 4.1 8B/10B Data symbol mapping

Name	EDCBA	abcdei (RD−)	abcdei (RD+)	Name	EDCBA	abcdei (RD−)	abcdei (RD+)
D00	00000	100111	011000	D16	10000	011011	100100
D01	00001	011101	100010	D17	10001	100011	100011
D02	00010	101101	010010	D18	10010	010011	010011
D03	00011	110101	001010	D19	10011	110010	110010
D04	00100	110101	001010	D20	10100	001011	001011
D05	00101	101001	101001	D21	10101	101010	101010
D06	00110	011001	011001	D22	10110	011010	011010
D07	00111	111000	000111	D23	10111	111010	000101
D08	01000	111001	000110	D24	11000	110011	001100
D09	01001	100101	100101	D25	11001	100110	100110
D10	01010	010101	010101	D26	11010	010110	010110
D11	01011	110100	110100	D27	11011	110110	001001
D12	01100	001101	00101	D28	11100	001110	001110
D13	01101	101100	101100	D29	11101	101110	010001
D14	01110	011100	011100	D30	11110	011110	100001
D15	01111	010111	101000	D31	11111	101011	010100
Name	HGF	fghj (RD−)	fghj (RD+)	Name	HGF	fghj (RD−)	fghj (RD+)
Dxx.0	000	1011	0100	Dxx.4	100	1101	0010
Dxx.1	001	1001	1001	Dxx.5	101	1010	1010
Dxx.2	010	0101	0101	Dxx.6	110	0110	0110
Dxx.3	011	1100	0011	Dxx.7	111	1110 or 0111	0001 or 1000

whether the byte is a data byte or a control code. The bits of this byte are grouped with the most significant three bits forming one group and the least significant five bits forming another group. A nomenclature for code symbols is created by using the notation "Dxx.y" for data bytes, or "Kxx.y" for control codes, where "xx" is the decimal equivalent of bits E down to A, and "yy" is the decimal equivalent of bits H down to F. These bit groupings (with the D/K flag) are then mapped into groups of bits to form the 10-bit symbol, as shown in Table 4.1. Bits "EDCBA" are mapped into 10-bit symbol bits "abcdei"; while bits "HGF" are mapped into bits "fghj." Note that there are two encodings shown for each mapping: a negative disparity mapping (RD−) which has at least as many 0's as 1's, and a positive disparity mapping (RD+) which has at least as many 1's as 0's.

Table 4.2 provides mapping for valid control symbols. Not all values of the "xx" bits are valid. Use of these control symbols varies somewhat depending on the protocol application.

Note that any valid mapping of the "abcdei" bits contains no more than four 1's or 0's, and any valid mapping of the "fghj" bits contains no more than three 1's or 0's. The characteristics of the code are such that the maximum run length that can ever be encountered is 5 bits.

A useful attribute of 8B/10B coding occurs because most (but not all) code words in the 8B/10B code have a Hamming distance of two from all other valid code words, and therefore most one-bit errors are detectable as a code word violation at the receiver. Additional one-bit errors may be detected as a violation of the disparity coding rules; receiving an RD– coding where an RD+ coding should have occurred, or vice versa, is likely the result of a bit error.

One drawback of 8B/10B codes is that the baud rate must be 25% higher than the data rate. This overhead factor becomes expensive at higher baud rates. For this reason, 64B/66B block codes have become more popular at higher baud rates. The 64B/66B block code adds two overhead bits to every 64 bits of data. These overhead bits are generated to ensure a data transition occurs and to equalize the number of 0 and 1 bits being transmitted. While the 8B/10B block code guaranteed a run length of 0's or 1's no greater than five, the 64B/66B only guarantees a transition every 66 bits, and does not have the error checking properties of the 8B/10B code. However, this is generally a reasonable trade-off given that the overhead of this code is only 3.125%.

IEEE 802.3 Clause 49 [6] defines 64B/66B coding for Ethernet network applications; this code is described in more detail in Sect. 5.3.4.2. The code is constructed by adding a two bit block tag to each 64-bit data block. The block tag is either "01" to indicate the 64-bit block contains data or "10" to indicate the 64-bit block is a control block. Control blocks may include control fields only, or a mixture of control fields and data. In either case, the block tag ensures that a data transition exists. Block tags of "00" and "11" are not valid.

Table 4.2 8B/10B Control symbol mapping

Name	EDCBA	abcdei (RD–)	abcdei (RD+)	Name	HGF	fghj (RD–)	fghj (RD+)
K23	10111	111010	000101	Kxx.0	000	1011	0100
K27	11011	110110	001001	Kxx.1	001	0110	1001
K28	11100	001111	110000	Kxx.2	010	1010	0101
K29	11101	101110	010001	Kxx.3	011	1100	0011
K30	11110	011110	100001	Dxx.4	100	1101	0010
				Kxx.5	101	0101	1010
				Kxx.6	110	1001	0110
				Kxx.7	111	0111	1000

4.2.2.2 Scrambling

Scrambling involves XOR'ing the output of a Linear Feedback Shift Register (LFSR) which generates a Pseudorandom Bit Sequence (PRBS) with the data being transmitted. The data is descrambled at the receiver using the inverse of the scrambling function. The purpose is to generate a more randomized content of 0's and 1's in the transmitted bit sequence, reducing the probability of a long run length of 0's or 1's occurring, and generating a frequency spectrum which more closely approximates a Gaussian distribution. Scrambling has no overhead, unlike block codes which always add overhead bits to the data.

Note that scrambling does not guarantee that a long run length of 0's or 1's does not occur. Given any scrambler starting at a given LFSR state, there is always a reverse scramble pattern which results in an all 0's or all 1's output. However, the probability of such a pattern occurring (provided it was not generated on purpose) is astronomically low, and can be discounted if the probability is well below the specified BER of the system. SONET and SDH, for example, are protocols which rely on scrambling to ensure data transitions occur. It is generally accepted that tolerating a maximum run length of 80 bits of all 0's or 1's is sufficient to ensure the system operates within the specified BER of the system (BER = 10^{-12}).

The scrambler algorithm is described using a shorthand polynomial notation. The polynomial for the scrambling used with 10 Gb Ethernet, 8 Gb Fibre Channel, etc., is:

$$G(x) = x^{58} + x^{39} + 1. \tag{4.1}$$

The serial implementation of the LFSR for this polynomial requires a 58-bit shift register (corresponding to the degree of the polynomial). The outputs of the 39th and 58th bit are XOR'd together to produce the feedback into the first bit of the shift register. The feedback term is also XOR'd with the data bit being transmitted (to scramble) or the data bit being received (to descramble). The LFSR is shifted by one bit for each bit transmitted or received.

The LFSR described above is part of the scrambler and descrambler implementations illustrated in Fig. 4.8. This figure illustrates both sidestream and self-synchronizing configurations. The sidestream scrambler and descrambler are implemented with identical logic. The feedback term loaded into the LFSR differs for the self-synchronizing scrambler and descrambler.

The LFSR contents of a sidestream scrambler and descrambler are not affected by the data being transmitted or received. In order to properly descramble the data, the state of the LFSR at the descrambler must match the state of the scrambler. In order to achieve this, the protocol must specify events upon which the LFSR at the scrambler and at the descrambler are reset to a seed value. One simple approach is to reset the scrambler and descrambler on the Start of Packet delimiter at the start of every packet, or on the framing pattern at the start of every SONET/SDH frame.

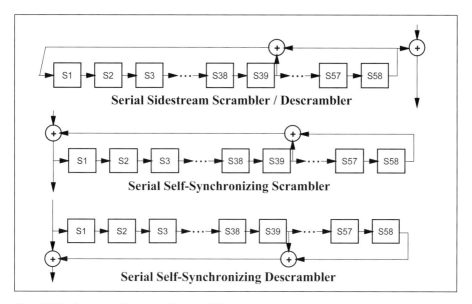

Fig. 4.8 Serial scrambler and descrambling architectures

The requirement for synchronizing the scrambler and descrambler can be eliminated by using a self-synchronizing scrambler. Fig. 4.8 illustrates these functions for the polynomial defined previously. The self-synchronizing scrambler differs from the sidestream scrambler in that the scrambled data bit is shifted into the LFSR rather using the feedback term alone. Likewise, the scrambled receive data is loaded into the LFSR at the descrambler. (The self-synchronizing scrambler and descrambler are not exactly the same design. They differ in terms of whether the data output or the data input are loaded into the LFSR.) For this architecture, the state of the scrambler and descrambler is entirely determined by the scrambled data. Regardless of the initial state of the descrambler, the descrambler state is synchronized after the first 58 bits are received, and correctly descrambles data thereafter.

Error propagation is one significant disadvantage of self-synchronizing descramblers. If the descrambler shown in Fig. 4.8 receives a bit which is in error, the incorrectly decoded bit alters the descrambler state. This error generates additional errors based on the feedback terms of the polynomial. Any error detection scheme must be robust enough to detect errors propagated by the characteristics of the scrambler scheme.

4.2.2.3 Parallel Scramblers

Although scrambler circuits are generally described in terms of their serial implementation, scrambling is generally not performed directly on the serial bit stream. Practical scramblers at baud rates used by HSS devices are implemented as part of the protocol logic and operate on the parallel data path. The logic equations necessary to implement the parallel implementation of a scrambler (or descrambler) are constructed through simple boolean manipulation of the scrambler polynomial. For example, assume a Sonet/SDH scrambler which uses the polynomial:

$$G(x) = x^7 + x^6 + 1. \tag{4.2}$$

Assume the scrambler (or descrambler) for (4.2) is implemented using seven flip-flops with the following initial states, where the number in the parenthesis indicates the advance of $n = 0$ bits of the sequence:

$$x^7(0) = x^7 \qquad\qquad x^3(0) = x^3$$
$$x^6(0) = x^6 \qquad\qquad x^2(0) = x^2$$
$$x^5(0) = x^5 \qquad\qquad x^1(0) = x^1$$
$$x^4(0) = x^4$$

After the sequence is advanced by one bit, the content of the serial scrambler state is shifted by one bit, and the feedback term is loaded into the first bit:

$$x^7(1) = x^6 \qquad\qquad x^3(1) = x^2$$
$$x^6(1) = x^5 \qquad\qquad x^2(1) = x^1$$
$$x^5(1) = x^4 \qquad\qquad x^1(1) = x^7 + x^6$$
$$x^4(1) = x^3$$

As the sequence continues to advance, after 8 bits of the sequence the scrambler state is:

$$x^7(8) = x^7 + x^6 \qquad\qquad x^3(8) = x^3 + x^2$$
$$x^6(8) = x^6 + x^5 \qquad\qquad x^2(8) = x^2 + x^1$$
$$x^5(8) = x^5 + x^4 \qquad\qquad x^1(8) = x^7 + x^6 + x^1$$
$$x^4(8) = x^4 + x^3$$

The above equations may be used to implement an 8-bit parallel sidestream scrambler for the polynomial in (4.2). Given a 7-bit scrambler state, these equations determine the next state for a parallel scrambler implementation. Keeping in mind that the bit order for transmission of SONET/SDH is most significant bit to least significant bit, the 8-bit data is scrambled using the following equations:

$$d_{7\text{scrambled}} = x^7 + d_7 \qquad\qquad d_{2\text{scrambled}} = x^2 + d_2$$
$$d_{6\text{scrambled}} = x^6 + d_6 \qquad\qquad d_{1\text{scrambled}} = x^1 + d_1$$
$$d_{5\text{scrambled}} = x^5 + d_5 \qquad\qquad d_{0\text{scrambled}} = x^7 + x^6 + d_0$$
$$d_{4\text{scrambled}} = x^4 + d_4$$
$$d_{3\text{scrambled}} = x^3 + d_3$$

where d_7 is the most significant bit and d_0 is the least significant bit. Likewise, at the receiver data is descrambled with these same equations.

Another useful example is a self-synchronizing scrambler using the polynomial defined in (4.1). This scrambler contains 58 flip-flops with the following initial states at $n = 0$:

$$x^{58}(0) = x^{58}$$
$$\vdots$$
$$x^1(0) = x^1$$

Note that the "⋮" is used to avoid writing out all 58 equations, however the equations for the intermediate bits should be obvious. The first bit transmitted is scrambled by the feedback term:

$$d_{scrambled} = x^{58} + x^{39} + d$$

and the scrambler state is advanced by one bit as follows:

$$x^{58}(1) = x^{57}$$

$$\vdots$$

$$x^2(1) = x^1$$
$$x^1(1) = x^{58} + x^{39} + d$$

Assuming a 32-bit data bus where d_{31} is the first bit transmitted, and d_0 is the last bit transmitted, the data is scrambled using the following equations:

$$d_{31scrambled} = x^{58} + x^{39} + d_{31}$$
$$d_{30scrambled} = x^{57} + x^{38} + d_{30}$$

$$\vdots$$

$$d_{1scrambled} = x^{28} + x^9 + d_1$$
$$d_{0scrambled} = x^{27} + x^8 + d_0$$

and the scrambler state is advanced by 32 bits using the following equations:

$$x^{58}(32) = x^{26}$$

$$\vdots$$

$$x^{33}(32) = x^1$$
$$x^{32}(32) = x^{58} + x^{39} + d_{31}$$

$$\vdots$$

$$x^1(32) = x^{27} + x^8 + d_0$$

It is left as an exercise to the reader to adapt the above equations for the self-synchronizing descrambler.

4.2.3 Error Detection and Correction

Any serial data link is prone to occasional bit errors, hopefully at a rate below the specified BER for the system. Most protocol applications include some means of detecting bit errors so that higher layers of the protocol may take appropriate corrective or reporting actions. This section describes the basics of error detection and correction schemes.

4.2.3.1 Parity Bits

Parity generation and checking is a simple form of error detection. Parity is calculated by XOR'ing all of the bits of the transmission using one of the following equations:

$$P_{even} = d_0 + d_1 + d_2 + d_3 + \ldots + d_n \tag{4.3}$$
$$P_{odd} = d_0 + d_1 + d_2 + d_3 + \ldots + d_n + 1. \tag{4.4}$$

The parity bit generated by the selected equation is appended to the transmission. Parity is again calculated at the receiver, this time with the transmitted parity bit included in the calculation. Assuming no errors, the check result for even parity is always "0," and the check result for odd parity is always "1." A bit error causes the parity check to be incorrect, thereby detecting the error.

Parity is guaranteed to detect any one-bit error in the transmission, but it does not detect the error if two bits are corrupted. Bit errors in many transmission media tend to occur in bursts, and parity does not guarantee error detection in such systems. As a result, parity is not commonly used to detect errors in serial link applications.

4.2.3.2 Bit Interleaved Parity

Bit interleaved parity (BIP) is an extension of a parity calculation where the data is broken into *n*-bit symbols, and the symbols are XOR'd together to form an *n*-bit parity symbol. The notation BIP-*n* indicates a BIP calculation performed on *n*-bit symbols.

SONET/SDH protocols use BIP-*n* for error detection. All bytes in the frame are XOR'd together, and the resulting parity byte is transmitted in the BIP-8 SOH field of the next frame. At the receiver, the BIP-8 byte is recalculated and compared with the BIP-8 field, and an error is flagged if a mismatch occurs.

Any parity error detection scheme is prone to multiple bit errors cancelling such that the error is not detected. A BIP-*n* calculation is such that a burst of bit errors up to *n* bits in length is detected by the BIP-*n* scheme. However, 2-bit errors separated by *n* bit positions cancel. The BIP-*n* scheme is far from robust, but larger values of *n* significantly reduce the probability of error cancellation. BIP-*n* schemes have proven sufficient for the SONET/SDH environment, where the only intent is to flag links which have degraded to an unacceptable BER level so that corrective maintenance actions can be taken. This application is not intended to ensure quality of the data being delivered.

4.2.3.3 Cyclic Redundancy Check (CRC)

Packet protocols often calculate a Cycle Redundancy Check (CRC) word during transmission and reception of the packet. The remainder of the CRC calculation is transmitted at the end of the packet. When the CRC is calculated at the receiver, and this remainder is included, the result should always be a fixed value. The calculation of the CRC word can be viewed mathematically as the division of polynomials:

$$L(x) / G(x) = C(x), \tag{4.5}$$

where $L(x)$ is a polynomial containing all n terms of an n-bit wide calculation:

$$L(x) = x^n + x^{n-1} + ... + x^2 + x^1 + 1 \tag{4.6}$$

and $G(x)$ is the standard generator polynomial. Several protocols including Fibre Channel use the following $n = 32$ generator polynomial:

$$G(x) = x^{32} + x^{26} + x^{23} + x^{22} + x^{16} + x^{12} +$$
$$x^{11} + x^{10} + x^8 + x^7 + x^5 + x^4 + x^2 + x + 1. \tag{4.7}$$

The resulting $C(x)$ polynomial remainder is therefore:

$$C(x) = x^{31} + x^{30} + x^{26} + x^{25} + x^{24} + x^{18} + x^{15} + x^{14} +$$
$$x^{12} + x^{11} + x^{10} + x^8 + x^6 + x^5 + x^4 + x^3 + x + 1. \tag{4.8}$$

A serial implementation of this CRC generator (or checker) consists of a 32-bit LFSR implementing the $C(X)$ polynomial. The LFSR is reset to a seed value at the start of the packet, and the serial data is XOR'd with the feedback terms and shifted into the LFSR as each bit is transmitted. The remainder is appended to the packet. At the receiver, the same LFSR implementation is used, however, the CRC word at the end of the packet is included in the CRC calculation at the receiver. The remainder in the LFSR after including this CRC word should always be a fixed value.

Although multiple bit errors can cancel such that a CRC check does not flag the error, this is extremely unlikely. CRC checks are very robust at flagging bit errors, including multiple bit errors, bursts of errors, etc. However, CRC checks have no ability to correct bit errors when they occur. Higher layer protocols must take responsibility for recovering from any error events.

4.2.3.4 Error Correction

Protocol standards sometimes incorporate error correction schemes at high baud rates. Use of these schemes is generally optional, and hardware implementations are not widely deployed. This may change as baud rates for serial link protocols continue to increase.

Error correction is used to improve the BER performance of an interface. However, any error correction implementation inherently requires some amount of buffering. Such buffering adds latency to the system which is considered undesirable in many applications. For this reason, other methods of improving system performance are generally pursued in the system design before any error correction schemes are considered.

Forward Error Correction schemes used in protocol standards of interest to serial links are generally based on some form of a cyclic code. At the transmitter, data is broken into fixed length blocks. The error correction code words for each block are calculated as data is being transmitted, and then are appended to the block. At the receiver, the error correction code words are recalculated and compared to the received values in much the same manner as error detection code words are processed. If values do not match, then an error has occurred.

Error correction schemes differ from error detection in that it is possible to deduce which bits are in error from the recalculated error correction code words. (Methods of doing this are beyond the scope of this text.) Given that the bits which are in error can be identified, the values of these bits can be corrected (by inverting the bits) and downstream logic processes the corrected data. Note that it is not possible to identify which bits are in error until the entire data block has been received, and the data bits which must be corrected may be anywhere in the block. Therefore, the protocol logic must buffer the data block until it is possible to determine whether corrections are needed. The data block is then forwarded to downstream logic with any corrections.

The size of the data block and the number of bits in the error correction code words therefore becomes a trade-off. Larger block sizes require larger buffers and introduce more latency. Smaller block sizes introduce higher overhead for error correction code words. The block size and complexity of the error correction code words also impact the *strength* of the code. Error correction strength is an indication of how many bit errors may occur in the data block before the error correction scheme cannot correctly identify which bits are incorrect. Bit errors often occur in bursts, and the ability to correct such bursts is also an important characteristic of the error correction code.

Cyclic error correction codes are partially described by the notation (n,m) where n is the total number of symbols in the data block, and m is the number of those symbols which are payload symbols. The difference $n - m$ indicates the number of symbols in the error correction code word. Often each symbol is one bit, but the notation can be applied to any symbol size.

The IEEE 802.3ap Backplane Ethernet standard [7] defines a Forward Error Correction scheme in clause 74 which is summarized in Sect. 5.3.5.3. This scheme is optionally used as part of the 10GBASE-R physical coding sublayer. This is a (2112,2080) code which is constructed by shortening the cycle code (42987,42955). Data and transcode bits are mapped as shown in Fig. 5.21 and the 32-bit error correction code word, calculated using (5.5), is appended to the block. This code is capable of correcting a burst of up to 11-bit errors in the data block.

The Optical Internetworking Forum has defined a set of electrical link standards for 5–6 Gb and 8–11 Gb baud rate ranges. The Common Electrical I/O (CEI) Implementation Agreements are inherently designed for a 10^{-15} BER. The Common Electrical I/O Protocol (CEI-P) Implementation Agreement [5], described in Sect. 5.2.4, defines a physical layer protocol for use with CEI electrical links. This protocol optionally includes an error correction scheme based on 1,584 bit data blocks as illustrated in Fig. 5.11. The data block includes 1,560 payload bits, 4 status bits, and a 20-bit error correction control word calculated using the Fire Code polynomial specified in (5.2). Using this error correction scheme, the expected BER of the link is extended beyond 10^{18}.

The ITU-T G709 protocol [7] implements an overhead shell around a SONET or SDH protocol. The additional overhead bytes transmitted for G709 requires a 7% higher baud rate than for the native SONET/SDH protocol. This overhead shell includes various control functions, and includes Reed-Solomon error correction code words. Although a higher baud rate is employed, the error correction improves the Signal-to-Noise of the system such that optical links can run for longer distances between repeater units. This is a significant cost reduction for long-lines telecom carriers.

This error correction scheme uses Reed-Solomon (255,239) code (where symbols of this code are bytes rather than bits). The transmission is broken into data blocks of 255 bytes, where every 4,080 bytes are byte interleaved among 16 independent Reed-Solomon codecs. A 16-byte error correction code word is produced for each block. This scheme is capable of correcting errors in up to 8 bytes of each 255 byte data block.

4.2.4 Parallel Data Interface

Serdes transmitter and receiver channels derive the parallel data clocks from the baud rate clocks generated by the PLL slice in the Serdes core. The clocks for each parallel interface are produced independently for each transmitter and receiver channel, and are used to clock parallel data to the transmitter, and to capture parallel data from the receiver. Although the system clock on the chip may be derived from the same reference as is supplied to the HSS cores, the phase relationship between the system clock and the parallel data clocks on the Serdes cannot be guaranteed. This section describes techniques used in protocol logic to interface to the parallel data interface of the HSS core.

4.2.4.1 Transmitter Parallel Data Interface

Figure 4.9 illustrates the parallel data interface to two or more HSS transmitter channels. Transmit data originates on the chip synchronous to a system clock, and is supplied in parallel to the logic associated with multiple HSS transmitter channels. Each HSS transmitter has an associated parallel data input and parallel data clock output. The system clock is assumed to be frequency locked to the reference clock for the HSS cores, however, the phase relationship between the system clock and the parallel data clocks cannot be known or guaranteed. In this example, it is also assumed that the phase relationship between parallel data clocks of different transmitters is also indeterminate.

In Fig. 4.9, elastic FIFOs are used to synchronize the data in the system clock domain to the parallel data interface for each individual transmitter channel. The n-deep elastic FIFO consists of n registers, each with a bit width corresponding to the width of the parallel data interface. Data is written into one of these registers on each system clock as determined by the current value of the write address counter. Data is read from one of these registers on each parallel data clock as determined by the current value of the read address counter. The write address counter and the read address counter are reset to address values with a difference of $n / 2$. Assuming the system clock and parallel data clock remain frequency locked, and assuming an appropriate value for FIFO depth n, the FIFO operates indefinitely after reset without the possibility of an overflow or underflow.

Note that only a small number of flops are clocked by the parallel data clock. This allows this clock domain to be implemented with minimal clock latency on the chip, aiding in timing closure on this timing critical interface.

The depth of the FIFO is selected to take into account any wander that may occur between the system clock and the parallel data clock. Wander results from differing delays of the system clock distribution path relative to the clock path through the HSS core. The delay of these paths varies over time due to process, voltage, and temperature variation in different regions of the chip. As the two clocks drift relative to each other, the FIFO must have sufficient depth to absorb this variation without an overflow or underflow occurring.

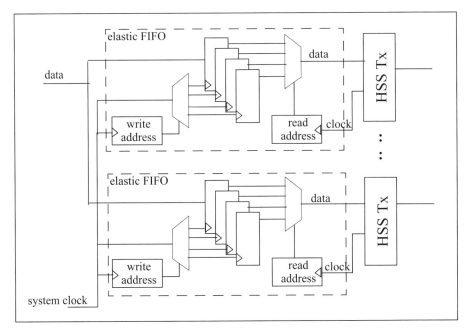

Fig. 4.9 Serdes transmitter elastic buffers

If the application is a multibit interface, and the various transmitter channels must meet a skew specification at the output of the chip, then additional considerations apply. Any phase difference between parallel data clocks, or any difference in when FIFO counters begin to increment after a reset, results in additional skew which may exceed the specification. Therefore, implementations of multibit interfaces must ensure the following:

- The parallel data clocks must be initialized such that they are in phase. The HSS EX10 core provides a *resynchronization* feature which resets clock dividers, resulting in TXxDCLKs being approximately in phase.

- The elastic FIFOs must be initialized such that they exit reset approximately in phase. Figure 4.10 illustrates a timing example where the elastic FIFOs have not been properly reset, and skew is induced as a result.

Assuming the above conditions are met, skew contributors are limited to propagation delay differences in the routing of the reference clock distribution network to the HSS cores, time of flight differences for serial data signals through the package, and the channel-to-channel skew specifications for the HSS cores as defined in the core databook.

An alternative approach is shown in Fig. 4.11. This scheme assumes that the various transmitter channels have been resynchronized, and therefore parallel data clocks are approximately in phase. This implementation uses one of these parallel data clocks to read data from all FIFOs; other parallel data clocks are not used.

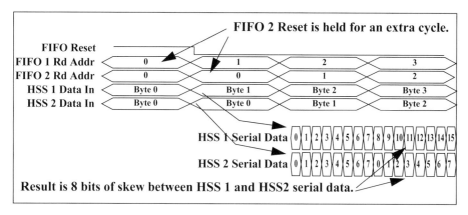

Fig. 4.10 Skew induced by out-of-sync elastic FIFO read pointers

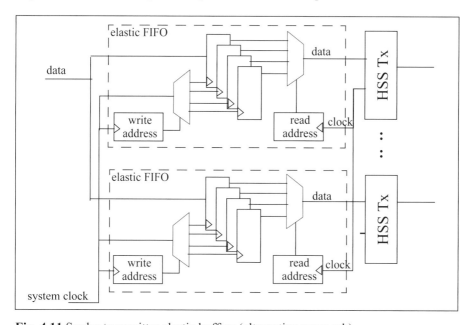

Fig. 4.11 Serdes transmitter elastic buffers (alternative approach)

This approach solves the problem of ensuring all FIFOs exit reset in phase with each other. However, this approach does introduce additional difficulty in closing timing on the parallel data interface:

- Setup and hold times of parallel data inputs must be adjusted to take into account channel-to-channel, and in some cases core-to-core, specifications for skew and wander. This has the effect of increasing both the setup and hold times.

- The FIFO read address is distributed to more logic, which may increase propagation delay. These signals are in the critical timing path.

4.2.4.2 Receiver Parallel Data Interface

Figure 4.12 illustrates the parallel data interface to two or more HSS receiver channels. Serial data is received by each HSS receiver channel, is deserialized, and then is driven onto the parallel data outputs of the core. Each receiver channel has a parallel data clock which should be used to sample the data output. The frequency of this clock is recovered from the bit transitions on the incoming data, and in applications using plesiosynchronous clocks this clock frequency is not the same as the local reference clock.

In this example, elastic FIFOs are used to synchronize the data in the local RXxDCLK domains to the system clock domain. The n-deep elastic FIFO consists of n registers, with a bit width corresponding to the parallel data width. Data is written into one of these registers on each RXxDCLK as determined by the current value of the write address counter. Data is read from one of these registers on each system clock as determined by the current value of the read address counter. The write address counter and the read address counter are reset to address values that have a difference of $n / 2$. In plesiosynchronous applications, the FIFO read and write pointers eventually catch up with each and a FIFO overflow or underflow may occur. The protocol must provide a mechanism to add or drop bytes (or symbols) to correct for differences in the clock rate. The depth required for the FIFO depends on the exact mechanism used and the frequency tolerance specification for the reference clock.

An example of calculating the minimum required FIFO depth follows: Assume a packet protocol application where data is transmitted as packets of up to 32,768 bytes in length. Between packets there are idle symbols which are ignored by the protocol. These symbols may be duplicated or dropped if needed to recenter the FIFO pointers. Assume the protocol specifies that a 200 ppm frequency difference is allowed between the reference clocks for the transmitter and the receiver of the link. Therefore:

- Frequency difference = 200 / 1,000,000 = 0.0002
- 1 UI slip occurs every 1 / 0.0002 = 5,000 bits
- Packet length of 32,768 bytes = 262,144 bits
- The total slip that occurs during a maximum length packet is up to: 262,144 / 5,000 = 53 bits (rounding up) = 7 bytes (rounding up)

The elastic FIFO must therefore tolerate the write and read pointers drifting by up to ± 7 FIFO locations relative to each other during the reception of a frame. The minimum depth of this FIFO is therefore 15 bytes (a range of ± 7 locations from the current location). In addition, the designer may want to increase the FIFO depth in order to avoid having to handle boundary conditions when FIFO pointers are at their extreme limits.

The elastic FIFO must be designed with some mechanism to recenter the read and/or write pointers while idles are being received between frames. Many possible design approaches are possible for this problem. The selected solution is often dictated by the needs of the application, and specific approaches are beyond the scope of this text.

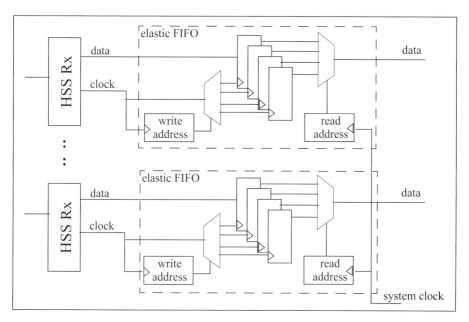

Fig. 4.12 Serdes receiver elastic buffers

The parallel data clocks from each of the receiver channels have no guaranteed phase relationship to either the system clock (even if frequency lock can be assumed) or to each other. If the various receiver channels represent multiple serial data bits on a multibit interface where all the bits are being driven by the same far-end chip, then the channels can be assumed to be frequency locked to each other. It may be possible, depending on the HSS core design, to assume the parallel data clocks of the various receiver channels are in phase immediately after a reset or resynchronization of the channels. However, as soon as reset completes the clock recovery function in each channel makes independent decisions as to when to update the data sampling point. Therefore, the phase relationships between the parallel data clocks quickly drift relative to each other, and no assumptions as to phase relationship should ever be assumed. Routing of clocks within the core contributes an additional skew factor, and variations in voltage and temperature contribute to wander, causing further phase differences.

Figure 4.12 assumes a system where each HSS channel is receiving an independent serial bit data bit stream. If the various receiver channels represent multiple serial data bits on a multibit interface, then there may be skew between the various bits of the data as it arrives at the various receiver serial data inputs. Deskew logic is the subject of a separate discussion in Scct. 4.2.6.

It is important to note that because a phase relationship between the parallel data clocks cannot be assumed, this potential phase difference contributes additional skew and wander which impacts the required deskew range.

4.2.5 Bit Alignment

The protocol logic inherently drives an integral number of bytes (or 10 bit symbols) of data into the parallel data inputs of the HSS transmitter. This data is serialized by the HSS transmitter, driven across the channel, and deserialized by the HSS receiver. The byte (or symbol) boundaries of the bit stream are not inherently known to the HSS receiver, and therefore the parallel data output of the receiver channel may be misaligned. For example, an 8-bit output may contain the latter portion of one byte and the first portion of the byte after that.

An example of this is shown in Fig. 4.13 where parallel data is transmitted least significant bit first, and then is deserialized starting at a bit position that is three bits out of sync with the transmitter. The resulting parallel data at the receiver appears significantly different from that which was transmitted. (It is not actually different – the bits are all there but just out of alignment.)

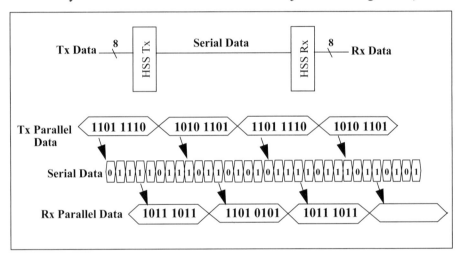

Fig. 4.13 Arbitrary bit alignment of received data

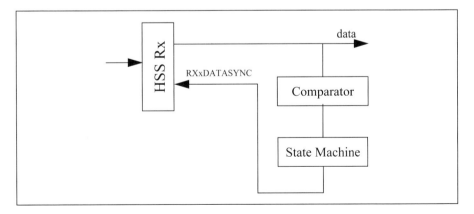

Fig. 4.14 Bit alignment using datasync

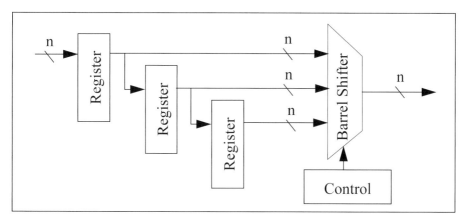

Fig. 4.15 Generic deskew logic

Most Serdes cores provide a mechanism to adjust the bit alignment until it falls on the desired byte (or symbol) boundary. The RXxDATASYNC feature on the HSS EX10 core was described in Chap. 2. Use of this feature requires that the protocol have a method of determining the proper byte (or symbol) alignment. One common method uses a symbol that occurs in the data stream, such as the idle symbols occurring between packets, to align the receiver. Figure 4.14 provides a block diagram of the logic that would perform bit alignment. The data output of the receiver is monitored by a comparator circuit to determine whether or not the data is bit aligned. (The data can be monitored either before or after the elastic FIFO, although monitoring prior to the FIFO might be necessary for some FIFO designs.) If the data is not aligned, then a state machine pulses the RXxDATASYNC input. This causes the data to "slip" by one bit. If the data remains unaligned, the state machine repeats this process until the data is aligned on the desired boundary. Once the data is aligned, it should remain aligned unless the interface is reinitialized.

The bit alignment feature is used by both serial interfaces and multibit interfaces. This feature is also related to deskew, as discussed in Sect. 4.2.6. It is not necessary to use the RXxDATASYNC feature to perform bit alignment; this function can also be performed by downstream logic in the datapath without interaction with the HSS core.

4.2.6 Deskewing Multiple Serial Data Links

Any interface implemented using multiple serial data links requires that the output of the various links be deskewed to correct for any skew introduced by the transmitter chip, the channel, and the HSS receiver. In many applications several UI of skew may exist, and the deskew logic must have a sufficient range to correct for this.

A generic example of the deskew logic for a single serial data link is shown in Fig. 4.15. This example assumes the parallel data output of the HSS receiver is n-bits wide. The n-bit parallel data is retimed to the system clock domain, and then is connected to the input of this deskew logic. The deskew logic shown includes three registers which contain the last $3n$ bits of data. The barrel shifter stage is a $3n : n$ multiplexor which can select any n consecutive bits out of the $3n$ bit input, thus providing a total deskew range of $3n$ bits.

The select input for the barrel shifter is controlled by unspecified control logic in this example. This control logic determines the correct setting for the barrel shifter associated with each serial data link. The goal is for the outputs of all of the barrel shifters to be aligned such the combined output is a single k by n bit parallel data bus (where k is the number of serial data links). Many schemes exist in protocol standards for providing a reference for control logic to use to perform deskew. Some of these are discussed below.

4.2.6.1 Deskew Schemes

This section describes several deskew approaches used by various protocol standards to provide a reference for aligning multiple serial data lanes of an interface. More detailed descriptions of these deskew methods are provided, where applicable, in Chap. 5.

Link Synchronization Patterns. Some protocol standards specify a link synchronization or training pattern which is transmitted regularly. The deskew function of the XGMII Extended Sublayer (XGXS) Physical Coding Sublayer (PCS) associated with IEEE 802.3 10Gb Ethernet [6] is an example of an interface which uses a specified pattern for deskew. The XGXS PCS is a four-lane packet-based protocol which uses 8B/10B block encoding. As described in Sect. 5.3.3.2, various control symbols are transmitted on the link between packets; these include *align symbols* (K28.3) which are transmitted at intervals of 16–31 symbol times. Align symbols are transmitted on all links simultaneously, and are used by the receiver to deskew the links. Given the interval at which these symbols are transmitted, deskew of up to ±15 symbol times (150 UI) is possible (although achieving this might require reception of several align symbols with various spacings).

The K28.3 symbol is chosen for performing deskew because its bit sequence is unique. This minimizes the possibility of falsely detecting the symbol in a data stream, and incorrectly deskewing the interface as a result. Such false detection, called *aliasing*, can occur for some training patterns used for alignment if the there is not a robust mix of "1" and "0" values on each link, and if the data immediately preceding or following the training pattern coincidentally has the right value.

PCI Express, another packet-based protocol which uses 8B/10B block code, does not explicitly specify how the receiver is to perform deskew. However, the *skip ordered set* which is transmitted between packets is typically used by the receiver to perform this function in a similar manner to the XGXS PCS application. The skip ordered set consists of a *COM* symbol (K28.5) followed by three *SKP* symbols (K28.0), and is transmitted on all lanes simultaneously. (Note that intermediary nodes may add or detract *SKP* symbols, so the receiver may see a *COM* followed by up to six *SKP* symbols.)

Framing References. The Scalable System packet Interface (SPI-S) [4] defined by the Optical Internetworking Forum (OIF) uses a variation on the alignment pattern scheme. As described in Sect. 5.2.3.3, deskew is performed across multiple lanes by leveraging the framing position of the individual links. The tag bits of the 66-bit code words of IEEE 64B/66B code as defined by IEEE 802.3 Clause 49 [6] are always "01" or "10." Framing is performed on individual links by examining bits for several consecutive code words, and searching for a framing reference where the tag bits are always valid (i.e., never "00" or "11"). Once a frame reference is established for each link, deskew may be performed across links by comparing the relative frame reference positions. Using this method, deskew of up to ± 32 UI is possible.

In general, any method used to establish a framing reference for a link may be leveraged to deskew multiple links. As an alternative to 64B/66B bit code words, SPI-S may also carry data coded and scrambled as described by the OIF Common Electrical I/O Protocol (CEI-P) Implementation Agreement [5]. CEI-P payload is mapped into frames as shown in Fig. 5.11, and the receiver searches for a frame reference where the FEC parity bits are consistently valid. Once a frame reference is established for each link, deskew is performed across links in a similar manner to the previous case.

Deskew Channel. The OIF Serdes-Framer Interface Generation 5 (SFI-5) protocol [2] uses a *deskew channel* to transmit a reference for alignment of the data channels. The deskew channel transmits a framing pattern, followed by 64 bits from each data channel in a round robin fashion. Deskew logic in the receiver uses the framing pattern as a reference point and aligns each data channel to corresponding data on the deskew channel.

The OIF SFI-5.2 Implementation Agreement [3] also uses a deskew channel, but with the contents described in Sect. 5.2.2.2. The deskew channel transmits a 10-bit frame consisting of bits from each of the four data channels and even/odd parity bits corresponding to the data channels. Deskew logic uses parity calculations to determine reference points for the 10-bit deskew frame, and then aligns data channels to the bits on the deskew channel.

4.2.6.2 Optimizing the Deskew Implementation

The reader may note that the generic deskew circuit in Fig. 4.15 was essentially an *n*-bit wide shift register and a barrel shifter used to insert a selectable delay into the data pipeline. There are two opportunities in the receiver logic prior to reaching this circuit where similar delays can be inserted to perform a similar function. This presents an opportunity to distribute the deskew function so that the shift register in Fig. 4.15 is reduced in size or even eliminated.

The first opportunity for optimization utilizes the RXxDATASYNC function. Previously, this function was described in the context of correcting byte (or symbol) alignment on a byte-striped serial data link. Given that each serial data link is byte (or symbol) aligned, any remaining skew present between data links at the deskew stage is in units of bytes or symbols. Other arbitrary bit alignments are not possible. This reduces the complexity of any alignment comparisons for the control logic, as well as reducing the complexity of the barrel shifter.

Also, for bit-interleaved applications, the RXxDATASYNC function is available to perform some of the deskew. For an *n*-bit wide parallel data path output of the HSS receiver, the RXxDATASYNC can be used to perform deskew within an *n*-bit range. The circuit in Fig. 4.15 can be used to perform additional deskew if this range is not sufficient.

The next opportunity for optimization utilizes the elastic FIFO function. The FIFO already implements the equivalent of the shift register in Fig. 4.15, and the read pointer and output multiplexor implement a limited barrel shifter function. Assuming a byte-striped application where RXxDATASYNC has been used to achieve bit alignment at the input to the FIFO, deskew can be implemented by adjusting the read pointer address. This potentially eliminates any requirement for additional downstream logic. Note, however, that the FIFO depth must be sufficient to permit deskew adjustments without resulting in FIFO overruns or underruns.

4.2.6.3 Skew Budget

The required range of the deskew logic is determined by the specification for skew and relative wander at the receiver input pins, plus any additional skew introduced by the receiver circuits.

Table 4.3 provides selected specifications for skew and relative wander for the Optical Internetworking Forum SFI-5.2 interface as described in [3]. As shown in this table, the transmitter must meet skew and relative wander requirements as measured on the serial data output pins. Skew may result due to several factors in the transmitter chip, including: differences in signal routing of the reference clock to the individual HSS cores, differences in signal routing internal to the HSS cores, and time of flight differences in the routing of the serial data signals through the chip package. The total contribution of all these factors must not exceed 5.50 UI peak-to-peak as measured at the output pins of the chip package.

In addition to skew, if the transmitter uses multiple HSS cores to implement the interface then phase differences may exist between the individual PLL slices in these cores. These phase differences may vary over time as temperature and voltage conditions change. This results in relative wander as specified in Table 4.3. Because relative wander is below the cutoff frequency of the CDR, the elastic FIFO and deskew logic must compensate for this variation.

Table 4.3 also specifies skew and relative wander at the receiver input. The difference between the skew specifications for the transmitter and the receiver reflects the skew that is allowed to be introduced by routing differences in the channel. Pattern-dependent distortions introduced by the channel may also result in slightly more relative wander as seen at the receiver. The total skew plus relative wander as seen at the receiver is of significance to the design of the deskew logic.

Table 4.3 Selected specifications for OIF SFI-5.2 skew and relative wander

Parameter	System points		Units
	Tx Output	Rx Input	
Skew	5.50	11.00	UI peak-to-peak
Relative wander	1.30	1.50	UI peak-to-peak
Total Skew + Wander at Receiver		12.50	UI peak-to-peak

When determining the range for the deskew logic, the designer starts with the specification for skew and wander at the receiver input, and then adds the additional skew contributors within the receiver. An example is shown in Table 4.4, although note any calculation is implementation dependent.

The first contributor that is considered in this table results from signal routing differences within the receiver chip, including: differences in signal routing of the reference clock to the individual HSS cores, differences in signal routing internal to the HSS cores, and time of flight differences in the routing of the serial data signals through the chip package. The magnitude of these contributors is implementation dependent, but it is reasonable to assume that the values are similar to the corresponding values for the transmitter. Table 4.4 assumes these contributors result in an addition 5.50 UIpp of skew.

The other contributor to skew in the receiver results from the phase differences between the RXxDCLK outputs of the various receiver slices. Unlike the transmitter, where synchronizing the phases of the TXxDCLK clocks is critical to meet the interface skew specifications, there is usually no attempt to synchronize RXxDCLKs. Even if these clocks were to be resynchronized, the CDR circuits in the individual receiver slices operate independently and the RXxDCLK phases would diverge over time. If the receiver slices are part of different HSS cores, then phase differences between the individual PLLs in each of these cores also contribute to changes to the RXxDCLK phase. If the interface uses RXxDATASYNC to perform a symbol alignment function, this changes the RXxDCLK phase by design. The only safe assumption for the deskew logic designer is that the phase relationships of the RXxDCLK clocks is arbitrary and any relationship is possible. Assuming an n-bit parallel data path, the cumulative effect of these factors can result in up to n UI of skew. This is reflected in Table 4.4 assuming $n = 8$.

As can be seen from the above example, a substantial deskew range may be required. The total skew budget for all contributors in Table 4.4 is 26 UIpp. Often one of the receive channels is chosen as a reference, and other channels are deskewed to align data with the reference channel. In most applications it is possible for the reference channel to be at the extreme limit of the skew specification, and therefore the deskew logic on the other channels in this example requires a total range of ± 26 UIpp.

Table 4.4 Skew budget for sample SFI-5.2 implementation

Parameter	Rx Input	Notes
Total skew + wander at receiver	12.50 UIpp	Per interface specification
Skew introduced in receiver due to signal routing differences	5.50 UIpp	Sum of following: • Time-of-flight differences in routing of serial data signals through receiver package. • Differences in signal routing internal to the HSS Rx cores. • Differences in signal routing of the reference clock to individual HSS cores in the receiver chip.
Skew introduced in receiver due to RXxDCLK phase differences.	8.00 UIpp	Assumes parallel data bus width = 8 bits
Total skew budget for deskew logic	26.00 UIpp	UI peak-to-peak

4.3 References and Additional Reading

A comprehensive list of interface standards documents for various network protocols can be found in Sect. 5.6. Refer to that list for more information on standards mentioned in this chapter.

In addition, the following interface standards documents are referenced in this chapter:

1. "Common Electrical I/O (CEI) – Electrical and Jitter Interoperability agreements for 6G+ bps and 11G+ bps I/O", OIF-CEI-02.0, Optical Internetworking Forum, Feb. 28 2005.

2. "Serdes Framer Interface Level 5 (SFI-5): Implementation Agreement for 40Gb/s Interface for Physical Layer Devices", OIF-SFI5-01.0, Optical Internetworking Forum, Jan. 29 2002.

3. "Serdes Framer Interface Level 5 Phase 2 (SFI-5.2): Implementation Agreement for 40Gb/s Interface for Physical Layer Devices", OIF-SFI5-02.0, Optical Internetworking Forum, Oct. 2 2006.

4. "Scalable System Packet Interface (SPI-S) Implementation Agreement: System Packet Interface Capable of Operating as an Adaption Layer for Serial Data Links", OIF-SPI-S-01.0, Optical Internetworking Forum, Nov. 17 2006.

5. "Common Electrical I/O – Protocol (CEI-P) – Implementation Agreement", OIF-CEI-P-01.0, Optical Internetworking Forum, March 2005.

6. "IEEE Standard for Information Technology – Telecommunications and Information Exchange Between Systems – Local and Metropolitan Area Networks – Carrier Sense Multiple Access with Collision Detection (CSMA/CD) Access Method and Physical Layer Specifications", IEEE 802.3-2005, Institute of Electrical and Electronic Engineers, Dec. 12 2005.

7. "Amendment: Ethernet Operation over Electrical Backplanes", IEEE P802.3ap, Draft 3.3, Institute of Electrical and Electronic Engineers, Jan. 26 2007.

8. "ITU-T G.709 – Series G: Transmission Systems and Media, Digital Systems and Networks, Digital Terminal Equipment – General, Interface for the Optical Transport Network (OTN)", International Telecommunications Union, 2001.

4.4 Exercises

1. Draw an eye mask similar to that shown in Fig. 4.2 for each set of parameters below. Label all indices and indicate the minimum eye width and eye height:

   (a) $T\_X1 = 0.1$ UI, $T\_X2 = 0.3$ UI, $T\_Y1 = 400$ mV, $T\_Y2 = 600$ mV

   (b) $T\_X1 = 0.25$ UI, $T\_X2 = 0.50$ UI, $T\_Y1 = 200$ mV, $T\_Y2 = 400$ mV

   (c) $T\_X1 = 1$ ps, $T\_X2 = 2$ ps, $T\_Y1 = 400$ mV,
   $T\_Y2 = 600$ mV, 4 ps = 1 UI

2. Some sample signal detect threshold settings for the HSS EX10 receiver were described in Table 2.18. Assume the amplitude of a serial data signal is initially 50 mVppd, and slowly ramps up to 400 mVppd, and then back down to 50 mVppd. Show the corresponding RXxSIGDET waveform for each of the threshold settings in Table 2.18, and show the signal amplitudes at which RXxSIGDET is "0," "1," or "X" (indeterminate). Assume the loss of signal response time is zero.

3. Deskew logic for a particular application is expected to be set based on a training pattern when the interface is initialized, and then is expected to receive data continuously thereafter:

   (a) One proposal for this interface specifies a relatively large value for the skew between data lanes. Other than affecting the elastic buffer sizes and training pattern selection, are there any issues created by the large skew value? Why?

   (b) Another proposal for this interface specifies a relatively large value for the wander between data lanes. Other than affecting the elastic buffer sizes and training pattern selection, are there any issues created by the large wander value? Why?

4. Protocol logic which uses a 128-bit datapath (with bits labelled DX127 to DX0) is to be connected in a bit-interleaved fashion to eight serial data links for transmission. Two HSS EX10 cores are used to implement this interface; the Tx slices of these cores are labelled D7 to D0. (Bits DX0 and D0 are the least significant bits for these interfaces.) The most significant byte of the 128-bit datapath is transmitted first on the D7 to D0 serial data channels, followed by the next most significant byte, etc., until all 16 bytes have been transmitted. Specify the connections of the DX[127:0] to TXxD[15:0] pins on the transmitter slices that implement this bit transmission order.

5. Two HSS EX10 cores (with eight receiver slices labelled D7 to D0) are to be connected to protocol logic using a 128-bit datapath with bits labelled DX127 to DX0. (Bits DX0 and D0 are the least significant bits for these interfaces.) The data is byte-striped across the interface from least significant byte (transmitted on D0) to most significant byte (transmitted on D7) when it is transmitted, and the least significant bit of each byte is transmitted first. Specify the connections of the RXxD[15:0] pins on the receiver slices to the DX[127:0] bits corresponding to this bit transmission order.

6. The following data bytes are to be coded using 8B/10B block code. Determine the corresponding Dxx.y nomenclature for each of these data bytes:

   (a) "13"h　　　(b) "22"h　　　(c) "5B"h.　　　(d) "E6"h

   (e) "F1"h　　　(f) "30"h　　　(g) "77"h.　　　(h) "8A"h

7. For each of the data bytes in Exercise 6 specify the corresponding 10-bit codeword (bits a to j) for both positive and negative disparity.

8. The following 10-bit codewords are coded using 8B/10B block code. Determine the corresponding Dxx.y or Kxx.y nomenclature for these codewords, and whether the codeword uses positive or negative disparity.

   (a) "346"h   (b) "0FC"h   (c) "241"h.   (d) "225"h

   (e) "218"h   (f) "3AA"h   (g) "0BE"h.   (h) "184"h

9. For each of the codewords in Exercise 8 specify the corresponding D/K bit and data byte value.

10. What is the longest run length of 0's or 1's that can be formed by concatenating one 8B/10B codeword with another codeword of the opposite disparity? Give an example of two codewords which result in this maximum run length.

11. Provide eight examples of Dxx.y symbols which result in the same codeword for both positive and negative disparity.

12. Design logic (Verilog or VHDL) for a self-synchronizing serial scrambler and descrambler which uses the following scrambler polynomial: $G(x) = x^{58} + x^{39} + 1$.

13. Design logic (Verilog or VHDL) for a self-synchronizing 8-bit parallel scrambler and descrambler which uses the scrambler polynomial specified in Exercise 12.

14. Design logic (Verilog or VHDL) for a sidestream serial scrambler (or descrambler) which uses the following scrambler polynomial: $G(x) = x^7 + x^6 + 1$.

15. Design logic (Verilog or VHDL) for a sidestream 16-bit parallel scrambler (or descrambler) which uses the scrambler polynomial specified in Exercise 14.

16. Modify the logic for the scrambler in Exercise 13 as follows: Add a D/K bit to the data path. When the D/K bit = 0, the data is scrambled and the scrambler state is advanced. When the D/K bit = 1, the data is not scrambled and the scrambler state remains unchanged.

17. Modify the logic for the scrambler in Exercise 15 as follows: The scrambler searches the unscrambled data input for the value 'F628'h. When this data value occurs, the data is not scrambled and the scrambler state is reset to all 1's for the next cycle clock cycle.

18. Design logic (Verilog or VHDL) for a BIP-8 generator. This logic has a 16-bit input which is passed through to the output. When the 16-bit input data is "F628"h, the BIP-8 accumulator is reset to the value determined by XOR'ing the "F6"h and "28"h bytes. This 16-bit data value marks the beginning of the frame and occurs every 405 clock cycles. Each byte for the next 404 clock cycles is XOR'd with the accumulator state. At the end

of each frame, the accumulated BIP-8 value is saved in a register, and a new BIP-8 calculation begins.

19. Modify the logic for the BIP-8 generator in Exercise 18 as follows: Add a 16-bit data output for the circuit driven by multiplexors which normally redrive the 16-bit data input onto the data output. The multiplexors alternatively allow the most significant 8-bits of the data output to be driven from the BIP-8 value that was saved in a register. Add state machine logic such that the BIP-8 value from the previous frame (which is saved in the register at the end of each frame) is inserted into the 90th byte position of the frame (the most significant byte of the 45th clock cycle after the beginning of the frame).

20. Design logic (Verilog or VHDL) for a serial implementation of the 32-bit CRC generator described by (4.8). Assume the CRC seed value is all 1's. The logic should have a *packet_gate* input and a *crc_reset* input. The *crc_reset* input is asserted (to "1") for one clock cycle while *packet_gate* = "0" to reset the LFSR to the seed value. The LFSR state is updated based on the data input for every clock cycle that *packet_gate* = "1."

21. One method of implementing a CRC checker is to calculate the CRC for the received frame, including the CRC field. The remainder of this calculation (if there are no bit errors) is always the same value. For the CRC implementation in Exercise 20, what would this value be?

22. Design logic (Verilog or VHDL) for an 8-bit parallel implementation of the 32-bit CRC generator described by (4.8). The CRC seed value and control inputs should be the same as specified in Exercise 20.

23. Modify the logic for the CRC generator in Exercise 22 as follows: Add an 8-bit data output for the circuit driven by multiplexors which normally redrive the 8-bit data input onto the data output. The multiplexors alternatively allow each byte of the CRC remainder to be driven onto the data output. Add state machine logic such that each byte of the CRC remainder is driven to the output on the four consecutive clock cycles after *packet_gate* transitions from "1" to "0." (This implementation assumes that *packet_gate* = "1" continuously for transmission of each data packet, and returns to "0" between packets.)

24. Design logic (Verilog or VHDL) for a single instance of an elastic buffer in Fig. 4.9. Assume the elastic buffer is 10-bits wide and 4-symbols deep. Include a *reset* input to initialize the read and write addresses.

25. Design logic (Verilog or VHDL) which instantiates four of the elastic buffers designed in Exercise 24, and connects these in the configuration shown in Fig. 4.11.

26. If the HSS transmitters shown in Fig. 4.11 are not resynchronized, explain how the resulting behavior of the interface can add skew between data lanes. Illustrate this with appropriate timing diagrams.

27. A novice engineer proposes that all of the logic on the chip associated with the transmit interface shown in Fig. 4.11 can be clocked by the TXxDCLK output of one of the HSS cores, and that this eliminates the need for the elastic FIFOs. There are several thousand flip-flops in the logic block that are associated with this transmit interface. Why is this not a good idea?

28. Design logic (Verilog or VHDL) which instantiates four of the elastic buffers designed in Exercise 24, and connects these in the configuration shown in Fig. 4.9. For purposes of this exercise, synchronization issues associated with resetting the FIFO read pointers can be ignored.

29. Design logic to drive the *reset* inputs to the elastic buffers in Exercise 28 such that the FIFO read pointers will exit reset in phase. (Hint: An approach similar to the HSSRESYNC logic described in Sect. 2.4.7 is needed.)

30. Modify the logic for the elastic buffer designed in Exercise 24 as follows:
    - Add control logic in the write clock domain which compares the read and write address. If a K28.0 8B/10B symbol is being written to the buffer, and the read address is lagging the write address by 3 FIFO words (i.e., write address + 1 = read address), then the symbol is not written into the FIFO and the write address is not updated.
    - Add control logic in the read clock domain which compares the read and write address. If a K28.0 8B/10B symbol is being read from the buffer, and the read address is lagging the write address by 1 FIFO word (i.e., read address + 1 = write address), then the read address is not updated in this cycle. This causes the K28.0 symbol to be read twice.

31. Given that comparing the read and write addresses of the elastic buffer in Exercise 30 required the comparison of values from two different asynchronous clock domains, how was this handled in your design? If your design did not handle the asynchronous domain crossing such that the resulting design will operate without bugs, then suggest possible approaches for fixing this problem.

32. The elastic buffer in Exercise 30 uses K28.0 symbols to adjust the FIFO read and write addresses and thus avoid overflow and underflow events. This elastic buffer is used in the receiver of a plesiosynchronous system for which a frequency tolerance of 400 ppm has been specified. However, if the K28.0 symbols do not occur with sufficient frequency, overflows or underflows may still occur. What is the maximum spacing of the K28.0 symbols that can be allowed in the system?

33. Repeat Exercise 32 assuming the frequency tolerance is 50 ppm.

34. Design logic (Verilog or VHDL) for a bit alignment circuit as described by Fig. 4.13. The circuit should have a 10-bit data input, and should include comparators which can detect reception of a K28.5 8B/10B symbol regardless of the bit alignment of this symbol. If a K28.5 symbol is received but it is not aligned on the 10-bit input, then the logic should create an RXxDATASYNC pulse.

35. Assume the K28.5 symbols received by the circuit described in Exercise 34 are initially misaligned by two bits. Draw a timing diagram illustrating operation of the circuit in Exercise 34 until such time as the K28.5 symbols are properly aligned.

36. Design logic (Verilog or VHDL) for a barrel shifter circuit as described by Fig. 4.15. The circuit should have a 10-bit data input and a 10-bit data output, and should be able to realign data to any 10-bit boundary. For purposes of this exercise, assume the multiplexor select is an input to the circuit.

37. Modify the logic for the elastic buffer designed in Exercise 36 as follows: Add comparators which can detect reception of a K28.5 8B/10B symbol regardless of the bit alignment of this symbol. Add control logic which sets the multiplexor selection based on the alignment of the received K28.5 symbol such that the K28.5 is aligned on the 10-bit boundary of the data output.

38. Design logic (Verilog or VHDL) which instantiates two elastic buffers as described in Exercise 30, and two bit alignment circuits as described in Exercise 34. In addition, the logic is needed to implement the following functionality:
    - The 10-bit output of one of the elastic buffers (the reference lane) is always delayed by one clock cycle.
    - The 10-bit output of the other elastic buffer is connected to a barrel shifter circuit which has the capability for delaying the data by 0–3 clock cycles.
    - A control circuit monitors the reference lane output, and adjusts the barrel shifter such that a K28.5 symbol appears on the output of this circuit in the same clock cycle as a K28.5 symbol on the reference lane. (The protocol always sends K28.5 symbols simultaneously on all lanes.)

39. Is the logic implemented in Exercise 38 sufficient for the skew budget in Table 4.4? Why or why not?

Chapter 5
Overview of Protocol Standards

Older protocol standards defined electrical interfaces which used source synchronous bus approaches that were described in Chap. 1. As interface standards evolved to higher and higher data rates, most standards turned to specification of high-speed signals implemented with HSS cores. This chapter is by no means a complete summary of all of the protocol standards for which HSS cores are used, nor does this chapter contain sufficient information to design the logic associated with any given protocol standard. The goal is simply to provide the reader with a basic knowledge of some of the more popular standards.

Some of the protocol standards that are popular and relevant as of the time of this writing are listed below:

1. **Telecom Standards**

    (a) *International Standards*:

    - *SONET / SDH*: Telcordia GR-253-Core

        SONET (Synchronous Optical NETwork) is a serial protocol for optical telecom networks used within North America; SDH (Synchronous Digital Hierarchy) is the equivalent protocol used on other continents. The protocol is based on continuously sending frames containing multiple connection clients at the rate of 8,192 frames per second. Relevant baud rates are: 2.48832 Gbps (OC-48), 9.95328 Gbps (OC-192), and 39.81312 Gbps (OC-768).

    (b) *Optical Internetworking Forum* (OIF):

    OIF is an industry forum which develops *Implementation Agreements* that define interfaces internal to a system that are not within the scope of other standards bodies.

    - *OIF-SxI-5-01.0*: System Interface Level 5 (SxI-5): Common Electrical Characteristics for 2.488–3.125 Gbps Parallel Interfaces

        SxI-5 is an electrical layer standard incorporated by reference in the SFI-5 and SPI-5 protocols. Relevant baud rates are from 2.488 Gbps up to 3.125 Gbps.

    - *OIF-SFI5-01.0*: Serdes Framer Interface Level 5 (SFI-5): 40 Gbps Interface for Physical Layer Devices

        SFI-5 is a protocol layer standard defining the chip-to-chip or chip-to-module interface between a SONET/SDH Framer chip and an OC-768 40-Gbps Serdes chip. It is also used to interface between a Framer or Serdes chip and a Forward Error Correction (FEC) chip. Protocol uses

D. R. Stauffer et al., *High Speed Serdes Devices and Applications,*
© Springer 2008

16 SxI-5 electrical links at 2.48832 Gbps per link to achieve 39.81312 Gbps, plus an additional link used as a deskew reference. Baud rates are higher on links between a FEC chip and a Serdes chip; baud rate in this application depends on the FEC implementation.

- *OIF-SPI5-01.1*: System Packet Interface Level 5 (SPI-5): OC-768 System Interface for Physical and Link Layer Devices

  SPI-5 is a protocol layer standard defining chip-to-chip or backplane interfaces between SONET/SDH Framer chips and other Network Processing Elements (NPE) of an OC-768 system. The protocol uses 16 SxI-5 electrical links to implement an interface with a total bandwidth of 39.81312–50.000 Gbps. Additional links in the opposite direction are used to communicate queue status.

- *OIF-CEI-02.0*: Common Electrical I/O (CEI): Electrical and Jitter Interoperability agreements for 6G+ bps and 11G+ bps I/O

  CEI is an electrical layer standard for interfaces with baud rates in the range of 4.976–6.375 Gbps, and in the range of 9.95–11.10 Gbps. Multiple variants are defined to target various reach objectives (short reach chip-to-chip, and long reach backplane). This electrical layer specification is incorporated by reference in the CEI-P and SFI-5.2 protocols.

- *OIF-CEI-P-01.0*: Common Electrical I/O: Protocol (CEI-P): Implementation Agreement

  CEI-P is a protocol standard for a Physical coding sublayer for CEI electrical interfaces, specifying overhead bits using a variation on 64B/66B block encoding with data scrambling for encoding data on CEI electrical links. The specification includes FEC codes used for error checking, and optionally for error correction. This lower layer protocol is incorporated by reference in the SPI-S protocol.

- *OIF-SFI5-02.0*: Serdes Framer Interface Level 5 Phase 2 (SFI-5.2): Implementation Agreement for 40 Gbps Interface for Physical Layer Devices

  SFI-5.2 is a protocol layer standard defining the chip-to-chip or chip-to-module interface with similar application to SFI-5. The protocol uses four CEI-11G-SR electrical links at 9.95328 Gbps per link to achieve 39.81312 Gbps, plus an additional link used as a deskew reference.

- *OIF-SPI-S-01.0*: Scalable System Packet Interface Implementation Agreement: System Packet Interface Capable of Operating as an Adaption Layer for Serial Data Links

  The SPI-S protocol layer standard defines chip-to-chip or backplane interfaces between SONET/SDH Framer chips and other Network Processing Elements (NPE) of an OC-768 system. The protocol extends SPI-5 and other prior protocols defined by OIF, specifying a scalable interface (both in terms of baud rate and in terms of bit width) capable of achieving any bandwidth. SPI-S builds on top of the CEI-P sublayer, although it may also be used with other 64B/66B encoded protocols.

2. Data Networking Standards

(a) *IEEE Std-802.3 Ethernet*: Ethernet is a packet protocol for data networks. The standard is written as a clause-based document which has expanded over time to include a multitude of interface variants. A sampling of relevant variants are listed below.

- *IEEE Std 802.3, 2000 Edition, Clause 38*: 1000Base-SX, LX

  This clause defines a Physical Medium Dependent (PMD) layer for 1,000 Mbps Ethernet. This protocol uses 8B/10B encoding, so the actual baud rate is 1.25 Gbps.

- *IEEE Std 802.3ae, 2002 Edition, Clause 47*: XGMII Extended Sublayer (XGXS) and 10-Gb Attachment Unit Interface (XAUI) Specification

  This clause defines a Physical Medium Dependent (PMD) layer for a 10-Gbps Ethernet interface between two chips, and is sometimes used across a backplane. The XAUI interface is specified as four links using 8B/10B encoding and running at a baud rate of 3.125 Gbps.

- *IEEE Std 802.3ak, 2004 Edition, Clause 54*: 10GBASE-CX4 Specification

  This clause defines a physical medium dependent (PMD) layer for a 10-Gbps Ethernet interface over an IBT cable. The protocol is similar to the XAUI specification in clause 47.

- *IEEE Std 802.3ap, Clauses 70, 71, and 72*: Backplane Ethernet

  These clauses define a physical medium dependent (PMD) layers for using the Ethernet protocol across a backplane. Clause 70 specifies a 1,000 Mbps PMD layer similar to clause 38. Clause 71 specifies a four link at 3.125 Gbps PMD layer similar to clause 47. Clause 72 specifies a serial 10.3 Gbps PMD layer.

3. Storage Networking Standards

(a) *INCITS T11 Fibre Channel*: Fibre Channel is a packet protocol for storage networks. Fibre Channel is used both for communication between systems in storage area networks and between host and disk drive devices. Fibre Channel is specified by a collection of documents which specify the physical layer, protocol, and other aspects of systems. Key documents are listed below.

- *FC-PI-4*: Fibre-Channel: Physical Interfaces: 4

  This standard specifies electrical and optical physical layer Fibre Channel variants defined by the T11.2 Task Group. The more popular serial variants utilize a single link at a baud rate of 1.0625, 2.125, 4.250, or 8.500 Gbps. There is also a 10 Gbps variant specified in a separate document which utilizes an interface similar to XAUI (see prior Ethernet descriptions).

- *FC-FS-2*: Fibre-Channel: Framing and Signaling: 2

   This standard specifies the Fibre Channel protocol for use over any of the physical layer variants specified in FC-PI-x and is defined by the T11.3 Task Group. The protocol utilizes 8B/10B block encoding. Serial variants with baud rates of 8.500 Gbps also scramble data prior to the block encoding stage.

(b) *INCITS T10 Serial Attached SCSI*: 2 (SAS-2) *T10/1760*: SAS is a packet protocol for storage networks. SAS is used for communication between host and disk drive devices. Current supported baud rates are 1.5 and 3.0 Gbps; a 6.0-Gbps baud rate is expected in the future. The protocol utilizes 8B/10B block encoding with data scrambled prior to the block encoding stage.

(c) *Serial ATA International Organization: Serial ATA Revision 2.5* (SATA): SATA is a packet protocol for storage networks. SATA is used for communication between host and disk drive devices. Current supported baud rates are 1.5 and 3.0 Gbps; a 6.0-Gbps baud rate is expected in the future. The protocol utilizes 8B/10B block encoding.

4. **Transaction Protocols**

(a) *Peripheral Component Interconnect Special Interest Group* (PCI-SIG): *PCI Express Baseline Specification, Version 2.0*: PCI Express is a transaction protocol for interconnecting peripheral devices in computing and communication platforms. Current supported baud rates are 2.5 and 5.0 Gbps. Ports use 1, 2, 4, 8, 12, 16, or 32 links in parallel to scale bandwidth as needed. The protocol utilizes 8B/10B block encoding with data scrambled prior to the block encoding stage.

(b) *InfiniBand Trade Association* (IBTA) *Infiniband Architecture Specification Volume 2, Release 1.2*: Infiniband is a transaction protocol for interconnecting host and peripheral devices, generally through a switch fabric. Current supported baud rates are 2.5, 5.0, and 10.0 Gbps. Ports use 1, 4, or 12 links in parallel to scale bandwidth as needed. The protocol utilizes 8B/10B block encoding.

A selection of the above standards are described in more detail in the remainder of this chapter. Detailed references for these standards documents are provided in Sect. 5.6.

5.1 SONET/SDH Networks

The original work on the Synchronous Optical NETwork (SONET) standard was performed by Bell Communications Research (Bellcore), now Telcordia, which was founded as the result of the breakup of American Telephone and Telegraph (AT&T) in 1984. Prior to SONET, higher bandwidth telecommunications equipment tended to use proprietary specifications and interoperability of equipment made by different vendors was not generally possible. Bellcore proposed SONET to the American National Standards Institute (ANSI) in 1985 as a network solution for fiber standardization.

The International Telecommunications Union (ITU-T) began working on the Synchronous Digital Hierarchy (SDH) standard in 1986. This was in response to progress in the United States on SONET, which was not compatible with existing needs of Europe. The United States officially proposed SONET to ITU-T for consideration in the SDH standard in 1986. (The ITU-T is sponsored by the United Nations and the official representative to this body is the United States Department of State.) For political reasons the direction of the SDH standard diverged from the original proposals. Intense work in the ANSI T1X1.4 committee resulted in publication of a revised proposal in February, 1988 that adopted for the most part the bit rates and formats in the emerging SDH standard. ITU-T published the SDH specifications, known as G.707, G.708, and G.709, in June 1988. SONET and SDH differs in terminology, and in the definition and use of certain overhead fields and alarm conditions, however, they are otherwise fully compatible.

Currently, SONET is the dominant standard used for telecommunications in North America, and SDH is the dominant standard on other continents. Most telecommunications equipment can be configured to comply with either standard.

This section describes the system reference model and frame formats used by SONET/SDH. Overhead bytes in the frame format are described to the extent that they are relevant to the HSS core carrying SONET/SDH traffic. Clock requirements are also discussed, since these are especially relevant to SONET/SDH networks.

5.1.1 System Reference Model

The SONET system reference model is shown in Fig. 5.1. There are four protocol layers defined by SONET: the photonic layer, the section layer, the line layer, and the path layer. These layers are defined below:

Photonic layer. This layer consists of the transmission path consisting of an electrical to optical conversion, optical interconnection, and optical to electrical conversion. There are a variety of photonic layers that have been employed for SONET and SDH networks.

Section layer. This layer manages transport of SONET frames across the physical path using the photonic layer. Section Overhead (SOH) bytes of the SONET frame are generated and monitored by the section layer regenerators. These overhead bytes are used for framing, scrambling, and error monitoring. Regenerators receive and retransmit the SONET frames, and do not modify the payload contents of the frame or overhead bytes for other layers. Regenerators must use the timing of the received signal as a reference for transmission since there is no provision in the section layer to add or drop bits of the frame.

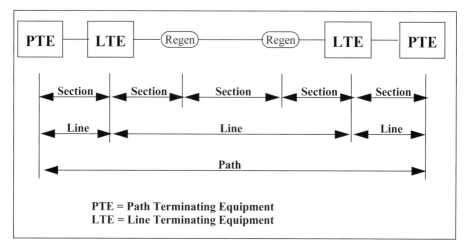

Fig. 5.1 SONET architecture

Line layer. The line layer manages the transport of the SONET payload across the physical medium. Functions of this layer include multiplexing and synchronization functions. LTE units can multiplex multiple lower speed SONET lines onto a higher speed SONET line, and vice versa. LTE units also perform pointer management to allow bytes to be added or dropped to adjust the frequency of the received data to the local clock reference. Line Overhead (LOH) bytes of the SONET frame are generated and monitored by the LTE unit to perform these functions.

Path layer. This layer maps network service components into a format that the line layer requires, and manages the end-to-end transport of these services. Path Overhead (POH) bytes of the SONET frame are generated and monitored by the PTE unit to perform these functions.

It may also be interesting to the reader to understand the scale of the above layers. The physical length of a section can vary from hundreds of meters to many kilometers depending on the type of photonic layer being used. LTE units break the network into maintenance spans, and the line span may be anywhere from several kilometers to hundreds of kilometers in length. The path termination can be in different cities or on different continents.

5.1.2 STS-1 Frame Format

The basic frame format of SONET is the Synchronous Transport Signal Level 1 (STS-1) frame shown in Fig. 5.2. This frame is viewed as a two-dimensional array of bytes, consisting of columns 1–90, and rows 1–9. Row 1 is transmitted from left to right, followed by row 2, etc. In order to achieve transmission of one frame every 125 μ s, the frame must be sent at a line rate of 51.840 Mbps.

Columns 4–90 inclusive contain payload data, including any path management overhead. Columns 1, 2, and 3 contain SOH and LOH overhead bytes as shown in the figure. These are defined further below.

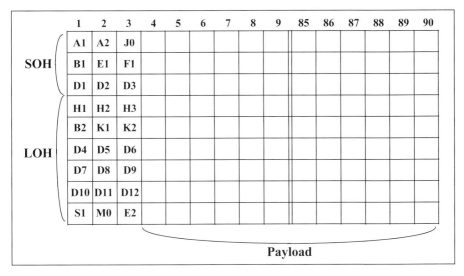

Fig. 5.2 STS-1 Frame

5.1.2.1 SONET Scrambler

Except for the bytes in row 1, columns 1–3, the SONET frame is scrambled with a sidestream scrambler. The row 1 overhead bytes are not scrambled so that they can be reliably used as a framing pattern. The scrambler resets after the J0 byte of each frame and uses the polynomial:

$$G(x) = x^7 + x^6 + 1. \tag{5.1}$$

SONET does not use any block coding. Scrambling provides a high probability that bit transitions will occur in the serial data. However, it is possible for payload data to contain a pattern which, when scrambled, produces a long run length of "1" or "0" bits. It is generally accepted that designing hardware to handle a run length of 80 is sufficient; the probability of longer run lengths is less than the specified 10^{-12} BER.

5.1.2.2 Section Overhead (SOH) Bytes

The nine bytes located in rows 1, 2, and 3, and columns 1, 2, and 3, are used by section regenerator equipment to perform section functions. The following bytes are of particular relevance to the topics in this text:

Framing Bytes (A1, A2): The A1/A2 bytes are a hardcoded 16-bit hexadecimal value of "F628"h, and are not scrambled. The receive logic in the section regenerator looks for the "F628"h data pattern in order to determine where the frame begins, and aligns row/column states based on this. Note that this data pattern may also occur in other overhead bytes or in payload data (called *aliasing*). The receive logic looks for this pattern to repeat at the proper interval for several frames before declaring that frames are being received correctly.

B1: The B1 byte is a BIP-8 even parity calculation generated by XOR'ing all the scrambled bytes in the frame together, and then inserting the result in the B1 byte position in the following frame. The B1 byte is inserted in the frame before scrambling. This is the primary method of error checking used by the SONET section layer.

Other SOH bytes provide management functions which, while important, are beyond the scope of this text.

5.1.2.3 Line Overhead (LOH) Bytes

The 18 bytes located in rows 4–9, and columns 1, 2, and 3, are used by LTE units to perform line functions. The following bytes are of particular relevance to the topics in this text:

Pointer Bytes (H1, H2, H3): The payload frame, called the synchronous payload envelope (SPE), may begin at any position within the payload portion of the STS-1 frame. The H1/H2 bytes provide a row/column position for the first byte of the SPE as shown in Fig. 5.3. The SPE frames continue from this position for the next 783 payload bytes, stretching into the next frame and the start of the next SPE.

From time to time the LTE unit may adjust the start of the SPE in order to compensate for clock frequency differences. Figure 5.3 illustrates an example of *positive stuffing* to compensate for the SPE source running slower than the line transmitter in the LTE. When this occurs, an extra byte of nondata is stuffed in the payload position following the H3 byte, and the payload pointer is incremented so that the next SPE starts one byte later in the STS-1 frame.

A pointer adjustment can also be made to adjust for the SPE source running faster than the line transmitter in the LTE. This is called *negative stuffing*. When this occurs, an extra byte of payload data is transmitted as the H3 byte, and the payload pointer is decremented so that the next SPE starts one byte earlier in the STS-1 frame. The H3 byte does not contain useful information except when used for negative stuffing.

The format of the H1/H2 bytes and the protocol for the LTE generation and response for these bytes is beyond the scope of this text. However, note that the maximum rate at which pointers may be adjusted is once every four STS-1 frames, and pointers can only be adjusted by one byte at a time. This places stringent limits on the accuracy of clock sources for telecom equipment.

B2: The B2 byte is a BIP-8 even parity calculation generated by XOR'ing all the bytes in the frame together prior to scrambling, except for the SOH bytes. The result is inserted as the B2 byte before scrambling the following frame. This is the primary method of error checking used by the SONET line layer.

Note that B2 differs from B1 in that (1) B2 is calculated prior to scrambling while B1 is calculated using scrambled data and (2) B2 excludes the SOH bytes from the calculation. This is consistent with the SONET layer definitions, since the SOH contents and the scrambled version of the data are not accessible to the line layer.

Other LOH bytes provide management and status functions which, while important, are beyond the scope of this text.

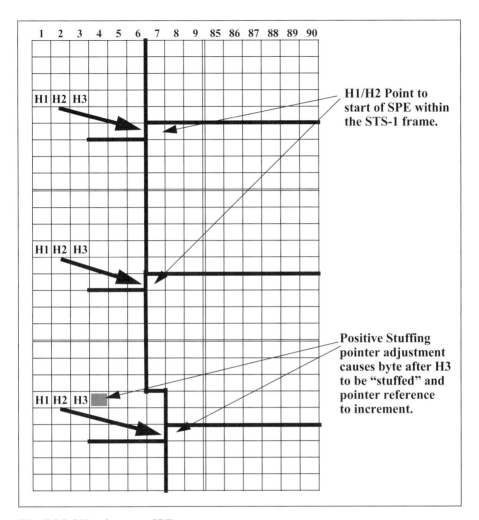

Fig. 5.3 LOH pointers to SPE

5.1.2.4 Synchronous Payload Envelope (SPE) and Path Overhead (POH)

The SPE format in Fig. 5.4 consists of 9 rows by 87 columns. Of these, the first column contains the path overhead (POH) bytes, columns 30 and 59 contain *fixed stuff* which is not useful information, and the remaining columns contain the payload. Note that the column numbers in the SPE are relative to the start of the SPE in the STS-1 frame as determined by the H1/H2 bytes.

POH bytes provide management and status functions which, while important, are beyond the scope of this text. However, the following byte is significant to this discussion:

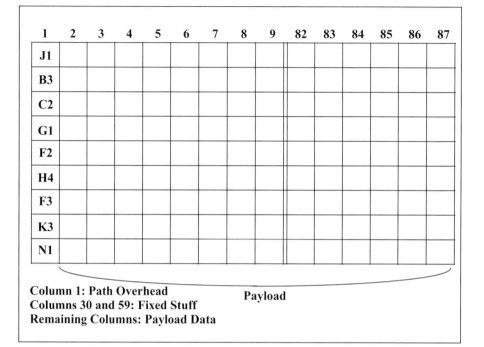

1	2	3	4	5	6	7	8	9	82	83	84	85	86	87
J1														
B3														
C2														
G1														
F2														
H4														
F3														
K3														
N1														

Column 1: Path Overhead   **Payload**
Columns 30 and 59: Fixed Stuff
Remaining Columns: Payload Data

Fig. 5.4 Synchronous payload envelope (SPE) format

B3: The B3 byte is a BIP-8 even parity calculation generated by XOR'ing all the bytes in the SPE together prior to scrambling. The result is inserted as the B3 byte before scrambling the following SPE. This is the primary method of error checking used by the SONET path layer.

5.1.3 STS-N Frame Format

SONET evolved from the proprietary methods of the telecommunications industry for multiplexing lower speed lines onto higher speed lines. As a result, a key feature of SONET is that some number n of STS-1 frames may be multiplexed together to form an STS-n frame. When STS-1 frames are stacked in this manner, the transmission order becomes: send the row 1 column 1 byte of each STS-1 frame in turn, followed by the row 1 column 2 byte of each STS-1 frame in turn, and so forth. When lower speed STS-1 frames are multiplexed into higher speed STS-n frames, the bit rate increases by n. Regardless of the multiplexing factor, there are always 8,000 STS-n frames sent per second, or one frame per 125 µs.

Note that this multiplexing is performed at the SONET line layer. Multiple SPE sources are mapped into STS-1 frames, and these STS-1 frames are multiplexed together to form an STS-n frame. Each STS-1 frame has its own LOH and its own SPE (including POH). The H1/H2 bytes for each STS-1 frame do *not* have to be the same, and SPEs for each STS-1 frame may start at arbitrary different positions.

SOH bytes are added in the SONET section layer after line layer multiplexing has already occurred. For this reason, only the SOH bytes associated with the first STS-1 frame are defined, and the SOH byte positions in the other STS-1 frames are not used by the protocol. The exception to this is:

Framing Bytes (A1, A2): These bytes are fixed values for all STS-1 frames in the STS-n frame. Receiving equipment uses the transition between the A1 and A2 bytes to determine the frame alignment.

J0/Z0: The J0 byte is associated with the first STS-1 frame in the STS-n frame. The corresponding bytes in the remaining frames are designated as Z0 bytes, and are reserved for future definition.

The above description of multiplexing independent STS-1 frames into an STS-n frame is the classical definition of SONET multiplexing. Other concatenated formats are also defined, which permit an STS-n frame ($n > 1$) to have a single LOH and a concatenated payload. Details of these definitions are beyond the scope of this text. When such formats are used, a "c" is appended to the notation. For example, an STS-3c frame is a concatenated STS-3 frame.

Although in theory any STS-n level is allowed, only certain values of n are commonly used. The common levels and their corresponding line rates are shown in Table 5.1. Note that although $n = 1$ is the base value for frame definition, it is *not* a commonly used line rate. The lowest SONET line rate in common use corresponds to multiplexing together three STS-1 frames to form an STS-3 frame. This corresponds to the SDH definition of an STM-1 frame. Commonly used line rates increase by factors of four thereafter.

Another point that should be made is that the row 1 SOH bytes are not scrambled for line rates up to STS-192/OC-192. For STS-192, there are 576 consecutive unscrambled bytes in the frame, which begins to affect the spectral characteristics and DC balance of the signal. Depending on the J0/Z0 byte values, excessive run lengths are also possible. Scrambling is therefore redefined for OC-768 such that only the last 64 A1 bytes and first 64 A2 bytes are unscrambled. This was deemed acceptable since framing functions only look for the transition between the A1 and A2 bytes.

Table 5.1 Commonly used SONET/SDH speeds

Optical level terminology	Electrical level terminology	Line rate	SDH equivalent terminology
OC-3	STS-3	155.520 Mbps	STM-1
OC-12	STS-12	601.344 Mbps	STM-4
OC-48	STS-48	2.48832 Gbps	STM-16
OC-192	STS-192	9.95328 Gbps	STM-64
OC-768	STS-768	39.81312 Gbps	STM-256

5.1.4 Clock Distribution and Stratum Clocks

A traditional packet switching network is only concerned with switching the source and destination of data on a per packet basis. Packets are not subdivided with various bytes fed to different destinations. Therefore, the need for network synchronization may not be obvious to readers experienced with IP networks and the Internet.

In SONET and SDH the payload envelope (SPE) is subdivided. When the SONET path is used for classic telephone voice connections, each payload byte of the SPE is associated with a different telephone call and has its own independent source and destination. Each telephone conversation is allocated one byte in each STS-1 frame, providing a total bandwidth of 8,000 bytes per second, and generating a voice bandwidth of 4 kHz. Switching systems must be able to separate out individual bytes in the SPE to add telephone connections originating locally, or drop telephone connections terminating locally.

In order for all of this to work, the SONET/SDH network must operate synchronously. Furthermore, to facilitate intercommunication between networks owned by different carriers, all of the networks in the world must be synchronized. Clock distribution therefore becomes a major concern.

The primary reference clock (PRC) used for SONET systems is the atomic master clock maintained by the United States government in Boulder, Colorado. This clock reference is distributed using the Global Positioning System (GPS). Prior to GPS, the Long-Range Navigation System (LORAN) was used. This clock is distributed to major sites containing SONET network equipment as the *Stratum 1* primary clock reference. *Stratum 2* clocks are connected to this stratum 1 clock; *Stratum 3* clocks connect to stratum 2 clocks, etc. When clock distribution fails, then a local oscillator must provide the clock reference within a defined frequency tolerance.

Table 5.2 describes the four stratum levels for clocks. The stratum 1 clock requires either a Cesium or Rubidium clock reference to back up the clock distribution from the GPS source. This is then distributed to toll switches, which have a backup clock source requiring stratum 2 accuracy. The stratum 2 clock is sufficiently accurate so that the toll switch should continue to operate with fewer than 255 errors in 86 days after loosing the connection to the stratum 1 clock reference.

Stratum 2 clocks are distributed across stratum 3 distribution to Local Switches in local telephone exchanges and Digital Cross-Connect Systems (DCS). These devices must have local clock backups with stratum 3 accuracy that are sufficient so that the unit can continue to operate with fewer than 255 errors in 24 hours after loosing the connection to the stratum 2 clock reference.

Rather than using expensive clock connections in the SONET network, many lower-level devices use *loop timing* to receive and retransmit data. Stratum 4 user equipment, section regenerators, etc. will often provide the capability to use the clock derived from the receive data to generate the reference clock for transmission. In order to support loop timing, HSS receivers must provide a suitable clock output for the received data clock which can be used by a PLL to supply a reference clock for the HSS transmitter. This clock must not be gated, and must minimize jitter as much as possible.

Table 5.2 Commonly used SONET/SDH speeds

Stratum level	Accuracy	Connected equipment
Stratum 1	0.00001 ppm (1 s / 300,000 yr)	Primary clock reference
Stratum 2	0.016 ppm	Toll switches
Stratum 3	4.6 ppm	Local switches, Digital cross-connect systems
Stratum 4	No requirement	PBX systems, T1 muxes

5.2 OIF Protocols

The Optical Internetworking Forum (OIF) is an industry forum in which member companies develop *Implementation Agreements* (IAs) which standardize interfaces within telecom systems that are not within the scope of internationally recognized standards bodies. OIF develops implementation agreements for both optical and electrical interfaces; the electrical interface applications are of interest to this text.

This section describes the system reference model used by OIF, and describes the relationship of this model to a number of OIF implementation agreements and to SONET/SDH. HSS devices are used in the implementation of electrical interfaces with baud rates of 2.488 Gbps and above. A selection of recently published protocol layer implementation agreements, and the underlying electrical layer implementation agreements, are described in some detail. Note that descriptions of these interfaces primarily focuses on details relevant to Serdes cores, as well as details related to the approach of the interface to the issues described in Chap. 4. The reader is referred to the references at the end of this chapter for a more complete description of these interfaces.

5.2.1 System Reference Model

The system reference model used by OIF is shown in Fig. 5.5. This model describes a line card as consisting of a Serdes device connected to a PHY device by a Serdes-Framer Interface (SFI), with the PHY device then connected to Link Layer devices by a System Packet Interface (SPI). The dataflow from the Link device through to the Serdes device is the *transmit interface* from the perspective of the line card; the dataflow from the Serdes device through to the Link device is the *receive interface*. Note that each electrical interface in this path has HSS transmitters and receivers associated with it; therefore the context of these terms is important. Sometimes these paths are referred to as the *egress* and *ingress* path (respectively) to avoid confusion.

The Serdes device in this figure is generally one or more chips implementing a SONET/SDH OC-192 (10 Gbps) or OC-768 (40 Gbps) interface, although other possibilities also exist. Very often the Serdes is implemented in a Silicon–Germanium (SiGe) or Indium Phosphide (InP) technology, where interface complexity is a significant cost and power driver. SFI protocols therefore strive to minimize the complexity of this interface.

The other end of the SFI connects to a PHY chip. For SONET/SDH applications, this is generally a SONET/SDH Framer chip or a Forward Error Correction (FEC) chip. SFI is used to connect Serdes devices to FEC or Framer devices; it is also used to connect the FEC device to the Framer device. The receive side of the Framer chip processes the various overhead layers of the SONET/SDH protocol, and extracts data for the link layer. This data is formatted into packets and sent to the receive link layer device using an SPI protocol. The transmit side of the Framer chip receives packets from the transmit link layer device and formats this into SONET/SDH frames for transmission by the Serdes device.

Figure 5.6 extends this system reference model into the context of a broader system. The link layer devices in Fig. 5.5 are represented generically as *network processing elements* (NPEs) in Fig. 5.6. SPI connects the SONET/SDH Framer to the link layer NPEs, and is also used to connect NPEs to other NPEs and to the switch fabric. This packet processing environment is independent of the protocol implemented by the line card, and SPI has also been used in Ethernet systems. Note that Fig. 5.6 distinguishes the SPI packet traffic from the transaction-based look-aside interfaces in the system, which require different protocols.

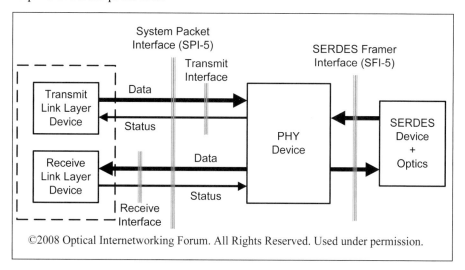

Fig. 5.5 OIF System reference model

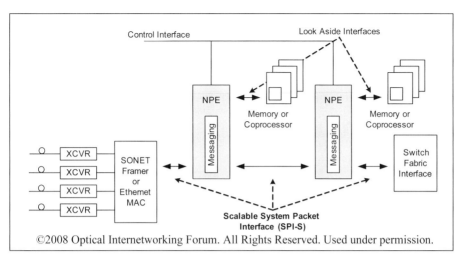

Fig. 5.6 OIF system reference model (extended)

Table 5.3 OIF implementation agreements

SONET rate	Interface type	Protocol layer IA	Electrical layer IA	General notes
OC-48	SPI	OIF-SPI3-01.0 (SPI-3)		104 Mbps x 8 lanes + status
OC-192	SPI	OIF-SPI4-01.0 (SPI-4.1)		311+ Mbps x 32 lanes + status. Limited industry use
		OIF-SPI4-02.01 (SPI-4.2)		622+ Mbps x 16 lanes + status. Widely used in industry
	SFI	OIF-SFI4-01.0 (SFI-4)		622+ Mbps x 16 lanes. Widely used in industry
		OIF-SFI4-02.0 (SFI-4.2)	OIF-SxI5-01.0 (SxI-5)	2.5 Gbps x 4 lanes
OC-768	SPI	OIF-SPI5-01.1 (SPI-5)	OIF-SxI5-01.0 (SxI-5)	2.500–3.125 Gbps x 16 lanes + status
	SFI	OIF-SFI5-01.0 (SFI-5)	OIF-SxI5-01.0 (SxI-5)	2.500–3.125 Gbps x 16 lanes + deskew
		OIF-SFI5-02.0 (SFI-5.2)	OIF-CEI-02.0 Clause 8	9.952+ Gbps x 4 lanes + deskew
OC-xxx	SPI	OIF-SPI-S-01.0 (SPI-S) OIF-CEI-P-01.0 (CEI-P)	OIF-CEI-02.0	Scalable to any width and link speed

Table 5.3 relates various OIF implementation agreements to the system reference model and to generations of SONET/SDH. OIF protocols associated with OC-48 are designated as *level 3* and contain a "-3" in the common name; OC-192 protocols are indicated by a "-4," and OC-768 protocols with a "-5." OIF protocols are additionally correlated to the reference model as either SFI or SPI protocols. (There are also OIF protocols associated with Time Domain Multiplexor (TDM) systems which are not discussed in this text.) From time to time, as newer HSS technologies come into existence, OIF has published an implementation agreement which is intended to replace a previous implementation agreement with an interface definition that uses fewer signals. For instance, SFI-5.2 implements a 4 x 10 Gbps datapath to replace the SFI-5 which used a 16 x 2.5 Gbps interface.

Earlier OIF implementation agreements defined the electrical layer and the protocol layer of the interface in the same document. Starting with protocols based on 2.5-Gbps HSS technologies, OIF began defining the electrical layer in a separate document, and using this layer in multiple protocols. Starting with SPI-S, OIF has also moved to scalable protocols rather than redefining the protocol for every SONET/SDH generation.

To provide a representative and hopefully relevant sampling of the various protocols in Table 5.3, this text focuses on descriptions of SFI-5.2 and SPI-S. These protocols are the most recent generations of the SFI and SPI protocols, respectively. This text also describes the corresponding implementation agreements which provide lower level protocol and electrical layer building blocks for these protocols.

5.2.2 SFI-5.2 Implementation Agreement

The SFI-5.2 Implementation Agreement targets OC-768 SONET/SDH applications using a 4 lane by 9.952 Gbps datapath. SFI-5.2 incorporates the electrical layer defined by CEI Clause 8 by reference, which supports baud rates in the range of 9.95–11.1 Gbps to allow for telecom applications using FEC protocols. The SFI-5.2 protocol does not require data to be encoded in any specific way (as long as data meets criteria defined by CEI), but does contain specific features to facilitate transmission of bit-interleaved SONET/SDH frames.

5.2.2.1 Reference Model Description

The SFI-5.2 reference model is shown in Fig. 5.7. As shown in this figure, the SFI-5.2 interface is used between Serdes and FEC devices, and between FEC and Framer devices. For systems which do not include a FEC device, SFI-5.2 is used to connect the Serdes device directly to the Framer device. Each interface consists of a 4 lane wide datapath plus an additional lane used for deskew. The deskew channel uses pattern matching to align each data lane with the deskew channel (and thereby align data lanes with each other). This is described more later.

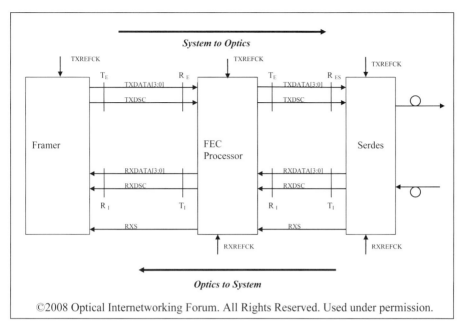

©2008 Optical Internetworking Forum. All Rights Reserved. Used under permission.

Fig. 5.7 Serdes–Framer interface reference model

Compliance points in the system are identified at the various device pins using the designations T_I, R_I, T_E, R_E, and R_{ES}. These designations differentiate the various compliance points for the electrical transmitters (T) and receivers (R) on the egress (E) and ingress (I) datapaths. Note that the Serdes input R_{ES} is differentiated from other egress receivers. Jitter budgets are described for each of these combinations of compliance points in more detail later.

SFI-5.2 signals are described in Table 5.4.

5.2.2.2 Deskew Description

The contents of the deskew channel are generated as described in Fig. 5.8. A 10-bit frame is generated as follows:

1. One bit from each of the data lanes is transmitted in a round-robin fashion on the deskew channel.

2. The odd parity calculation for the bits in step 1 is transmitted on the deskew channel.

3. Step 1 is repeated.

4. The even parity calculation for the bits in step 3 is transmitted on the deskew channel.

Table 5.4 SFI-5.2 pin descriptions

Pin name	Direction	Description
Receive signals		
RXDATA[3:0]	Optics to system	Receive Data. Data is bit-interleaved with RXDATA[3] being first data bit received from serial optics. Baud rate is in range 9.95–11.1 Gbps
RXDSC	Optics to system	Receive Deskew Channel. Continuously transmits 10-bit reference frames as described below
RXREFCK	Optics to system	Receive Reference Clock. The frequency of RXREFCK is 1/16th the baud rate of RXDATA/RXDSC. This clock is recovered by the optics from the serial data. Signal must be driven; use at the receiver is optional
RXS	Optics to system	Receive Status. LVCMOS active high alarm indicating data is invalid. Signal must be driven; downstream use is optional
Transmit signals		
TXDATA[3:0]	System to optics	Transmit Data. Data is bit-interleaved with TXDATA[3] being first data bit transmitted by serial optics. Baud rate is in range 9.95–11.1 Gbps
TXDSC	System to optics	Transmit Deskew Channel. Continuously transmits 10-bit reference frames as described below
TXREFCK	System to optics	Transmit Reference Clock. The frequency of TXREFCK is 1/16th the baud rate of TXDATA/TXDSC. At least one device in the transmit chain must use this clock

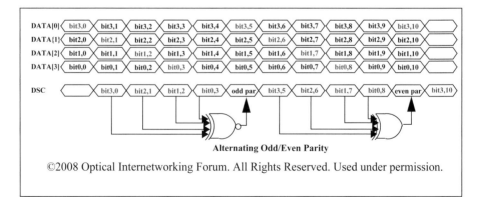

Alternating Odd/Even Parity

Fig. 5.8 Deskew channel generation

The receiver uses the deskew channel to align the data lanes as follows:

1. The parity bits are used to determine the frame alignment of the deskew channel. The deskew logic at the receiver calculates parity for each 4 bits and compares the results of this calculation to the received parity bit. If mismatches occur, then the frame alignment is not correct, and the deskew logic steps its reference point (possibly using RXxDATASYNC on the HSS core). When parity calculations consistently predict the parity bits on the deskew channel, then the deskew frame alignment is correct.

2. Once step 1 is complete, the deskew logic compares each data bit to the corresponding data bits on the deskew channel. If the bits do not match, the data lane is misaligned, and the deskew logic steps its reference point on the data lane (possibly using RXxDATASYNC on the HSS core, and additional downstream logic if more range is required). When data bits consistently match, the data lane is aligned.

3. When all data lanes are aligned, then the interface is aligned and ready to receive data correctly.

The deskew frame definition is considerably less complex than that used in the SFI-5 Implementation Agreement. This simplification was justified to optimize the cost and power of the Serdes devices, which are often implemented using non-CMOS technologies.

5.2.2.3 Nibble Inversion

SFI-5.2 is protocol agnostic and allows transmission of data encoded in any manner that is compliant with the CEI electrical layer. However, SFI-5.2 objectives specifically require the interface to be capable of carrying bit-interleaved SONET/SDH data. The framing pattern of SONET/SDH is not scrambled, and can lead to excessive run lengths and DC unbalance.

To eliminate these issues, SFI-5.2 requires data on the data lanes to be inverted for five nibbles out of every ten nibbles. This inversion is mandatory when the interface is carrying SONET/SDH data, and is optional otherwise. The five nibbles during which the data is to be inverted are correlated with the deskew channel frame, and correspond to the five data nibbles (five bits on each lane) leading up to and coinciding with transmission of the odd parity bit on the deskew channel.

5.2.2.4 Clock Architectures

SFI-5.2 defines reference clocks for the egress and ingress paths, but the connection of these clocks is left as an exercise for the system designer. In most SONET/SDH systems, reference clocks are produced by high-quality oscillators and driven directly to the Serdes device in order to optimize jitter performance. The Serdes device divides this reference clock and distributes it to other chips in the system.

Table 5.5 SFI-5.2 jitter / wander / skew budget

Parameter	Signal type	System points			Units
		Ti/Te	Ri/Re	Res	
Skew	Data	5.50	11.00	6.10	UI Peak
Correlated wander	All	5.00	7.00	7.00	UI peak to peak
Uncorrelated wander	All	0.65	0.75	0.75	UI peak to peak
Total wander	All	5.65	5.75	5.75	UI peak to peak
Relative wander	All	1.30	1.50	1.50	UI peak to peak
Skew + (rel. wander) / 2	All	6.15	11.75	6.85	UI peak to peak
Deterministic jitter	Data	0.15			UI peak to peak
Total jitter	Data	0.30	0.65	0.65	UI peak to peak

5.2.2.5 Skew Budget

Table 5.5 describes the Skew and Wander budget, and the Jitter budget for various compliance points in Fig. 5.7. The terms *skew*, *wander*, and *jitter* were defined in Sect. 4.1.2.5. The deskew logic in the receiver must have sufficient range to accommodate the sum of the skew and wander seen at the receiver input (plus any additional skew/wander introduced by the receiver). Note that the skew requirement at R_{ES} is significantly more stringent than for R_E; this is to minimize complexity of the Serdes device.

5.2.3 SPI-S Implementation Agreement

The *Scalable System Packet Interface* (SPI-S) defines an interface for transmitting packet traffic between *Network Processor Elements* (NPEs). Fig. 5.9 illustrates various system configurations. Each interface consists of some number of data lanes, and an optional status channel transmitting in the opposite direction. Unidirectional and bidirectional configurations are allowed; asymmetric unidirectional configurations include status channels in both directions.

The SPI-S protocol can be mapped onto an IEEE 64B/66B coding and scrambling layer, or onto a CEI-P coding and scrambling layer. Both of these layers are based on a data rate to baud rate ratio of 64:66. The protocol is independent of the electrical layer definition, however, is generally assumed to use one of the options defined in the CEI Implementation Agreement.

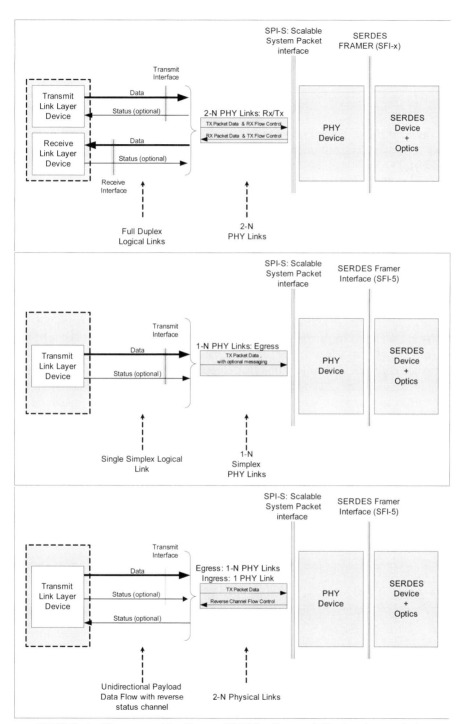

Fig. 5.9 SPI-S configurations

5.2.3.1 Data Path Operation

Data is transmitted over SPI-S in *bursts* with *idles* transmitted between bursts. The bursts may be complete *packets*, where the packet includes *control words* to indicate the start of the packet, end of packet, etc., and the packet length may be any integer number of bytes. Alternatively, the packet may be subdivided into *segments*, where the length of a segment is a provisionable constant, and where one or more segments may be transmitted in a burst.

Data within the packet is organized into 64-bit *blocks* as shown in Fig. 5.10. Each block consists either of a *control word* and four data bytes, or of eight data bytes. A *tag* field indicates which block format applies. When SPI-S is mapped in IEEE 64B/66B, the tag field selects whether the word sync is "01" or "10"; for CEI-P the tag is mapped to a dedicated bit in the CEI-P word. The control word consists of *flags* which indicate the control word type, an *address* specifying the destination of the packet, and a *CRC* field used for error detection. Detailed definitions for the control word are beyond the scope of this text.

Note that SPI-S does not define a baud rate or a data width. The protocol is scalable to any baud rate or data width.

5.2.3.2 Status Channel

SPI-S includes a status channel used to transmit queue status from the receiver backward to the transmitter source. The status channel is defined as two bits which indicate whether the queue is *satisfied*, *hungry*, or *starving*. This status channel must be transmitted by the receiver device at the same baud rate as the received data. If the HSS cores from both chips are provided with a reference clock from a common clock source, then this requires no special attention. If a common clock source is not guaranteed, then the HSS core transmitting the status channel must use loop timing and connect the HSSREFCLK input through a PLL to the recovered clock output of one of the receiver lanes on the HSS core receiving the SPI-S data.

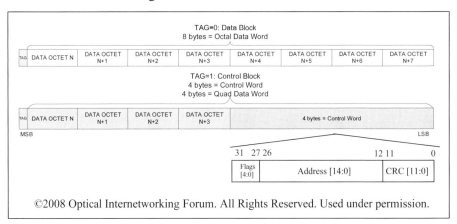

Fig. 5.10 CEI-P 64B/66B block formats

SPI-S defines the link state machines at the receiver and transmitter that generate and respond to the status channel as well as control word fields. SPI-S also defines provisionable parameters relevant to these link states. Link state and parameter definitions are beyond the scope of this text.

5.2.3.3 Framing Modes

As mentioned previously, SPI-S may be mapped onto an IEEE 64B/66B coding and scrambling layer, or onto a CEI-P coding and scrambling layer.

IEEE 64B/66B Coding:

IEEE 64B/66B coding and scrambling is defined in IEEE 802.3ae-2002 Clause 49. The SPI-S block tag is mapped into the IEEE 64B/66B framing field with a value of either a "01" or "10," and the remainder of the SPI-S block is mapped into the 64-bit payload field.

Framing is performed by receiver deskew logic by searching for the bit positions within the 66-bit symbol which are consistently either "01" or "10," and never "00" or "11." This provides a framing reference to distinguish tag fields from payload data. When the data width of the interface is 2 or more lanes, deskew is performed across lanes by aligning the tag field references.

CEI-P Coding:

The OIF CEI-P Implementation Agreement defines a coding and scrambling layer which provides an alternative to the IEEE 64B/66B code. The tag bit and 64-bit payload words of each block are mapped into CEI-P frames as shown in Fig. 5.11. Other fields in this frame, as well as framing procedures, will be described shortly for CEI-P.

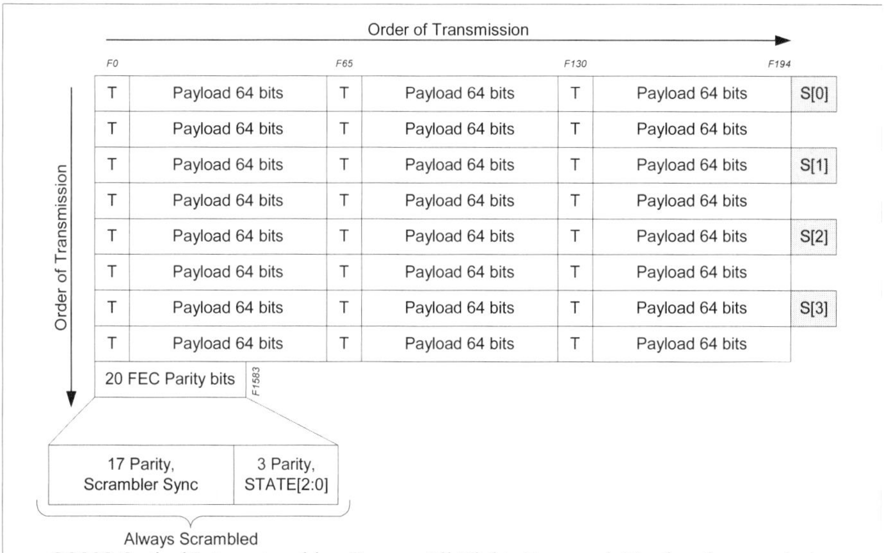

Fig. 5.11 CEI-P mapping

5.2.4 CEI-P Implementation Agreement

The CEI-P Implementation Agreement defines three sublayers of the protocol stack:

Adaption layer. This layer provides client signal alignment, performance monitoring, and mapping functions to convert from client protocol formats to CEI-P formats. This layer is application specific and is not specified in CEI-P.

Aggregation layer. This is an optional layer which multiplexes and de-multiplexes clients onto the CEI-P lane. This permits n lanes of a given baud rate to share one lane running at n times the baud rate. The purpose is to allow scalability of legacy protocols onto higher speed electrical lanes. For example, two OC-48 clients may share a CEI-6G lane using CEI-P. The detailed definition of this layer is beyond the scope of this text.

Framing layer. This layer defines the CEI-P framing, coding, scrambling, error detection and correction, etc. Key aspects of this are discussed in this section.

5.2.4.1 CEI-P Frame Format

The format of the 1584-bit CEI-P frame is defined in Fig. 5.11:

- Each 64-bit word of client data is mapped into the CEI-P payload words.
- The "T" bits are available for use by the client application; for SPI-S the T bits indicate the format of the payload word.
- The S[3:0] bits are available for use by an application-specific supervisory layer. Use of these bits is not defined in CEI-P.
- The FEC parity bits are generated by XOR'ing a 20-bit forward error correction (FEC) code with a 3-bit state code.

The FEC code used in CEI-P is a Fire Code with the following generator polynomial:

$$G(x) = (x^{13} + 1)(x^7 + x + 1). \tag{5.2}$$

Frames are scrambled using a free running scrambler with characteristic polynomial:

$$G(x) = x^{17} + x^{14} + 1. \tag{5.3}$$

All bits of the frame are scrambled. The receiver logic can determine the correct scrambler state by subtracting the calculated value of the FEC bits from received value of this field, and setting the descrambler state to be equal to the most significant 17 bits of this subtraction.

The state value is recovered at the receiver by XOR'ing the calculated value of the FEC bits with the received value of this field, and using the least significant 3 bits of this result as the state value. Note that any bit errors in the received data potentially cause the state value to be incorrectly decoded, and therefore the receiver logic must ignore transitory state transitions, and should only respond if the state value is consistent for several consecutive frames.

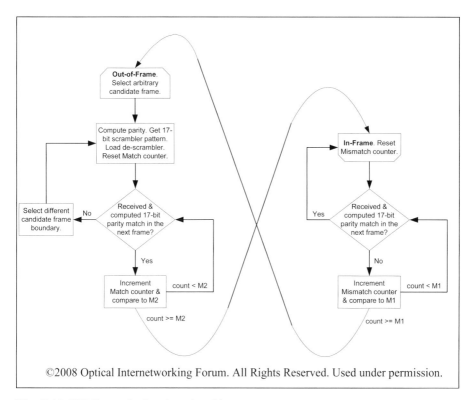

Fig. 5.12 CEI-P sample framing algorithm

5.2.4.2 CEI-P Link States

CEI-P links implement state machines which provide for the transmission of a specified training pattern. This training pattern may be used by the receiver to train the CDR and DFE circuits. Bidirectional interfaces pair each CEI-P link with a corresponding link in the reverse direction, and allow either chip to request that the other chip send the training pattern. CEI-P links on unidirectional interfaces only send training patterns when directed to do so by a supervisory function.

The state value in the CEI-P frame indicates the current link state of the transmitter. The receiver interprets the data on the link as either a training pattern or CEI-P frames based on this state. The receiver can request transmission of the training pattern by having the transmitter with which it is paired send the corresponding state request.

Detailed definitions of state values and state machine transitions are beyond the scope of this text.

5.2.4.3 CEI-P Framing

Logic in the receiver is required to determine the CEI-P frame reference. This is performed through a complex trial-and-error search utilizing the FEC parity calculation. Although other algorithms are possible, CEI-P provides the sample algorithm shown in Fig. 5.12. Parameters in this algorithm include:

Count: Frame match counter

M1: Constant number of mismatches required before the frame state transitions from *in-frame* to *out-of-frame*.

M2: Constant number of matches required before the frame state transitions from *out-of-frame* to *in-frame*.

The algorithm proceeds to find a frame reference in *out-of-frame* state as follows:

1. Assume an arbitrary frame reference.

2. Calculate the FEC code for the next frame and use this to initialize the descrambler state.

3. Calculate the FEC code for the next frame and compare to the received value. If the values do not match, try a new frame reference and go back to step 2. If the values match, then repeat this step for *M2* frames.

4. If the calculated and received FEC codes match for *M2* consecutive frames, the frame state transitions to *in-frame*.

5.2.5 Electrical Layer Implementation Agreements

OIF has published two electrical layer implementation agreements:

- *OIF-SxI-5-01.0*: System Interface Level 5 (SxI-5): Common Electrical Characteristics for 2.488–3.125 Gbps Parallel Interfaces: This implementation agreement defines an electrical layer for 2.488–3.125 Gbps link baud rates.

- *OIF-CEI-02.0*: Common Electrical I/O (CEI): Electrical and Jitter Interoperability agreements for 6G+ bps and 11G+ bps I/O: This implementation agreement is a clause-based document containing electrical layer definitions for 4.96–6.375 Gbps link baud rates, and for 9.95–11.10 Gbps link baud rates.

SxI-5 consists of normative transmitter and receiver specifications. CEI specifies a normative transmitter and a normative channel, and is described in further detail in this section.

5.2.5.1 CEI Variants

The OIF CEI Implementation Agreement is a clause-based document which specifies the variants shown in Table 5.6. For reference, SxI-5 is also compared in this table.

5.2.5.2 Transmitter Electrical Parameters

Basic electrical characteristics of the transmitter specification for various CEI variants are compared in Table 5.7.

Table 5.6 Common electrical I/O variants

Short name	Implementation agreement reference	Baud rate	Reach objective
SxI-5	OIF-SxI-5-01.0	2.488–3.125 Gbps	Capable of driving at least 8 in. of FR4 with 1 or 2 connectors
CEI-6G-SR	OIF-CEI-02.0 Clause 6	4.976–6.375 Gbps	Capable of driving 0–200 mm of PCB and up to 1 connector
CEI-6G-LR	OIF-CEI-02.0 Clause 7	4.976–6.375 Gbps	Capable of driving 0–1,000 mm of PCB and up to 2 connectors
CEI-11G-SR	OIF-CEI-02.0 Clause 8	9.95–11.1 Gbps	Capable of driving 0–200 mm of PCB and up to 1 connector
CEI-11G-MR	OIF-CEI-02.0 Clause 9	9.95–11.1 Gbps	Capable of driving 0–600 mm of PCB and up to 2 connectors for low power applications
CEI-11G-LR	OIF-CEI-02.0 Clause 9	9.95–11.1 Gbps	Capable of driving 0–1,000 mm of PCB and up to 2 connectors

The "Transmitter Reference Model" indicated in this table specifies the assumptions regarding the feed forward equalizer (FFE) in the transmitter device. There is no direct requirement specified for the transmitter in the implementation agreement; rather this requirement is implicitly derived from the channel compliance requirements. In some cases the description of the reference model may not be physically implementable; "infinite precision" for example is impossible to achieve in a digital design. This implies that the FFE designer must over-design in order to achieve an FFE that is at least as capable at compensating for the channel effects as the reference model specified.

The "Common Mode Voltage" specification in this table only applies if the transmitter supports DC Coupling. CEI is generally AC coupled, with DC coupling supported as an option.

Table 5.7 CEI transmitter electrical parameters

	CEI-6G-SR	CEI-6G-LR	CEI-11G-SR	CEI-11G-LR CEI-11G-MR
Transmitter reference model	2-tap FFE[a]	2-tap FFE[b]	No emphasis	3-tap FFE[c]
Output differential voltage	400–750 mVppd	800–1,200 mVppd	360–770 mVppd	800–1,200 mVppd
Common mode voltage (AC coupled load)	0–1.8 V	100–1,700 mV	0–3.55 V	100–1,700 mV
Common mode voltage (DC coupled load)	Range depends on Vtt at the receiver	630–1,100 mV	Range depends on Vtt at the receiver	630–1,100 mV
Rise/fall time	30 ps min.	30 ps min.	24 ps min.	24 ps min.
Uncorrelated bounded high probability jitter (T_UBHPJ)			0.15 UIpp	0.15 UIpp
Uncorrelated unbounded Gaussian jitter (T_UUGJ)			0.15 UIpp	0.15 UIpp
Uncorrelated high probability jitter (T_UHPJ)	0.15 UIpp	0.15 UIpp		
Duty cycle distortion (T_DCD)	0.05 UIpp	0.05 UIpp		0.05 UIpp
Total jitter (T_TJ)	0.30 UIpp	0.30 UIpp	0.30 UIpp	0.30 UIpp

[a]Single post tap transmitter, with ≤3dB of emphasis and infinite precision accuracy

[b]Either a single pretap or posttap transmitter, with ≤6dB of emphasis, with infinite precision accuracy

[c]Equalizing filter with 2 tap baud spaced emphasis no greater than a total of 6dB with finite resolution no better than 1.5dB

The jitter terminology in Table 5.7 is unique to the CEI Implementation Agreement. The detailed jitter budgets for each of the CEI clauses subdivided the deterministic and random jitter components in order to differentiate between contributors for which the FFE/DFE filters would or would not compensate. Those parameters of the jitter budget which are normative are shown in Table 5.7. The reader may generally assume as a first order approximation:

- T_UBHPJ = Unequalizable Deterministic Jitter (DJ).
- T_UHPJ = T_UBHPJ + T_DCD + other high probability jitter components.
- T_UUGJ = Unequalizable Random Jitter (RJ).

5.2.5.3 Receiver Electrical Parameters

Basic electrical characteristics of the receiver specification for various CEI variants are compared in Table 5.8. The "Receiver Reference Model" indicated in this table specifies the assumptions regarding the equalization in the receiver device. Where two possible reference models have been specified for a given CEI variant, the receiver designer may assume either one.

There is no direct requirement specified for the receiver equalization in the implementation agreement; rather this requirement is implicitly derived from the channel compliance requirements. DFE filters are assumed to have tap weights with infinite precision; this is not physically implementable, and implies the DFE designer must over-design in order to achieve equivalent performance.

The "Common Mode Voltage" specification in this table only applies if the transmitter supports DC Coupling. CEI is generally AC coupled, with DC coupling supported as an option.

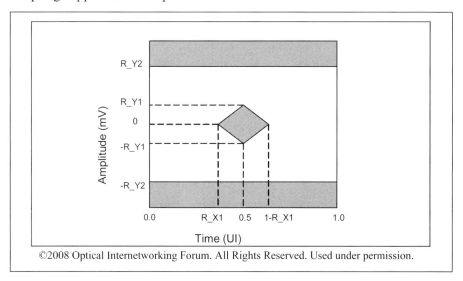

Fig. 5.13 Eye mask at output of receive equalizer for channel compliance

Table 5.8 CEI receiver electrical parameters

	CEI-6G-SR	CEI-6G-LR	CEI-11G-SR CEI-11G-MR	CEI-11G-LR
Receiver reference model	No equalization	5 tap DFE with constraints on tap weights	See Note (1)	4 tap DFE with constraints on tap weights
Input differential voltage	125–750 mVppd	1,200 mVppd max.	110–1,050 mVppd	1,200 mVppd max.
Common mode voltage (AC coupled load)	0–1.8 V	100–1,800 mV	0–3.60 V	100–1,800 mV
Common mode voltage (DC coupled load)	Range depends on Vtt at the receiver	595–(Vtt – 60) mV	Range depends on Vtt at the receiver	595–(Vtt – 60) mV

Note (1) Two reference receivers are defined:
- Reference Receiver A (CEI-11G-SR/MR): No equalization
- Reference Receiver B (CEI-11G-SR only): Single-zero single-pole filter

Table 5.9 Channel compliance parameters

	CEI-6G-SR	CEI-6G-LR	CEI-11G-SR	CEI-11G-LR
Transmitter reference model				
Baud rate	6.375 Gbps (See Note (1))	6.375 Gbps (See Note (1))	11.1 Gbps (See Note(1))	11.1 Gbps (See Note (1))
Equalization	See Note (2)	See Note (2)	None	See Note (2)
Amplitude	400 mVppd	800 mVppd	Both 360 and 770 mVppd (See Note 3)	800 mVppd
Jitter (T_UBHPJ)	0.15 UIpp	0.15 UIpp	0.15 UIpp	0.15 UIpp
Jitter (T_UUGJ)	0.15 UIpp	0.15 UIpp	0.15 UIpp	0.15 UIpp
Jitter (T_DCD)				0.05 UIpp

Table 5.9 Channel compliance parameters

	CEI-6G-SR	CEI-6G-LR	CEI-11G-SR	CEI-11G-LR
Tx edge rate filter	As specified in Note (4)	As specified in Note (4)	As specified in Note (4)	As specified in Note (5)
Return loss	RC Network model (Note (6))	RC Network model (Note (6))	RC Network model (Note (6))	RC Network model (Note (6))
Receiver reference model				
Equalization	None	See Note (7)	See Note (7)	See Note (7)
Return loss	RC Network model (Note (6))	RC Network model (Note (6))	RC Network model (Note (6))	RC Network model (Note (6))
Bit error rate (BER)	10^{-15}	10^{-15}	10^{-15}	10^{-15}
Sampling point	(See Note (8))	Not specified	(See Note (8))	Not specified
Eye mask parameters				
R_X1	0.30 UI	0.30 UI	0.35 UI / 0.25 UI (See Note (9))	0.2625 UI
R_Y1	62.5 mV	50 mV	55 mV	50 mV
R_Y2	375 mV	Not specified	525 mV	Not specified

Note (1) Lower of maximum baud rate of channel or baud rate specified in table

Note (2) See Table 5.7, "Transmitter Reference Model" entry, column for corresponding variant for specified transmitter equalization

Note (3) Channel compliance must be tested and pass for all specified amplitudes

Note (4) Transmitter edge filter is modeled as a simple 20 dB/dec low-pass filter at 75% of baud rate

Note (5) Transmitter edge filter is modeled as a simple 40 dB/dec low-pass filter at 75% of baud rate

Note (6) Return loss is modeled as an RC filter where R is the defined maximum allowed DC resistance of the interface and C is increased until the defined maximum return loss at the defined frequency is reached

Note (7) See Table 5.8, "Receiver Reference Model" entry, column for corresponding variant for specified receiver equalization

Note (8) Sampling point is defined as midpoint between average zero crossings of the differential signal

Note (9) First number is requirement when analyzing Reference Receiver A; second number is requirement when analyzing Reference Receiver B

Parameters regarding "Input Differential Voltage" are informative specifications. The receiver must receive any signal generated by a compliant transmitter driven through a compliant channel. For CEI variants with receiver reference models using DFEs, the signal amplitude at the receiver may approach zero for higher spectral components of the data.

Jitter parameters are also informative specifications with respect to the receiver. The individual CEI clauses provide informative values for jitter at the receiver which may be used as a guideline. The requirement is that the receiver must receive any signal generated by a compliant transmitter and driven through a compliant channel.

5.2.5.4 CEI Channel Compliance

All CEI variants include normative specifications for channel compliance. Channel compliance is determined by frequency domain analysis of the channel using the mathematical methods specified in the implementation agreement. This is analysis is performed as follows:

- A reference transmitter is assumed which incorporates certain assumptions regarding launch amplitude, jitter, equalizer capabilities, etc.

- The channel response is measured and 4-port S-parameters are generated for the channel.

- A reference receiver is assumed which incorporates certain assumptions regarding return loss, equalizer capabilities, etc. In some cases, more than one reference receiver is specified; the channel must work with all specified reference receivers.

- Statistical signal integrity analysis is performed with the above components to determine the width and height of the virtual eye at the output of the receiver equalizer. If there exists a set of filter coefficients for the transmitter and the receiver which are legal, and which produce a virtual eye that is of sufficient width and height for the target bit error rate (BER), then the channel is compliant.

Table 5.9 specifies the reference transmitter and receiver assumptions for channel compliance analysis of the various CEI variants. The table also specifies the eye mask parameters for the output of the receive equalizer which constitute the pass/fail condition; the eye mask is provided in Fig. 5.13. Note that:

1 If more than one Receiver Reference Model is specified, then the channel must pass using each reference model.

2. If more than one transmit amplitude is specified, then the channel must pass using each specified amplitude.

3. CEI-11G-MR channel compliance uses:
   - The transmitter reference model for CEI-11G-LR
   - The receiver reference model A for CEI-11G-SR
   - The eye mask parameters for CEI-11G-SR.

OIF has collaborated with the developers of an open source software tool called StatEye to provide an implementation of the mathematics specified in the CEI Implementation Agreement [12]. Certain releases of this tool include templates for the various CEI variants. StatEye performs the necessary statistical signal integrity analysis using these templates.

5.3 Ethernet Protocols

The Institute of Electrical and Electronics Engineers (IEEE) commissioned a committee to develop open network standards. Since this committee started work on this effort in February, 1980, the documents produced by this project were designated using the nomenclature IEEE 802.x. Several different network architectures were ultimately standardized, resulting in various ".x" extensions. The "Ethernet" system originally developed by Xerox was standardized and published as "IEEE 802.3 Carrier Sense Multiple Access with Collision Detection (CSMA/CD) Access Method and Physical Layer Specifications" in 1985. Although the IEEE document does not use the term, the "Ethernet" name continues to be commonly associated with this standard. While Ethernet was once one of several network standards with significant deployment (for example: the 802.5 Token Ring), over time it has emerged as the dominant network standard. The IEEE 802.3 document has gone through several revisions, adding additional clauses to support additional speeds, physical layers, etc., since its original publication.

The historical network topology for an Ethernet system consisted of multiple workstations connected to a common Ethernet bus using coaxial cable. Whenever one of the workstations wanted to send data, the network device would first listen to the bus to determine whether another workstation was using it. If the bus was in use, the device would wait. If the bus was not in use, the network device would transmit a data packet, while continuing to monitor the bus for any collisions resulting from two devices starting transmission at the same time. If a collision occurred, the device would abort the transmission and wait some amount of time before trying again. The CSMA/CD portion of the title for the IEEE 802.3 document results from this historical network access method.

Beginning in the early 1990s, "star-connected" network topologies became the network configuration of choice. In a "star-connected" topology, each workstation connects to a central network unit (either a network switch or a network hub), and each connection is a point-to-point link implemented using twisted-pair wire or optical fiber. Such topologies allowed full-duplex connections such that a network device can be both transmitting and receiving at the same time. Since each link had only one transmitter, collisions were no longer possible. Nevertheless, the CSMA/CD access methods were retained in the IEEE 802.3 standard through the 10 Mbps, 100 Mbps, and 1 Gbps generations. Clauses pertaining to 10 Gbps recognize that practical implementations are point-to-point links, and no longer support CSMA/CD access methods.

5.3.1 Physical Layer Reference Model

Figure 5.14 illustrates the Media Access Control (MAC) layer and the various components (with the associated Ethernet terminology) of the Physical Layer associated with 10-Gbps Ethernet. The Reconciliation Layer remaps signals between the 32-bit MAC/PLS service interface definition (clause 36) and the 10-Gbps Ethernet Media-Independent Interface (XGMII) definition. The MAC layer (via the Reconciliation Layer) and the Physical Layer are interconnected using the XGMII. The Physical Layer consists of the Physical Coding Sublayer (PCS), Physical Medium Attachment (PMA) sublayer, and the Physical Medium Dependent (PMD) sublayer. There are multiple clauses of the IEEE 802.3 document which apply to these sublayers, depending on the type of medium being used. HSS cores are used in the high-speed electrical implementation of these layers.

Table 5.10 describes the relationship between 10-Gbps Ethernet variants and normative clauses of the IEEE 802.3 document for the physical sublayers. The published clauses relevant to 10-Gb Ethernet are summarized below:

Clause 44. Introduction to the 10Gbps Baseband Network: This clause is a normative summary of applicable requirements for the 10-Gbps LAN/WAN variants described in Table 5.10. This clause references other 10Gbps clauses.

Clause 46. Reconciliation Sublayer (RS) and 10 Gigabit Media Independent Interface (XGMII): This clause defines both the Reconciliation Sublayer (RS) and the XGMII.

Clause 47. XGMII Extender Sublayer (XGXS) and 10 Gigabit Attachment Unit Interface (XAUI): This clause defines a method of extending the physical distance between the MAC layer device and the PCS/PMA layer device. An XGXS layer device is located near the MAC layer device (DTE XGXS), and a similar XGXS layer device is located near the PCS/PMA layer device (PHY XGXS). XAUI is the interface definition which interconnects the two XGXS layer devices.

Clause 48. Physical Coding Sublayer (PCS) and Physical Medium Attachment (PMA) sublayer, type 10GBASE-X: This clause defines the PCS and PMA layers for connecting an XGMII to a PMD using a 4 lanes by 3.125Gbps electrical interface. Block encoding/decoding (8B/10B), alignment, deskew, and clock rate compensation are components of this clause.

Clause 49. Physical Coding Sublayer (PCS) for 64b/66b, type 10GBASE-R: This clause defines the PCS layer for connecting an XGMII to a 10-Gbps serial PMA layer. Block encoding/decoding (64B/66B), alignment, and clock rate compensation are components of this clause.

Clause 50. WAN Interface Sublayer (WIS), type 10GBASE-W: This clause defines a sublayer which exists between the PCS and the PMA layer for Wide Area Network (WAN) variants. This sublayer maps Ethernet packets into the Synchronous Payload Envelope (SPE) of an STS-192c SONET frame.

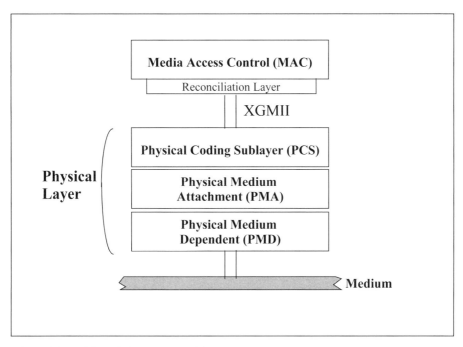

Fig. 5.14 LAN PHY/WAN PHY sublayers

Table 5.10 Relevant IEEE 802.3 clauses

Nomenclature	Type	PMD description	PCS layer clause	PMA layer clause	PMD layer clause	WIS (Clause 50)	Auto-Neg. (Clause 73)	FEC (Clause 74)
10GBASE-SR	LAN	850-nm serial optics	49	51	52			
10GBASE-LR	LAN	1,310-nm serial optics	49	51	52			
10GBASE-ER	LAN	1,550-nm serial optics	49	51	52			
10GBASE-LX4	LAN	4 x 1,310-nm CWDM optics	48	48	53			
10GBASE-CX4	LAN	4 x 3.125 Gbps electrical i/f	48	48	54			
10GBASE-SW	WAN	850-nm serial optics	49	51	52	X		
10GBASE-LW	WAN	1,310-nm serial optics	49	51	52	X		
10GBASE-EW	WAN	1,550-nm serial optics	49	51	52	X		
10GBASE-KX4	Backplane	4 x 3.125 Gbps electrical	48	48	71		X	
10GBASE-KR	Backplane	10 x Gbps serial electrical	49	51	72		X	O
Key: X = Mandatory, O = Optional								

Clause 51 Physical Medium Attachment (PMA) sublayer, serial: This clause defines a PMA sublayer for interfacing the PCS sublayer defined in clause 49 (or the WIS sublayer defined in clause 50) to a PMD using a serial 10.3125-Gbps electrical interface.

Clause 52 Physical Medium Dependent (PMD) sublayer and baseband medium, type 10GBASE-S (Short Wavelength Serial) and 10GBASE-L (Long Wavelength Serial) and 10GBASE-E (Extra Long Wavelength Serial): This clause defines the electrical-to-optical and optical-to-electrical PMD layer for 10-Gbps serial optical devices.

Clause 53 Physical Medium Dependent (PMD) sublayer and baseband medium, type 10GBASE-LX4: This clause defines the electrical-to-optical and optical-to-electrical PMD layer for a 10-Gbps optical device which employs Course Wave Division Multiplexing (CWDM) to transmit four 3.125 Gbps data streams on the fiber.

Clause 54 Physical Medium Dependent (PMD) sublayer and baseband medium, type 10GBASE-CX4: This clause defines the electrical PMD layer for a 10-Gbps electrical device which uses four 3.125-Gbps electrical differential signals to transmit data across a cable.

The Ethernet variants also exist for communication across backplanes within host systems. The clauses relevant to 10-Gb Backplane Ethernet are summarized below:

Clause 69. Introduction to Ethernet operation over electrical backplanes: This clause is a normative summary of applicable requirements for the 1 Gbps and 10 Gbps backplane variants described in Table 5.10. This clause references other clauses applicable to these variants.

Clause 71. Physical Medium Dependent Sublayer and Baseband Medium, Type 10GBASE-KX4: This clause defines the electrical PMD layer for a 10-Gbps electrical device which uses four 3.125-Gbps electrical differential signals to transmit data across a backplane.

Clause 72. Physical Medium Dependent Sublayer and Baseband Medium, Type 10GBASE-KR: This clause defines the electrical PMD layer for a 10-Gbps serial electrical device which uses a 10.3125-Gbps electrical differential signal to transmit data across a backplane. This clause additionally defines a training frame structure and training state machine. This training protocol monitors the signal metrics at the receiver during link initialization and updates FFE coefficients in the transmitter in order to optimize performance of the link.

Clause 73. Auto-Negotiation for Backplane Ethernet: This clause defines autonegotiation protocols which are mandatory for PCS/PMA layer implementations for Backplane Ethernet. This protocol allows devices to negotiate which Backplane Ethernet variant is to be used, whether FEC is to be used, and other parameters.

Clause 74. Forward Error Correction (FEC) sublayer for 10GBASE-R: This clause defines an optional FEC sublayer for enhancing the BER performance of 10GBASE-KR.

5.3.2 Media Access Control (MAC) Layer

In this section the basic format of an Ethernet frame is discussed, as well as the XGMII interface definition. Other details of MAC layer function are beyond the scope of this text.

5.3.2.1 Ethernet Packet Format

Ethernet is a packet delivery system. The basic format of an Ethernet packet is shown in Fig. 5.15. Each packet consists of a preamble and start frame delimiter (SFD) to indicate the beginning of a packet, destination and source addresses to identify the recipient and the sender, a packet length field indicating the number of data bytes in the packet, a variable length data field, and a frame check sequence (FCS) field to facilitate error detection.

Each Ethernet network interface card (NIC) is assigned a unique MAC address by its manufacturer. The first 24 bits of the 48-bit MAC address identify the manufacturer and are assigned by the IEEE Registration Authority. The remaining bits are assigned by the manufacturer and are generally programmed into the hardware such that they cannot be changed. This MAC address uniquely identifies the NIC, even if the hardware moves to another location.

The data portion of the MAC frame generally carries data for a higher layer protocol, which also requires some control information to be transmitted. The data portion of the frame in Fig. 5.15 often consists of a logical link control (LLC) layer header followed by payload data. If there are fewer than 46 bytes of payload data, then the payload data is padded to achieve this minimum length.

Preamble (7 bytes)	SFD (1)	Destination Address (6 bytes)	Source Address (6 bytes)	Length (2)	Data (Up to 1500 bytes)	FCS (4 bytes)

Preamble: Idles which set timing.
Start Frame Delimiter (SFD): 'AB'h indicating start of frame.
Destination Address: Six byte unique MAC address of recipient device.
Source Address: Six byte unique MAC address of sending device.
Length: Two bytes indicating length of data field.
Data: Layer 2 Payload Data field. Generally contains a Logical Link
Control (LLC) header and between 46 and 1500 bytes of payload data.
Data is padded if shorter than 46 bytes.
Frame Check Sequence (FCS): Cyclic Redundancy Check (CRC)
remainder for error detection.

Fig. 5.15 IEEE 802.3 ethernet frame

5.3.2.2 10-Gb Media Independent Interface (XGMII)

The XGMII interface consists of the following signals:

TXD[31:0]: This 32-bit bus is the transmit data bus driven by the MAC to the Physical Layer. Bit 31 is the most significant bit of this bus, and bit 0 is the least significant bit. Ethernet data is transmitted least significant bit first (i.e., TXD[0] first, then TXD[1], and so forth through TXD[31]).

TXC[3:0]: These control bits indicate whether the corresponding bytes of the TXD bus are data values or control values. The TXD[31:0] and TXC[3:0] signals are organized into "lanes" as is explained shortly.

TX_CLK: This signal is a 156.25-MHz transmit DDR clock and is synchronous to the TXD and TXC signals.

RXD[31:0]: This 32-bit bus is the receive data bus driven by the Physical Layer to the MAC. Bit 31 is the most significant bit of this bus, and bit 0 is the least significant bit. Ethernet data have received least significant bit first.

RXC[3:0]: These control bits indicate whether the corresponding bytes of the RXD bus are data values or control values. The RXD[31:0] and RXC[3:0] signals are organized into "lanes" as is explained shortly.

RX_CLK: This signal is a 156.25-MHz receive DDR clock and is synchronous to the RXD and RXC signals.

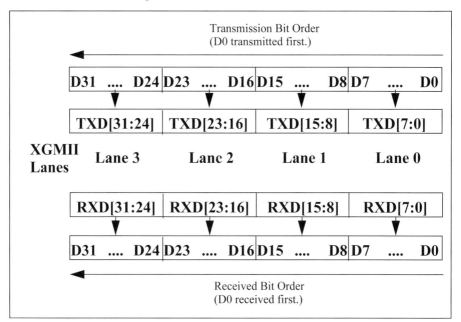

Fig. 5.16 MAC serial bit order and XGMII data lanes

Table 5.11 XGMII frame example

Contents	XGMII Lane 3		XGMII Lane 2		XGMII Lane 1		XGMII Lane 0	
	TXC3 or RXC3	TXD/ RXD [31:24]	TXC2 or RXC2	TXD/ RXD [23:16]	TXC1 or RXC1	TXD/ RXD [15:8]	TXC0 or RXC0	TXD/ RXD [7:0]
Preamble / start	0	"AA"h	0	"AA"h	0	"AA"h	1	"FB"h
SFD / preamble	0	"AB"h	0	"AA"h	0	"AA"h	0	"AA"h
Frame data	0	Byte 4	0	Byte 3	0	Byte 2	0	Byte 1
	⋮	⋮	⋮	⋮	⋮	⋮	⋮	⋮
End/frame data	1	"FD"h	0	Byte N	0	Byte N–1	0	Byte N–2

Table 5.12 XGMII control characters

	TXC/RXC	TXD/RXD	8B/10B Code
MAC data	0	"00"h to "FF"h	Dx.y
Idle	1	"07"h	
Sequence (only valid in lane 0)	1	"9C"h	K28.4
Start (only valid in lane 0)	1	"FB"h	K27.7
Terminate	1	"FD"h	K29.7
Transmit error propagation or receive error	1	"FE"h	K30.7

The TXD[31:0] and TXC[3:0] signals are organized into lanes as follows:

Lane 0 consists of TXD[7:0] and TXC[0].
Lane 1 consists of TXD[15:8] and TXC[1].
Lane 2 consists of TXD[23:16] and TXC[2].
Lane 3 consists of TXD[31:24] and TXC[3].

The RXD[31:0] and RXC[3:0] signals are organized into lanes in a similar manner. The order in which MAC bits are mapped onto the XGMII signals is illustrated in Fig. 5.16.

Table 5.11 illustrates the format of a packet on the XGMII bus. The first preamble character generated by the MAC layer is replaced by the XGMII "Start" control symbol (see control character definitions in Table 5.12), which must be aligned in Lane 0 of the XGMII. The Start character is followed by six additional preamble bytes, and the MAC Start Frame Delimiter. From the perspective of the XGMII, all characters after the Start control character are data bytes.

The next N data bytes correspond to the MAC frame header, data, and Frame Check Sequence as defined previously. After the last FCS byte, a "Terminate" control character is sent on the XGMII. Note that the length of the MAC frame does not need to be divisible by four, and therefore the frame may end (and the "Terminate" character may occur) on any lane of the XGMII. The MAC layer may also abort the frame using an End character ('FE'h). This character is interpreted at the receiver as either MAC propagation of an error on the transmitter or as a receive error detected by the Physical Layer.

In between frames, the Idle characters or Sequence characters are sent.

5.3.3 XGMII Extender Sublayer (XGXS)

The XGMII Extended Sublayer (XGXS) extends the reach the XGMII interface across a 10-Gigabit Attachment Unit Interface (XAUI). This extension across XAUI, with an XGXS at either end, is shown in Fig. 5.17.

The XGXS encodes and serializes the XGMII transmit data (TXD/TXC signals) for transmission on XAUI. Each XGMII lane is 8B/10B encoded, serialized, and transmitted at a baud rate of 3.125 Gbps. XAUI serial data signals received by the XGXS are deserialized, 8B/10B decoded, and then driven to the MAC or PHY on the XGMII receive data (RXD/RXC signals). XAUI is a full duplex interface consisting of four transmit and four receive serial data signals, corresponding to the four transmit and four receive lanes of XGMII.

It should be noted that the functions for the PCS/PMA sublayers defined in clause 48 of the 802.3 specification are identical to the function of the XGXS as specified in clause 47. The logic implementation is therefore similar. These clauses differ in that clause 47 defines the electrical parameters for the serial data signals that are applicable to XAUI. When used as the PCS/PMA layer for 10GBASE-X variants, the electrical characteristics of the signals between the PMA and PMD sublayers are implementation dependent.

The XGXS function is also similar for backplane applications designed to clause 71. Electrical parameters which apply for backplane applications are contained in clause 71.

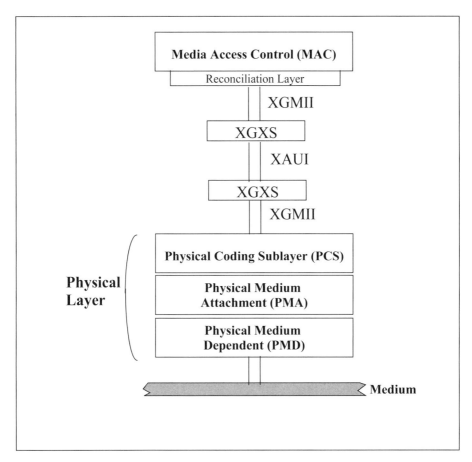

Fig. 5.17 XAUI/XGXS relationship to Ethernet layers

5.3.3.1 XGXS Function Overview

The PMA sublayer of an XGXS is implemented with HSS cores. The PCS sublayer requires additional logic to implement the following functions:

- Transmitter pseudorandom idle generation between packets.
- Clock compensation between the HSS core clock domains for each lane and the XGMII clock domain (through insertion or deletion of Idles between packets).
- 8B/10B encoding of XGMII transmit data to the PMA; 8B/10B decoding of PMA receive data to the XGMII.
- Receiver detection of synchronization idles.
- Generation of bit alignment pulses to the HSS core to perform symbol alignment of 8B/10B symbols received by the HSS core.
- Receiver detection of align idles and deskew of all lanes.

The HSS core implements the following XGXS PMA sublayer functions:

- Generation of XGXS transmit clocks (for data between the PCS and PMA sublayers) based on the local reference clock.

- Serialization of PMA transmit data and transmission over XAUI.

- Reception of XAUI data, clock data recovery, and generation of XGXS receive clocks (for data between the PMA and PCS sublayers).

- Deserialization of PMA receive data for the PMA interface to the PCS sublayer.

- Implementation of a bit alignment feature (as implemented on the HSS EX10 core by the RXxDATASYNC input) to facilitate bit-shifting and alignment of receive data on the interface between the PMA and PCS sublayer.

- Continuous status indication of valid clocks and data to the PCS sublayer.

The signals which interface between the PCS and PMA sublayers are implementation dependent. The following signal descriptions are based on IBM cores, but are typical of what would be generically required:

PMA_DATA_OUT_LANEx[9:0] (x = 0,1,2,3): These busses are the transmit data for each PMA lane. There are four sets of signals corresponding to each of the four lanes. Data for each lane is 10-bits wide corresponding to the 10-bit symbol produced by 8B/10B encoding.

PMA_TXR_CLK_IN[3:0]: These are the 312.5 MHz transmit symbol clock outputs of the HSS core and used by the XGXS PCS layer logic. There is one clock per lane. The resynchronization function of the HSS core must be used to minimize skew between transmit clocks.

PMA_DATA_IN_LANEx[9:0] (x = 0,1,2,3): These busses are the receive data for each PMA lane. There are four sets of signals corresponding to each of the four lanes. Data for each lane is 10-bits wide corresponding to the 10-bit symbol produced by 8B/10B encoding.

PMA_RCVR_CLK_IN[3:0]: These are the 312.5-MHz receive symbol clock outputs of the HSS core and used by the XGXS PCS layer logic. There is one clock per lane.

PMA_DATA_SYNC_LANE[3:0]: These outputs of the PCS sublayer are connected to the RXxDATASYNC inputs of the HSS core for each of the respective lanes.

PMA_CLKS_READY: Status input from the HSS core to the PCS sublayer indicating that the PLL is locked, clocks are stable, and the HSS core is ready to receive and transmit data. The proper function of this signal depends on the defined initialization sequence for the HSS core being used.

One of the functions of the XGXS PCS receive logic is to compensate for frequency differences between PMA_RCVR_CLK_IN (as determined by the far-end transmitter's clock reference) and the XGMII clocks (derived from the local clock reference). There is a 100 ppm tolerance between these clocks.

The XGXS PCS has sufficient buffer capability such that this frequency difference is tolerated, and Idles are added or dropped between packets to compensate. When PMA_CLKS_READY is not asserted, it indicates that clocks may be beyond this specification, and therefore no attempt is made to receive data or compensate for differences in clock frequencies.

5.3.3.2 Ordered Sets and Special Code Groups

The XGXS PCS sublayer performs 8B/10B encoding and decoding of MAC frames. Each byte of data maps to a corresponding 10-bit symbol. In addition, the XGMII control characters in Table 5.12, with the exception of the Idle character, directly map to corresponding 10-bit symbols. This mapping is shown in Table 5.12 using the symbol nomenclature for 8B/10B codes. (The 8B/10B block code was described in Sect. 4.2.2.1.)

XGMII Idles are randomly converted to one of the following 10-bit codes: *Sync* (K28.5), *Skip* (K28.0), or *Align* (K28.3). A LFSR is used to randomly either generate *Sync* or *Skip* symbols. *Align* symbols are inserted at randomized intervals in the range of 16–31 symbols, with this interval also determined by a LFSR. The same symbol is transmitted on all lanes.

Align symbols are used by the receive PCS logic to perform deskew across all lanes of the interface such that the XGMII RXD[31:0] output is aligned to a common clock domain. The *Align* symbol is unique and does not otherwise occur in the protocol. Since it is transmitted on all lanes simultaneously, and is only transmitted every 16–31 symbols, the lanes can be deskewed by aligning the occurrence of this symbol.

Skip symbols are used by the receive PCS logic to adjust for frequency differences between the recovered receive clock and the local XGMII clock. *Skip* symbols are inserted or dropped as needed.

The *Sequence* symbol (K28.4) is used on lane 0, in conjunction with Dxx.y symbols on the other lanes, to convey management information. Such information is inserted by the MAC layer between frames in place of Idles, and is conveyed through the XGXS accordingly.

The *Sync* symbol does not have any special use.

5.3.4 10-Gb Serial Electrical Interface

The 10-Gb Serial Electrical Interface (XFI) was developed as part of the 10-Gigabit Small Form Factor Pluggable (XFP) Multi Source Agreement (MSA). This MSA was developed by a consortium of optics module vendors who agreed to source optics modules meeting the XFP requirements. The XFI is a 10-Gbps serial electrical interface for communications between the XFP module and a SONET framer chip, or an Ethernet PCS/PMA layer. The PCS and PMA sublayers defined in clauses 49 and 51, respectively, define the necessary functions to interface between XGMII and XFI, and are used with all 10-Gb Ethernet serial variants (including backplane applications using the PMD defined in clause 72). Although the XFI acronym does not appear anywhere in the IEEE 802.3 standard, for historical reasons implementations of clause 49 and 51 are commonly called XFI.

The XFI PCS encodes and serializes the XGMII transmit data (TXD/TXC signals) for transmission on XFI. Each XGMII lane is 64B/66B encoded, scrambled, serialized, and transmitted at a baud rate of 10.3125 Gbps. XFI serial data signals received by the XFI PCS are deserialized, descrambled, 64B/66B decoded, and then driven to the MAC or PHY on the XGMII receive data (RXD/RXC signals). XFI is a full duplex interface consisting of one transmit and one receive serial data signals.

An optional 16-bit bus is defined to interface between the PCS sublayer and the PMA sublayer. While this is the reference interface between the XFI PCS and the HSS core implementing the PMA sublayer as defined by clause 49, it does require data to be transferred on the parallel bus of the HSS core at 644.53 Mtransfers per second. Implementations may choose HSS core options which utilize more bits of parallel data at a slower transfer rate. (Available IBM cores use a 32-bit interface.) The definition of this interface has significance for two reasons:

- The 66-bit code word blocks produced by 64B/66B encoding of the XGMII data are not an integer multiple of the width of the parallel data bus on most HSS cores. Data must be steered from cycle to cycle by "gearbox" logic in order to map the 66-bit blocks onto the 16-bit (or 32-bit) parallel data bus of the HSS core.

- 10GBASE-W Wide Area Network (WAN) variants are intended to inter-operate with other network nodes on a SONET/SDH network. These variants define a WAN Interface Sublayer (WIS) in clause 50 which maps the Ethernet packets into a SONET SPE, and then wraps this data with SONET/SDH overhead bytes with the level of support defined in clause 50. WIS is sandwiched between the PCS and PMA sublayers for 10GBASE-W variants.

5.3.4.1 XFI Function Overview

The PMA sublayer of an XFI is implemented with HSS cores. The PCS sublayer requires additional logic to implement the following functions:

- Mapping XGMII data into block payloads.

- Clock compensation between the HSS core clock domain and the XGMII clock domain (through insertion or deletion of Idles between packets).

- 64B/66B encoding and scrambling of XGMII transmit data to the PMA; 64B/66B decoding and descrambling of PMA receive data to the XGMII.

- Gearbox data steering to map 66-bit code words onto the HSS core parallel data bus for transmission, and similarly unmap received data.

- Receiver detection of sync headers in received data stream.

- Block alignment using sync headers to form 66-bit blocks from incoming data.

The HSS core implements the following XFI PMA sublayer functions:

- Generation of XFI transmit clock (for data between the PCS and PMA sublayers) based on the local reference clock.

- Serialization of PMA transmit data and transmission over XFI.

- Reception of XFI data, clock data recovery, and generation of XFI receive clock (for data between the PMA and PCS sublayers).

- Deserialization of PMA receive data for the PMA interface to the PCS sublayer.

- Continuous status indication of valid clocks and data to the PCS sublayer.

The signals which interface between the PCS and PMA sublayers are implementation dependent. A 16-bit PMA Service Interface (XSBI) is defined for reference in clause 51, but is not required. Some variation from the XSBI definition exists in most XFI implementations. The following signal descriptions are based on IBM cores, but are typical of what would be generically required:

TX_DATA_OUT[31:0]: This is the transmit data parallel data bus connection from the XFI PCS to the HSS core.

PMA_TX_CLK: This is the 322.265-MHz transmit clock output of the HSS core and is used by the XFI PCS sublayer logic.

RX_DATA_IN[31:0]: This is the receive data parallel data bus connection from the HSS core to the XFI PCS.

PMA_RX_CLK: This is the 322.265-MHz receive clock output of the HSS core and is used by the XFI PCS sublayer logic.

PMA_TX_READY: Status input from the HSS core to the XFI PCS sublayer indicating that the PLL is locked, clocks are stable, and the HSS core is ready to transmit data. The proper function of this signal depends on the defined initialization sequence for the HSS core being used.

PMA_RX_READY: Status input from the HSS core to the XFI PCS sublayer indicating that the PLL is locked, clocks are stable, and the HSS core is ready to receive data. The proper function of this signal depends on the defined initialization sequence for the HSS core being used.

One of the functions of the XFI PCS receive logic is to compensate for frequency differences between PMA_RCVR_CLK_IN (as determined by the far-end transmitter's clock reference) and the XGMII clocks (derived from the local clock reference). There is a 100 ppm tolerance between these clocks. The XFI PCS has sufficient buffer capability such that this frequency difference is tolerated, and Idles are added or dropped between packets to compensate. When PMA_TX_READY or PMA_RX_READY is not asserted, it indicates that clocks may be beyond this specification, and therefore no attempt is made to receive data or compensate for differences in clock frequencies.

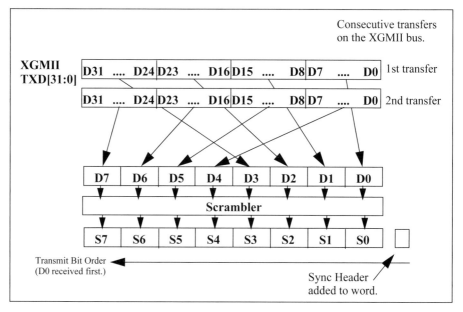

Fig. 5.18 MAC XGMII data mapping into 64B/66B code word

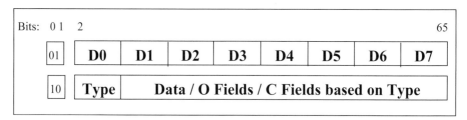

Fig. 5.19 64B/66B data blocks and control blocks

5.3.4.2 64B/66B Encoding and Scrambling

The XFI PCS maps MAC data on the XGMII into 66-bit code words for transmission in the manner shown in Fig. 5.18. The reverse of this mapping is performed at the receiver. Two data transfers on the XGMII bus are mapped into a single 64-bit block. This block is scrambled with a self-synchronizing scrambler implementing the polynomial:

$$G(x) = x^{58} + x^{39} + 1. \tag{5.4}$$

A two-bit Sync Header (either "10" or "01") is then added to the block to form a 66-bit code word. The code word is transmitted with the Sync Header bits transmitted first, followed by data starting with the least significant bit and proceeding to the most significant bit.

The Sync Header is not scrambled and is always "10" or "01," limiting the data run length such that at least one transition occurs every 66 bits. The XFI PCS logic at the receiver determines the alignment of the 66-bit block by looking for an alignment where the sync header is consistently "10" or "01."

Table 5.13 Control block formats

XGMII contents	Block type	Remaining bytes
Idle	"1E"h	Control characters
Sequence	"4B"h	Ordered Set mapped from first XGMII transfer followed by control bytes
	"2D"h	Control characters followed by Ordered Set mapped from second XGMII transfer
	"55"h	Two Ordered Sets mapped from consecutive XGMII transfers
Start	"78"h	First seven bytes of packet data
	"33"h	Four control characters followed by first three bytes of packet data
Combined sequence / start	"66"h	Ordered set mapped from first XGMII transfer and first three bytes of packet data mapped from second XGMII transfer
Terminate	"87"h	Control bytes
	"99"h	Last 1 bytes of packet data followed by control characters
	"AA"h	Last 2 bytes of packet data followed by control characters
	"B4"h	Last 3 bytes of packet data followed by control characters
	"CC"h	Last 4 bytes of packet data followed by control characters
	"D2"h	Last 5 bytes of packet data followed by control characters
	"E1"h	Last 6 bytes of packet data followed by control characters
	"FF"h	Last 7 bytes of packet data

Figure 5.18 illustrates the code words with the most significant bit toward the left and the least significant bit toward the right. It is more convenient when discussing block encoding to picture the code with the first bit to be transmitted on the left and the last bit to be transmitted on the right, as in Fig. 5.19.

There are two code word formats shown in Fig. 5.19. The first is a *data block* formed from two consecutive transfers of XGMII data. The sync header field is "01" for this block format when viewed with the first bit transmitted on the left. A sync header field of "10" indicates a *control block*. A control block contains either all control information, or a mix of control information and data. The first byte of the control block is the *block type field* and determines the format of the remainder of the block.

Block types are defined in Table 5.13. Block types are used to denote the start and end of packets, and the transmission of ordered sets. Control characters are 7-bit fields; eight control characters plus the 8-bit block type field can be transmitted in a single control block. Ordered sets include a 4-bit O code and three data bytes; two ordered sets plus an 8-bit block type field can be transmitted in a single control block. The precise bit mapping for control characters, ordered sets, and data into the block formats in Table 5.13 is beyond the scope of this text. It should be noted that some of these formats have unused bits as the result of how field sizes combine.

The following control bytes are defined for use in control blocks:

- Idle is encoded as the seven bit control character: "0000000."

- Error is encoded as the seven bit control character: "0011110."

Various other control character codes are either reserved or illegal.

5.3.4.3 WAN Interface Sublayer (WIS)

10GBASE-W Wide Area Network (WAN) variants interoperate with other network nodes on a SONET/SDH network. The WAN Interface Sublayer (WIS) defined in clause 50 maps the Ethernet packets into a SONET SPE, and then wraps this data with SONET/SDH overhead bytes. WIS is sandwiched between the PCS and PMA sublayers for 10GBASE-W variants.

The frequency of operation of the PMA sublayer when used with a 10GBASE-W PMD must match the STS-192 bit rate. Therefore, the following bit rates and clock rates apply when WIS is used:

- The PMA sublayer operates at 9.95328 Gbaud for 10GBASE-W variants.

- The clocks for a 16-bit XSBI interface to this PMA sublayer operate at 622.08 MHz. The clocks for a 32-bit equivalent interface (as used by IBM cores) operate at 311.04 MHz.

- The maximum transfer rate on a 16-bit XSBI interface from the PCS sublayer into the WIS is 599.04 Mtransfers per second to allow for SONET/SDH overhead added by the WIS. The maximum transfer rate on an equivalent 32-bit interface (as used by IBM cores) is 299.52 Mtransfers per second.

To properly limit the transfer rate through the PCS layer to the WIS, the MAC layer must be provisioned to insert extra idles between frames. WIS drops Idle characters to make room for the SONET/SDH overhead bytes.

Figure 5.20 illustrates the SONET/SDH overhead bytes which are implemented by WIS.

Section overhead (SOH). Framing pattern A1/A2 bytes and the B1 byte are supported. A fixed path trace value is generated and checked for the J0 byte.

Line overhead (LOH). The B2 and M1 (STS-N extension of M0) bytes are supported. H1/H2 pointer bytes are set to a constant value corresponding to a fixed position of the SPE in the STS-192 frame. K1/K2 are set to fixed values.

Path overhead (POH). The B3 and G1 bytes are supported. J1 and C2 are set to fixed values.

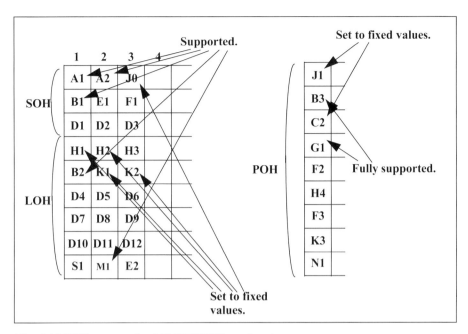

Fig. 5.20 WIS support for SONET/SDH overhead

Overhead bytes not listed above are not supported. They are set to fixed values and are not checked.

One last note on the WIS sublayer is in regard to the orientation of most significant bit and least significant bits on the interface between this layer and other sublayers. Ethernet bits are transmitted least significant bit first, while SONET/SDH octets are transmitted most significant bit first. For this reason, bit/octet ordering is interposed between the PCS and the PMA sublayers by the WIS sublayer.

5.3.5 Backplane Ethernet

There are three significant features associated with Backplane Ethernet which are not used in other Ethernet variants. These features are:

- Training protocol (defined in Clause 72 for 10GBASE-KR)
- Autonegotiation (defined in Clause 73 for all Backplane Ethernet variants)
- Forward error correction (FEC) (defined in Clause 74 for optional use with 10GBASE-KR)

5.3.5.1 Training Protocol

The training protocol is included in the IEEE 802.3 clause 72 specification for the 10GBASE-KR variant and must be executed upon start-up to initialize the PMD for this variant. This protocol continuously exchanges fixed-length training frames between the PMD and the link partner. The training frames contain both a pseudorandom training pattern and an FFE coefficient control field. During the training process, each PMD monitors the quality of the received training pattern and generates FFE coefficient control commands to the PMD at the other end of the link. At the same time the PMD monitors received FFE coefficient control commands and updates the HSS configuration accordingly. It is important to note that Ethernet assumes full-duplex communication between PMDs, making it possible to perform two-way communication to negotiate link parameters.

Training Frames:

The training frame is a fixed length frame containing the following fields:

- *Frame Marker.* A four byte unique marker pattern denoting the start of the training frame. The marker pattern is "FFFF0000" hexadecimal. This marker is transmitted at 10.3125 Gbps.

- *Coefficient Update.* A 16-bit coefficient update field which allows the PMD to request changes to the FFE coefficients being used by the PMD at the other end of the link. This 16-bit coefficient field is transmitted using Manchester coding at a baud rate which is one-eighth of the normal 10.3125 Gbps baud rate. Transmission of this 16-bit field is the equivalent of transmitting 16 bytes at 10.3125 Gbps.

- *Status Report.* A 16-bit status field which indicates whether the PMD has finished its training process, as well as the status of any requested FFE coefficient updates. This 16-bit coefficient field is transmitted using Manchester coding at a baud rate which is one-eighth of the normal 10.3125 Gbps baud rate. Transmission of this 16-bit field is the equivalent of transmitting 16 bytes at 10.3125 Gbps.

- *Training Pattern.* Pseudorandom data generated with the polynomial:
  $G(x) = 1 + x^9 + x^{11}$
  This field is 512 bytes in length (4,094 bits PRBS bits followed by two zero bits), transmitted at 10.3125 Gbps.

The purpose of using low-speed Manchester encoding for the *Coefficient Update* and *Status Report* fields is to ensure reliable information is communicated even when the link performance is not optimized. These fields can be implemented by transmitting the parallel data byte "00000000" or "11111111" to represent a Manchester encoded "0" bit, and transmitting the parallel data byte "00001111" or "11110000" to represent a Manchester encoded "1" bit. (The byte pattern used to denote a Manchester encoded "0" or "1" is selected to ensure that a transition always occurs between bits.)

The *Coefficient Update* field contains the following subfields:

- *Preset.* Forces all FFE coefficients to be set to a state where equalization is turned off (i.e., coefficients for the main FFE tap is set to its maximum value, and coefficients for precursor and postcursor taps are set to zero).

- *Initialize.* Forces FFE to be turned on and coefficients to be set to predefined initialization values.

- *Coefficient* (+1) *Update.* Directs that the postcursor FFE coefficient shall be incremented, decremented, or held at current value. .
- *Coefficient* (0) *Update.* Directs that the main FFE coefficient shall be incremented, decremented, or held at current value.
- *Coefficient* (-1) *Update.* Directs that the precursor FFE coefficient shall be incremented, decremented, or held at current value.

Update requests to increment or decrement coefficients should only be sent if the received *Status Report* field corresponding to the coefficient indicates the coefficient is in *not updated* state.

The *Status Report* field contains the following subfields:

- *Receiver Ready.* Indicates that reporting node has finished training its receiver and is ready to receive data.
- *Coefficient* (+1) *Status.* Reports that the postcursor FFE coefficient is *not updated*, is *updated*, is at *maximum* value, or is at *minimum* value. The PMD initially reports all coefficient status as *not updated*. After receiving a Coefficient Update request to update the coefficient, the PMD updates the coefficient and then reports either *updated*, *maximum*, or *minimum* status. After receiving a Coefficient Update request to hold the previous value, the PMD returns status to *not updated*.
- *Coefficient* (0) *Status.* Reports that the main FFE coefficient is *not updated*, is *updated*, is at *maximum* value, or is at *minimum* value. Status transitions are similar to the description for the postcursor FFE coefficient.
- *Coefficient* (−1) *Status.* Reports that the precursor FFE coefficient is *not updated*, is *updated*, is at *maximum* value, or is at *minimum* value. Status transitions are similar to the description for the postcursor FFE coefficient.

The *coefficient status* and *coefficient update* fields associated with each FFE coefficient operate according to a simple two-way handshake as described above. Each PMD assesses the quality of the received signal (through analysis of the training pattern), and then instructs its partner node to update FFE coefficients in an attempt to improve the quality of the signal.

Training State Machine:

The training state machine is also defined by clause 72. When a node detects a signal on the link, it starts sending training frames and waits for reception of valid training frames from its link partner. Once training frames are being received, the node allows the receiver equalization to train, and determines whether updates are required to the transmitter equalization. FFE coefficient updates are requested via command bits in the transmitted training frame; the link partner's acknowledgement of these requests are communicated via the status bits in the received training frame. Simultaneously, the node monitors command bits in the received training frame to determine whether local FFE coefficient updates are being requested by the link partner, and if applicable executes these requests and returns an acknowledgement via the transmitted training frame.

It should be noted that clause 72 does *not* specify the algorithms which should be employed to determine optimal equalizer settings. These algorithms are left as an exercise to the designer, and some implementations may even choose to use preset FFE coefficient values. (Although such implementations may not work well in some system configurations.)

The methods of determining signal quality and criteria for exiting training are also *not* specified by clause 72. These methods and criteria may be based on the PRBS pattern in the training frame, or may use other proprietary features of the HSS core.

HSS Features:

Some HSS cores may implement features which aid in signal quality assessment and allow faster updates to FFE coefficient values. Features such as the following are helpful to the implementation of the clause 72 Training Protocol:

- FFE coefficient negotiation requires assessment of the eye quality of the signal being received by the receive PMD layer. This signal quality is not directly observable by measurements at higher sublayers. The HSS EX10 digital eye feature assesses eye quality and provides training state machines with quick access to these measurements. This is useful for implementing the training protocol.

- FFE coefficient updates by a partner link during the training process should be made in a timely manner. HSS core features to quickly reset, load, increment, or decrement FFE coefficients speed up training.

5.3.5.2 Autonegotiation

The autonegotiation protocol is defined in IEEE 802.3 clause 73. Backplane Ethernet PHYs are required to implement the autonegotiation protocol, however, use of the protocol is under the control of the management interface and is optional. The protocol allows a device to advertise the modes of operation it supports to another device at the remote end of a backplane Ethernet link, and to detect the corresponding operational modes being advertised by the other device. The objective is to allow the devices to agree to a configuration that maximizes the performance of the link. The protocol additionally ensures that the PHY device is attached to its link partner at the remote end of the link, and is not responding to a crosstalk signal.

Autonegotiation is performed using low-speed Manchester encoding of the information to be exchanged. Clause 73 describes a number of Differential Manchester Encoding (DME) pages which are exchanged by the protocol. Autonegotiation supports the following partial list of functions:

- Advertise which backplane Ethernet variants are supported and negotiate which variant should be selected.

- Advertise whether the interface supports FEC and negotiate whether to enable FEC.

- Communicate additional message information.

Clause 73 defines the DME page formats, management interface variables, and autonegotiation state machines that must be implemented by all Backplane Ethernet nodes.

5.3.5.3 Forward Error Correction (FEC)

The FEC sublayer is defined as being inserted between the PCS and PMA sublayers. However, it should be noted that some efficiency is gained by combining the PCS sublayer and FEC sublayer implementations. The FEC sublayer at the transmitter performs the following functions:

- The 66-bit code words generated by the PCS layer are unscrambled.
- Every 32 consecutive 66-bit code words are mapped into a FEC frame with the format shown in Fig. 5.21. Each 66-bit code word is mapped into a 65-bit word position in the FEC frame. This is performed by converting the two-bit Sync Header into a one-bit Transcoding bit.
- Parity bits are generated and appended to the FEC Frame.
- The resulting 2112 bit FEC Frame is scrambled and output to the PMA sublayer.

The FEC sublayer at the receiver performs the following functions:

- The 2112 bit FEC Frame is received from the PMA sublayer and unscrambled.
- The parity bits are generated and compared to the received parity bits. If a mismatch occurs, bit errors are corrected if possible.
- The 65-bit word positions in the FEC frame are encoded into 66-bit code words, scrambled, and output to the PCS sublayer.

Obviously, the cascaded scrambling/unscrambling logic in the PCS sublayer and FEC sublayer can be eliminated if the sublayers are combined. Some additional efficiency may also be gained in the conversion between Sync Header bits and Transcoding bits.

The FEC used to calculate parity bits is based on a (2112,2080) code which is constructed by shortening the cycle code (42987,42955). This code is calculated using the generator polynomial:

$$g(x) = x^{32} + x^{23} + x^{21} + x^{11} + x^2 + 1. \qquad (5.5)$$

Given a polynomial representation of the information bits $m(x)$, the code word $c(x)$ is calculated as follows:

$$p(x) = x^{32}\, m(x) \bmod g(x), \qquad (5.6)$$
$$c(x) = p(x) + x^{32}\, m(x). \qquad (5.7)$$

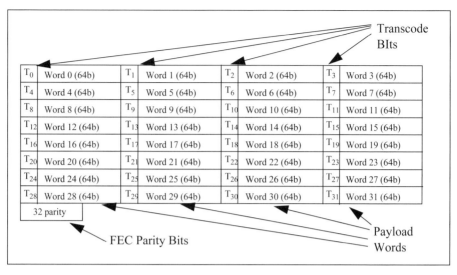

Fig. 5.21 Clause 74 FEC frame

5.3.6 PMD Sublayers for Electrical Variants

The IEEE 802.3 documents defines several PMD sublayer variants for 10-Gb Ethernet. Of these, many are optical interfaces which are not relevant to this text. This section summarizes the electrical variants which include 10GBASE-CX4, and Backplane Ethernet variants 10GBASE-KX4 and 10GBASE-KR. Also included in this section is the electrical definition of XAUI in clause 47.

The 10GBASE-CX4 specifies a normative cable assembly as the channel for this variant. Normative electrical parameters have been defined for both the transmitter and the receiver, however, jitter tolerance of the receiver is implicitly specified based on the transmitter jitter and normative channel (cable) characteristics.

Other electrical variants covered in this section provide normative specifications for the transmitter and receiver. An informative specification for Backplane Ethernet channels is provided in Annex 69B.

Transmitter electrical parameters are defined in Table 5.14. XAUI uses a traditional eye mask for describing limits for signal amplitude and total jitter. PMD clauses for other variants specify a time domain waveform template; the transmitter output waveform must fit within this template. The waveform template for 10GBASE-KR is parameterized based on FFE coefficient settings, which affect the shape of the transmit waveform.

Receiver electrical parameters are defined in Table 5.15. XAUI specifies that the receiver must tolerate 0.65 UIpp of jitter at its input, although no specification is provided for the type of jitter. 10GBASE-CX4 jitter tolerance is implied as discussed previously. The Backplane Ethernet variants use the Interference Tolerance Test defined in Annex 69.A to verify receiver jitter tolerance. The description of this test is beyond the scope of this text. Some of the jitter parameters used to perform this test are listed in Table 5.15 for the applicable variants.

Minimum signal amplitude at the receiver is not specified for any of the variants. A maximum amplitude is specified, although the actual amplitude may exceed this in a real system with an imperfect termination impedance. The minimum amplitude is implied by the BER specification; the signal amplitude must be sufficient for the receiver to receive the data and meet the specified BER.

Table 5.14 Ethernet PMD transmitter electrical parameters

	10GBASE-CX4 (Clause 54)	XAUI (Clause 47)	10GBASE-KX4 (Clause 71)	10GBASE-KR (Clause 72)
Number of lanes and baud rate	4 lanes x 3.125 Gbps ± 100 ppm	4 lanes x 3.125 Gbps ± 100 ppm	4 lanes x 3.125 Gbps ± 100 ppm	1 lane x 10.3125 Gbps ± 100 ppm
Transmitter waveform	Time domain waveform	Eye Mask	Time domain waveform	Time domain waveform as a function of FFE
Output differential voltage	800–1,200 mVppd	800–1,600 mVppd	800–1,200 mVppd	1,200 mVppd max
Common mode voltage (AC coupled)	−0.4 to 1.9 V	−0.4 to 2.3 V	−0.4 to 1.9 V	0–1.9 V
Rise/fall time	60–130 ps		60–130 ps	24–47 ps
Random jitter (RJ)	0.27 UIpp		0.27 UIpp	0.15 UIpp
Deterministic jitter (DJ)	0.17 UIpp	Near end: ± 0.085 UI Far end: ± 0.185 UI	0.17 UIpp	0.15 UIpp
Duty cycle distortion (DCD)				0.035 UIpp
Total jitter (TJ)	0.35 UIpp	Near end: ± 0.175 UI Far end: ± 0.275 UI	0.35 UIpp	0.28 UIpp

Table 5.15 Ethernet PMD receiver electrical parameters

	10GBASE-CX4 (Clause 54)	XAUI (Clause 47)	10GBASE-KX4 (Clause 71)	10GBASE-KR (Clause 72)
Number of lanes and baud rate	4 lanes x 3.125 Gbps ± 100 ppm	4 lanes x 3.125 Gbps ± 100 ppm	4 lanes x 3.125 Gbps ± 100 ppm	1 lane x 10.3125 Gbps ± 100 ppm
Receiver jitter tolerance (all units UIpp)	Implied	0.65 UIpp	Interference tolerance test with: SJ = 0.17 RJ = 0.18	Interference tolerance test with: SJ = 0.115 RJ = 0.130 DCD = 0.035
Input differential voltage	1,200 mVppd max	1,600 mVppd max	1,600 mVppd max	1,200 mVppd max
Bit error rate	10^{-12}	10^{-12}	10^{-12}	10^{-12}
Receiver coupling	AC	AC	AC	AC

5.4 Fibre Channel (FC) Storage Area Networks

The Technical Committee T11 (formerly X3T9.3) of the International Committee for Information Technology Standards (INCITS) began work on the Fibre Channel standard for Storage Area Networks (SANs) in 1988. INCITS coordinates with the American National Standards Institute (ANSI), and the initial versions of the Fibre Channel standard were approved by ANSI in 1994.

Hardware was deployed on a wide scale in 1998 for the 1.0625-Gbps serial baud rate in the initial Fibre Channel standard. Since then, additional serial variants have been added for 2.125 and 4.25 Gbps, and T11 is developing an 8.50-Gbps serial variant as of this writing. Variants have also been defined for 10.5-Gbps serial and an equivalent throughput variant implemented with four lanes by 3.1875-Gbps, however, these variants are less popular.

5.4.1 Storage Area Networks

Storage Area Networks (SANs) connect servers on the Internet Protocol (IP) network to storage devices. The SAN network can be configured in any number of ways. Fibre Channel supports two devices communicating directly across a point-to-point link. Some simple examples of more interesting config-urations are shown in Fig. 5.22, including a simple *fabric* topology where various servers on the IP Network connect to a Fibre Channel Switch. Any number of Tape, Disk, or Redundant Arrays of Independent Disk (RAID) systems are connected to the Fibre Channel switch, allowing the server to access storage without impacting traffic on the IP Network.

The SAN application virtualizes the storage connected to the fabric, in effect converting storage into one big disk. Files can be directed to any physical storage device based on available capacity, and mirroring of data can be employed for redundancy. Disk to Tape backup can be performed across the SAN without impacting traffic on the IP Network.

The SAN fabric configuration in Fig. 5.22 is limited by the number of ports available on the Fibre Channel Switch. A more complex SAN fabric may include multiple switches, thereby providing scalability to support any number of attached servers and devices. Also, Fibre Channel supports physical variants that can operate over optical links up to 100 km long. Therefore, the Fibre Channel switches can be physically distributed. This provides support for applications requiring remote mirroring of data at multiple sites, and backup to remote facilities (providing better disaster recovery).

Figure 5.22 also illustrates an *arbitrated loop* topology. Fibre Channel allows up to 127 devices to be connected in a loop without requiring a Fibre Channel switch. Devices arbitrate and gain control of the loop, and then communicate as if the link were point-to-point. This is not a token passing network; only one pair of devices can engage in the active exchange of frames at any given time. The bandwidth of the loop is shared among all devices on the loop. The arbitration methods for arbitrated loops are not addressed in this text; the reader is referred to the appropriate references for more information.

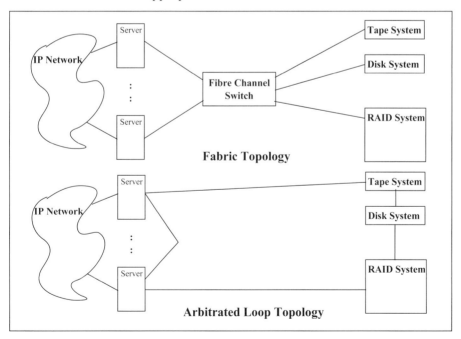

Fig. 5.22 Storage area network (SAN) configurations

5.4.2 Fibre Channel Protocol Layers

Fibre Channel defines the following levels of the Fibre Channel architecture and standards, and specifies these levels in various documents:

FC-0 Level.

This level defines the transmission media, transmitters, and receivers and their interfaces. Various "Fibre Channel – Physical Interfaces" documents, designated FC-PI-x, specify this level.

FC-1 Level.

This level defines serial encoding, decoding, and error control functions of the transmission protocol. The "Fibre Channel – Framing and Signaling–2" document, designated FC-FS-2, specify this level.

FC-2 Level.

This level defines the rules and mechanisms needed to transfer blocks of data between two devices. FC-FS-2 also defines this level.

FC-3 Level.

This level defines a set of services that are common across multiple ports of a node. FC-FS-2 partially defines this level. The "Fibre Channel – Link Services" document, designated FC-LS, specifies extended link services.

FC-4 Level.

This level defines mapping between lower levels of the Fibre Channel standards and Upper Level Protocols. Fibre Channel does not specify this level.

5.4.3 Framing and Signaling

The Fibre Channel framing and signaling layer are specified in "Fibre Channel –Framing and Signaling – 2" document, commonly referred to as "FC-FS-2." This document supersedes prior specifications of the FC-FS layer, and extends frame formats to add new features while maintaining backward compatibility. (Future FC-FS-x versions may supersede this document.) This section summarizes the framing and signaling features which are of some relevance to HSS.

5.4.3.1 Frame Format

Fibre Channel traffic is organized into 32-bit words of control and data information, which is subsequently 8B/10B encoded and transmitted across the physical interface. Although individual fields and payload data may be defined in terms of bytes, the organization into 32-bit words is applied throughout the FC-FS-2 document, and streamlines the implementation by consistently using a 32-bit parallel datapath.

Payload data is organized into frames as shown in Fig. 5.23, delineated by a Start of Frame (SOF) and End of Frame (EOF) delimiter. The SOF and EOF delimiters, as well as the Idles and other control information between frames, are examples of Ordered Sets. An *ordered set* is a 32-bit control word consisting of one 8B/10B control symbol, followed by three 8B/10B data symbols. There are multiple ordered sets defined for coding of SOF and EOF delimiters. Different

SOF codings are used to request and initiate a connection between two devices for a given service class, and to transmit payload data for the resulting connection. Different EOF codings are used by the sending device to indicate whether the frame payload data as sent is valid, corrupted, or truncated. The definition of the coding of the SOF and EOF ordered sets is beyond the scope of this text.

SOF (4 bytes)	Extended Header (optional)	Frame Header (24 bytes)	Data (0 to 2112 bytes)	CRC (4 bytes)	EOF (4 bytes)

Start Of Frame (SOF): Ordered Set denoting beginning of the frame.
Extended Header: Optional header extension supported for FC-2 frames.
Frame Header: Required frame header for FC-1 and FC-2 frames.
Data: Frame payload of anywhere from 0 to 2112 bytes in length.
Cyclic Redundancy Check (CRC): Frame check CRC for error detection.
End Of Frame (EOF): Ordered Set denoting end of the frame.

Fig. 5.23 Fibre channel frame

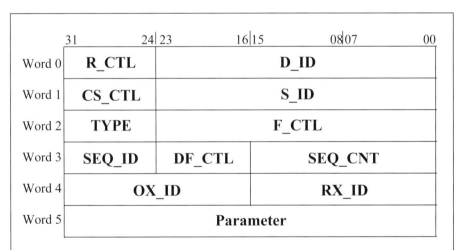

R_CTL: Routing Control
D_ID: Destination Address Identifier
S_ID: Source Address Identifier
CS_CTL: Class Specific Control / Priority
TYPE: Data Structure Type
F_CTL: Frame Control
DF_CTL: Data Field Control
SEQ_ID: Sequence Identifier
SEQ_CNT: Sequence Count
OX_ID: Originator Exchange Identifier
RX_ID: Recipient Exchange Identifier
Parameter: Definition depends on TYPE.

Fig. 5.24 Fibre channel header

IDLE ordered sets, and ordered sets denoting other control information, are sent between frames. The receiver can use these ordered sets in conjunction with the RXxDATASYNC feature on the HSS receiver to align the received data to a 32-bit word boundary in order to simplify downstream processing. Note that with the exception of the four lane variant of 10GFC (which will be discussed separately), Fibre Channel variants are serial data streams and links do not need to be deskewed relative to other links.

The frame content between the SOF and EOF delimiters includes the following fields:

Extended header.

FC-2 frames contain an optional header extension of variable length. Definition of the content of this field is beyond the scope of this text.

Frame header.

This is a field consisting of six words (24 bytes) which uniquely identify the frame sequence and routing. This field is discussed further below.

Data field.

Payload data consisting of 0–2112 bytes. The number of data bytes must be divisible by four (i.e. an integral number of 32-bit words).

Cyclic redundancy check (CRC).

Frame check remainder of a CRC calculation performed using the polynomial described in Sect. 4.2.3.3.

The Frame Header field consists of six words as shown in Fig. 5.24. General descriptions of fields in this header which are relevant to this text follow:

Destination Address Identifier (D_ID).

This field identifies the node(s) on the fabric which are the intended recipients of the frame. Each device on the fabric has a unique 24-bit address, and may also be assigned one or more alias IDs. This address is not fixed (as in the case of Ethernet MAC addresses); it is assigned dynamically by the fabric when the node initially connects (or "logs in") to the fabric.

Source Address Identifier (S_ID).

This field identifies the node on the fabric which originated the frame.

Sequence ID (SEQ_ID).

When a sequence is initiated for transmission between a pair of nodes on the fabric (identified by D_ID and S_ID), a sequence ID is assigned to this connection. All frames associated with this sequence are identified the sequence ID transmitted in the SEQ_ID field.

Sequence Count (SEQ_CNT).

For a given open sequence associated with a specific sequence ID, each frame transmitted for this sequence is assigned a two-byte sequence count value. This value is incremented for each frame transmitted. This allows the recipient to verify that all frames have been received, and the order of these frames.

Other fields serve various control functions. The Routing Control (R_CTL) identifies the type of routing service (basic or extended link services, device data, link control, video, etc.), and information type (solicited/unsolicited control/data). The CS_CTL field either specifies additional class specific control information, or a priority level, depending on the F_CTL field. The Data Structure Type (TYPE) field provides additional information specific to the routing service defined by R_CTL. The Frame Control (F_CTL) field provides various sequence control and status flags. The Data Field Control (DF_CTL) flags the presence of additional optional headers in the payload data. The Originator Exchange ID (OX_ID) and Recipient Exchange ID (RX_ID) identify exchange IDs assigned for exchanges of sequences. The definition of the PARAMETER field depends on the frame type. Detailed definitions of these fields are beyond the scope of this text. The reader is referred to the FC-FS-2 document for further information on these fields.

5.4.3.2 8B/10B Encoding and Scrambling

Figure 5.25 and 5.26 illustrate block diagrams of the datapath of the transmitter and receiver, respectively. These block diagrams show the relationship between the 8B/10B encoder/decoder, CRC generation/checking, and scrambling/descrambling functions. Scrambling is only relevant to 8GFC serial variants; lower baud rate serial variants are not scrambled. (The 10GFC variant is discussed later.)

Data words in the transmit datapath of the link layer shown in Fig. 5.25 are applied to the CRC generator, and the remainder is multiplexed onto the datapath at the end of the frame. Data is scrambled (for 8GFC only), and then multiplexed with Ordered Set primitives that are transmitted between frames or as SOF/EOF delimiters. Note that primitives are not scrambled and are not part of the CRC calculation. The physical layer 8B/10B encodes the data, and then uses an HSS core to transmit the data serially. Ordered Sets are aligned on word boundaries and defined such that only the first byte can be a Kxx.y control symbol (denoted prior to encoding in the figure by the Z bit) in the 8B/10B code.

Bytes (or the corresponding 8B/10B symbols) of each word are transmitted in the order of left to right. Symbols of the 8B/10B code are specified as 10-bit encoded values with bits labeled a through j. Bits within each symbol are transmitted serially in order starting with bit a and ending with bit j.

The physical layer of the receive datapath shown in Fig. 5.26 deserializes the data and decodes the 8B/10B symbols. Primitives are processed directly in the link layer. Data is descrambled (for 8GFC only) and propagated to downstream logic. CRC is calculated for the unscrambled data and the result is compared to the transmitted value.

The CRC Generator polynomial used by Fibre Channel is the same as was defined in Sect. 4.2.3.3. The CRC calculation is performed in a manner that is equivalent to a serial CRC calculation using the following byte/bit order: The byte order is the same as the order in which bytes (or corresponding symbols) are transmitted. The bit order within each byte is least significant bit through most significant bit. The CRC remainder is calculated across all words of the frame between the Start Of Frame (SOF) delimiter and the CRC field.

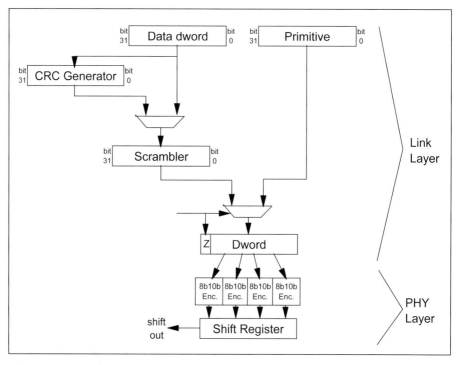

Fig. 5.25 Transmitter datapath block diagram

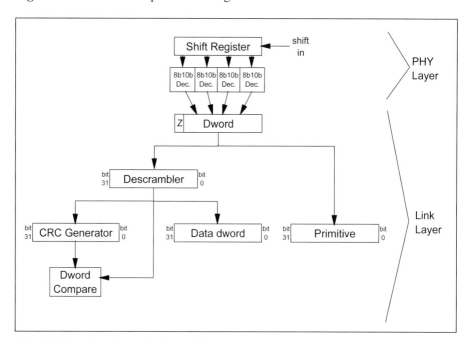

Fig. 5.26 Receiver datapath block diagram

The motivation for scrambling 8GFC serial variants is to avoid CDR biasing issues impacting the link jitter budget, as was discussed in Chap. 1. The 8GFC serial variants use a self-synchronizing scrambler with the following polynomial:

$$G(x) = x^{58} + x^{39} + 1. \tag{5.8}$$

The scrambler (and descrambler) are reset at the start of every frame to avoid any error propagation from one frame to the next. (Otherwise a received bit error at the end of one frame could corrupt the next frame as well.) By using a self-synchronizing scrambler, any variation in the content of the frame causes subsequent bytes to be scrambled differently. The SEQ_CNT field of the header is always guaranteed to change from frame to frame, even for a frame retransmission, and therefore the retransmitted frame is scrambled differently each time.

Note that the bit order for scrambling is not the same as the bit order defined for CRC calculation. Scrambling is performed equivalent to serially scrambling a bit stream consisting of each 32-bit word in the bit order of most significant bit to least significant bit. This convention simplifies implementation assuming a 32-bit datapath.

5.4.3.3 Speed Negotiation

One of the most common parameters that is negotiated by many protocols is the baud rate of the interface. The Fibre Channel standard, as an example, specifies a speed negotiation algorithm in the FC-FS-2 document which is designed to allow hardware to interoperate with legacy hardware supporting lower Fibre Channel speeds. It is important to note that Fibre Channel assumes full-duplex communication between PMDs, making it possible to perform two-way communication to negotiate link parameters. The following steps summarize operation of the Fibre Channel speed negotiation algorithm:

1. The port initially cycles through supported transmit speeds and transmits a specified data pattern until a response is received. Once a response (any response) is received, the port transitions to the next step.

2. The port starts negotiation by setting both the Serdes transmitter and receiver to the maximum speed supported by the port.

3. In this step the port performs a "master" role in the negotiation, potentially trying different transmit baud rates. This step exits when either it becomes obvious that the other port has assumed a "follow" role in the negotiation (indicating the other port is happy with the current transmit baud rate on this port), or the other port is performing a "master" roll at the same baud rate as this port (in which case this port relinquishes the "master" roll to the other port). This step proceeds as follows:

    (a) The port transmits a specified data pattern at the current transmit baud rate, and compares received data to the specified data pattern at the current receive baud rate. If no errors are detected, then the port proceeds to step 4.

(b) If errors are detected then the port tries other receive baud rates look-ing for one where the data pattern is received correctly. If there is a baud rate when the data pattern is received correctly, then the port proceeds to step 4.

(c) If no receive baud rates result in a correct data pattern, the port tries a lower baud rate for the transmitter and repeats the process.

4. In this step the port performs a "follow" role in the negotiation, and con-tinues this role until negotiation completes. This proceeds as follows:

(a) The port sets the transmit baud rate to match the receive baud rate.

(b) The port continues to transmit the specified data pattern, comparing received data to the expected data. If the other port is still assuming a "master" role and changes its transmit speed, then the received data miscompares. In this case the port tries another baud rate for the receiver, sets the transmitter baud rate to match, and repeats this step.

(c) If the port successfully compares the received data to the specified data pattern, the negotiation is complete.

The above negotiation also uses a watchdog timer and, optionally, monitors the RXxSIGDET signal for a Loss of Signal condition. In the event the watchdog timer expires, the negotiation process restarts at step 1. Loss of Signal causes the process to return to step 3.

The Fibre Channel speed negotiation sequence is generally implemented by a hardware state machine. The time spent monitoring the received data, and transitioning between steps, is constrained with minimum and maximum wait time specifications intended to ensure the follower port can try up to four receive baud rates before the master attempts to change the transmit baud rate.

To support this sequence, the HSS core must support changes of the baud rate between the speeds supported by Fibre Channel without requiring complete reinitialization of the core, or generating substantially periods where the receiver will detect a Loss of Signal. The HSS core must also support the transmitter and receiver baud rates being set independently. The transition between baud rates should not cause the HSS core to generate glitches on the TXxDCLK or RXxDCLK outputs. Such glitches could result in erroneous operation of logic being clocked by these clocks.

Currently defined Fibre Channel serial variant baud rates are 8.5 Gbps, 4.25 Gbps, 2.125 Gbps, and 1.0625 Gbps. Fibre Channel hardware is generally designed to support the latest baud rate and at least the next lower baud rate. The HSS EX10 core supports Fibre Channel baud rates by configuring the core to support the 8.5-Gbps baud rate, and using the rate select feature to select "Half-," "Quarter-," or "Eighth" rate modes to support legacy baud rates. Switching the selected rate does not require substantial reinitialization of the core, and if the proper sequence is executed this change of selection does not causes glitches on the parallel data clocks. Specific sequences to support this are described in Sect. 2.5.

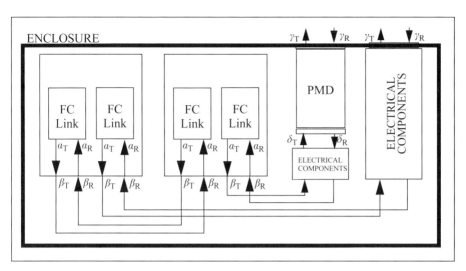

Fig. 5.27 Fibre channel interoperability points example

5.4.4 Physical Interfaces

The Fibre Channel physical interfaces layer variants are specified in "Fibre Channel – Physical Interfaces – 2" document, commonly referred to as "FC-PI-2." At the time of this writing, drafts exist for FC-PI-3 and FC-PI-4, which will eventually specify additional physical interfaces for 10GFC and 8GFC.

This section describes the compliance points for Fibre Channel physical interfaces. Fibre Channel uses the terminology "Interoperability Points" to refer to compliance points. This section also summarizes electrical specifications at the Interoperability Points for Fibre Channel physical variants. Optical interface specifications are important to Fibre Channel, but are beyond the scope of this text.

5.4.4.1 Interoperability Points

Fibre Channel defines electrical interoperability points at input/output pins of the chip, card connector, and internal PMD connector. Also, and possibily most important, the interoperability points defined at the input and output of the enclosure. Note, however, that these latter points may be electrical but are more often optical. The interoperability points defined by the FC-PI-x documents are as follows:

Alpha T (a_T) *and Alpha* R (a_R).

Interoperability points at the chip output pins of the transmitter and the chip input pins of the receiver for a Fibre Channel device or retiming element. This is an electrical interoperability point.

Beta T (β_T) *and Beta* R (β_R).

Interoperability points at the internal connector nearest to the alpha point. This may be a card backplane connector in systems where Fibre Channel traffic is carried across the backplane. In the event that the internal connector nearest to the alpha point meets the definition of a delta or gamma point, then the connector is a delta or gamma point and there is no beta point on the link. This is an electrical interoperability point.

Delta T (δ_T) and Delta R (δ_R).

Interoperability points at the internal connector of a removable Physical Medium Device (PMD). This is an electrical interoperability point.

Gamma T (γ_T) and Gamma R (γ_R).

Interoperability points at the external enclosure connector. This is usually an optical interoperability point. This is an electrical interoperability point for variants which transport Fibre Channel across an electrical cable.

Epsilon T(ε_T) and Epsilon R (ε_R).

Equivalent of δ_T and δ_R interoperability points for physical interface variants that assume an equalizer in the receiver.

Some examples of Fibre Channel connections and the associated interoperability points are shown in Fig. 5.27. This figure contains two cards plugged into a backplane which each have two Fibre Channel Tx/Rx links on them. The chip input/output pins are defined as a_T and a_R interoperability points and the connector pins nearest to these points are defined as the β_T and β_R interoperability points. The example illustrates cases where the β_T and β_R interoperability points are connected together across the backplane. This allows FC devices on separate cards within the same enclosure to communicate without requiring an intervening optical or electrical PMD.

The example in the figure also illustrates a removable PMD element with δ_T and δ_R interoperability points at the internal connector within the enclosure, and γ_T and γ_R interoperability points at the external connector. The specification allows for the possibility that unspecified active components may be required to convert between the electrical signals associated with the β_T and β_R points, and the electrical signals associated with δ_T and δ_R points.

The example also illustrates the case of β_T and β_R interoperability points interfacing directly to unspecified active components which provide γ_T and γ_R interoperability points at the external connector. This case, for example, may be represented by a PMD element which is not removable. In such cases, the FC-PI-x documents do not specify the function of the active components or the internal interface to those components.

5.4.4.2 Nomenclature and Types of Physical Interface Variants

Fibre Channel uses the following nomenclature to denote physical interface variants:

NNN-AA-BB-C,

Where the *NNN* field designates the speed of the interface. Serial variants of Fibre Channel support 100, 200, 400 MBytes per second as defined in FC-PI-2, and FC-PI-4 additionally defines an 800 MBytes per second variant. These speeds correspond to baud rates of 1.0625, 2.125, 4.25, and 8.50 Gbps, respectively.

Table 5.16 Optical physical interface variants

Variant	Specification	Baud rate (Gbps)	Laser (nm)	Receiver	Max. Range	Fiber type
100-SM-LL-V	FC-PI-2	1.0625	1,550	Limiting	50km	Single mode
100-SM-LC-L	FC-PI-2	1.0625	1,300	Limiting	10km	Single mode
100-M5-SN-I	FC-PI-2	1.0625	850	Limiting	500m	50 µm OM2 MM
100-M5E-SN-I	FC-PI-4	1.0625	780/850	Limiting	860m	50 µm OM3 MM
100-M6-SN-I	FC-PI-2	1.0625	780/850	Limiting	300m	62.5 µm OM1 MM
200-SM-LL-V	FC-PI-2	2.125	1,550	Limiting	50km	Single mode
200-SM-LC-L	FC-PI-2	2.125	1,300	Limiting	10km	Single mode
200-M5-SN-I	FC-PI-2	2.125	850	Limiting	300m	50 µm OM2 MM
200-M5E-SN-I	FC-PI-4	2.125	850	Limiting	500m	50 µm OM3 MM
200-M6-SN-I	FC-PI-2	2.125	850	Limiting	150m	62.5 µm OM1 MM
400-SM-LC-L	FC-PI-2	4.25	1,300	Limiting	10km	Single mode
400-SM-LC-M	FC-PI-2	4.25	1,300	Limiting	4km	Single mode
400-M5-SN-I	FC-PI-2	4.25	850	Limiting	150m	50 µm OM2 MM
400-M5E-SN-I	FC-PI-4	4.25	850	Limiting	380m	50 µm OM3 MM
400-M6-SN-I	FC-PI-2	4.25	850	Limiting	70m	62.5 µm OM1 MM
800-SM-LC-L	FC-PI-4	8.50	1,300	Limiting	10km	Single mode
800-SM-LC-I	FC-PI-4	8.50	1,300	Limiting	1.4km	Single mode
800-M5-SA-I	FC-PI-4	8.50	850	Linear	100m	50 µm OM2 MM
800-M5-SN-S	FC-PI-4	8.50	850	Limiting	50m	50 µm OM2 MM
800-M5E-SA-I	FC-PI-4	8.50	850	Linear	300m	50 µm OM3 MM
800-M5E-SN-I	FC-PI-4	8.50	850	Limiting	150m	50 µm OM3 MM
800-M6-SA-S	FC-PI-4	8.50	850	Linear	40m	62.5 µm OM1 MM
800-M6-SN-S	FC-PI-4	8.50	850	Limiting	21m	62.5 µm OM1 MM

Table 5.17 Electrical physical interface variants

Variant	Specification	Baud rate	Receiver	Signal type
100-SE-EL-S	FC-PI-2	1.0625 Gbps	No equalization	Single-ended
100-DF-EL-S	FC-PI-2	1.0625 Gbps	No equalization	Differential
200-SE-EL-S	FC-PI-2	2.125 Gbps	No equalization	Single-ended
200-DF-EL-S	FC-PI-2	2.125 Gbps	No equalization	Differential
400-DF-EL-S	FC-PI-2	4.25 Gbps	No equalization	Differential
800-DF-EL-S	FC-PI-4	8.50 Gbps	No equalization	Differential
800-DF-EA-S	FC-PI-4	8.50 Gbps	DFE	Differential

The AA field designates the type of media. Options include single mode optical fiber, various multimode optical fiber types, unbalanced (single-ended) electrical links, or balanced (differential) electrical links. Optical fiber media connect between γ_T and γ_R interoperability points; electrical links may interconnect between any interoperability points.

The BB field designates the type of transmitter and receiver device used on the link. The following transmitter devices are supported: 1,300 nm/1,550 nm long-wave laser devices, 1,300-nm long wave cost reduced laser devices, and 850-nm short wave laser devices. At baud rates of up to 4.25 Gbps, a limiter stage has typically been included in the optical receiver device. FC-PI-4 includes 8.50-Gbps variants for receivers containing a limiter stage, and also includes variants which use a linear optical receiver. An additional variant is defined in FC-PI-4 for an equalized electrical receiver.

The C field designates the distance range of the link. Maximum distances of 70 m, 2 km, 4 km, 10 km, and 50 km are defined.

Not all combinations of these designators are valid. Physical layer optical variants which have been defined (excluding 10GFC variants) are listed in Table 5.16. Although these are of interest to Fibre Channel users and developers, optical variants are not relevant to electrical HSS cores and these specifications are not summarized in this text.

Electrical physical interface variants defined in FC-PI-2, as well as variants which are defined in FC-PI-4, are listed in Table 5.17. All of these variants are defined as AC Coupled.

For 1GFC and 2GFC, electrical variants are defined which support both single-ended and differential interfaces, either within the enclosure or across an electrical cable between enclosures. Specifications are defined for these variants at β_T, β_R, δ_T, δ_R, γ_T, and γ_R interoperability points. Specifications for α_T and α_R interoperability points are application dependent.

At baud rates of 4.25 Gbps and above, electrical signaling using single-ended signals is not practical, nor is it practical to transmit electrical signals at these frequencies between enclosures. Therefore, only differential variants are defined for 4GFC and 8GFC, and only intraenclosure interoperability points are specified. Specifications are defined for these variants at β_T and β_R interoperability points only.

For 8GFC, variants are specified for receivers that have no equalization and for receivers that include a DFE. The 800-DF-EL-S (which assumes no equalization in the receiver) is used for relatively short electrical links within the enclosure, including links between FC devices and PMD devices, which implement optical variants that include a limiter in the optical receiver. The β_T, β_R, δ_T, and δ_R interoperability points apply to such interfaces.

The 800-DF-EA-S variant (which assumes a DFE circuit in the receiver) is used for longer electrical links within the enclosure, and for any ingress link from a PMD device which implements an optical variant that does not include a limiter in the optical receiver. This variant additionally defines ε_T and ε_R interoperability points. Note that it is assumed that ε_T is equivalent to β_T, however, there are differences in the specification for ε_R and β_R.

5.4.4.3 Specifications for Electrical Parameters

Transmitter electrical parameters for electrical variants defined in FC-PI-4 are summarized in Table 5.18. Jitter specifications for β_T, δ_T, and γ_T points are compliance measurements at the transmitter output with the transmitter connected directly to a termination. Jitter specifications for β_R, δ_R, and γ_R points are also compliance measurements for the transmitter output. These specifications are tested by connecting the transmitter to a reference channel as specified in FC-PI-4, and measuring the signal characteristics at the far end of the channel.

Electrical receiver devices for electrical variants defined in FC-PI-4 must receive a signal with a BER of 10^{-12} or better when the signal has the amplitude and jitter characteristics which are summarized in Table 5.19. The jitter waveform is generated by combining a Deterministic Jitter (DJ) component with a Sinusoidal Jitter (SJ) component. The amount of DJ is equivalent to the maximum DJ allowed at the transmitter output. The resulting signal should be constrained within the specified eye mask; these horizontal limits are specified by the Total Jitter (TJ) component.

Table 5.18 Fiber channel transmitter electrical parameters (FC-PI-4)

	100-DF-EL-S	200-DF-EL-S	400-DF-EL-S	800-DF-EL-S	800-DF-EA-S
Baud rate	1.0625 Gbps ± 100 ppm	2.125 Gbps ± 100 ppm	4.25 Gbps ± 100 ppm	8.50 Gbps ± 100 ppm	8.50 Gbps ± 100 ppm
Type	Differential	Differential	Differential	Differential	Differential
Transmitter waveform	Eye mask	Eye mask	Eye mask	Eye mask	Statistical Analysis
Output signal voltage measured at transmitter[a]	β_T: 600–2,000 δ_T: 650–2,000 γ_T: 1,100–2,000	β_T: 600–2,000 δ_T: 650–2,000 γ_T: 1,100–2,000	β_T: 310–1,600 δ_T: 310–1,600 γ_T: 650–1,600	δ_T: 180–700	β_T:665–1,200 ε_T: 535–1,200 δ_T: 180–700
Output signal voltage measured through a reference load (mVppd)	Not applicable	Not applicable	β_T: 276–1,600 γ_T: 276–1,600 δ_T: 600–1,600	Not applicable	Not applicable
Rise/fall time	100–385	75–192 ps	Not specified	Not specified	Not specified
Deterministic jitter (DJ) (Interenclosure, Units: UIpp)	δ_T: 0.12 γ_T: 0.13 γ_R: 0.35 δ_R: 0.36	δ_T: 0.14 γ_T: 0.16 γ_R: 0.37 δ_R: 0.39	δ_T: 0.14 γ_T: 0.37 γ_R: 0.37 δ_R: 0.39	δ_T: 0.17 δ_R: 0.42	δ_T: 0.17
Deterministic jitter (DJ) (Intraenclosure, Units: UIpp)	β_T: 0.11 β_R: 0.37	β_T: 0.20 β_R: 0.33	β_T: 0.33 β_R: 0.33	Not applicable	See note (b)
Total jitter (TJ) (Interenclosure, Units: UIpp)	δ_T: 0.25 γ_T: 0.27 γ_R: 0.54 δ_R: 0.56	δ_T: 0.26 γ_T: 0.30 γ_R: 0.57 δ_R: 0.59	δ_T: 0.26 γ_T: 0.57 γ_R: 0.57 δ_R: 0.59	δ_T: 0.31 δ_R: 0.71	δ_T: 0.31
Total jitter (TJ) (Intraenclosure, Units: UIpp)	β_T: 0.23 β_R: 0.58	β_T: 0.33 β_R: 0.52	β_T: 0.52 β_R: 0.52	Not applicable	See note (b)

[a]Units: mVpp for single-ended variants, mVppd for differential variants.
[b]Values of DDJ, BUJ, and RJ are specified for the β_T, ε_T, ε_R, and β_R points and are used in statistical signal integrity analysis to determine compliance.

Table 5.19 Fibre channel receiver electrical parameters (FC-PI-4)

	100-DF-EL-S	200-DF-EL-S	400-DF-EL-S	800-DF-EL-S	800-DF-EA-S
Baud rate	1.0625 Gbps ± 100 ppm	2.125 Gbps ± 100 ppm	4.25 Gbps ± 100 ppm	8.50 Gbps ± 100 ppm	8.50 Gbps ± 100 ppm
Type	Differential	Differential	Differential	Differential	Differential
Receiver waveform	Eye mask	Eye mask	Eye mask	Eye mask	Statistical Analysis
Input signal voltage[a]	β_R: 400–2,000 δ_R: 370–2,000 γ_R: 400–2,000	β_R: 400–2,000 δ_R: 370–2,000 γ_R: 400–2,000	β_R: 276–1,600 δ_R: 370–1,600 γ_R: 276–1,600	δ_R: 340–850	Not specified
Sinusoidal jitter (SJ) (Interenclosure, Units: UIpp)	δ_T,δ_R: 0.10 γ_T,γ_R: 0.10	δ_T,δ_R: 0.10 γ_T,γ_R: 0.10	δ_T,δ_R: 0.10 γ_T,γ_R: 0.10	δ_T,δ_R: 0.10 γ_T,γ_R: 0.10	Not applicable
Sinusoidal jitter (SJ) (Intraenclosure, Units: UIpp)	β_T,β_R: 0.10	β_T,β_R: 0.10	β_T,β_R: 0.10	β_T,β_R: 0.10	See note (b)
Deterministic jitter (DJ) (Interenclosure, Units: UIpp)	δ_T: 0.12 γ_T: 0.13 γ_R: 0.35 δ_R: 0.36	δ_T: 0.14 γ_T: 0.16 γ_R: 0.37 δ_R: 0.39	δ_T: 0.14 γ_T: 0.39 γ_R: 0.37 δ_R: 0.39	δ_R: 0.47	Not applicable
Deterministic jitter (DJ) (Intraenclosure, Units: UIpp)	β_T: 0.11 β_R: 0.37	β_T: 0.20 β_R: 0.33	β_T: 0.33 β_R: 0.33	Not applicable	See note (b)
Total jitter (TJ) (Interenclosure, Units: UIpp)	δ_T: 0.35 γ_T: 0.37 γ_R: 0.64 δ_R: 0.66	δ_T: 0.36 γ_T: 0.40 γ_R: 0.67 δ_R: 0.69	δ_T: 0.36 γ_T: 0.69 γ_R: 0.67 δ_R: 0.69	δ_R: 0.71	Not applicable
Total jitter (TJ) (Intraenclosure, Units: UIpp)	β_T: 0.33 β_R: 0.68	β_T: 0.43 β_R: 0.62	β_T: 0.62 β_R: 0.62	Not applicable	See note (b)

[a]Units: mVpp for single-ended variants, mVppd for differential variants.
[b]Values of DDJ, BUJ, and RJ are specified for the β_T, ε_T, ε_R, and β_R points and are used in statistical signal integrity analysis to determine compliance.

5.4.5 10-Gbps Fiber Channel

The Fibre Channel framing and signaling and physical interface layers for 10GFC are specified in the "Fibre Channel – 10 Gigabit" document, commonly referred to as "10GFC." This section summarizes the features which are of some relevance to HSS.

As was discussed previously, Fibre Channel uses a nomenclature of the following form to denote physical interface variants:

NNN-AA-BB-C,

Where the *NNN* field designates the speed of the interface. The speed for 10GFC is 1,200 MBytes per second.

The *AA* field designates the type of media. Options for 10GFC include: single-mode optical fiber and various multimode optical fiber types similar to those used with serial variants.

The *BB* field designates the type of transmitter and receiver device used on the link. Serial 10GFC variants use either 850-nm or 1,300-nm laser devices. Parallel 10GFC variants use either four lasers operating in parallel (850 nm), or a four-lane CWDM (850 nm or 1,300 nm) optical transmitter.

The *C* field designates the distance range of the link. Variants exist for both long and intermediate distances.

Valid physical interface variants for 10GFC are listed in Table 5.20. The base bit rate of 10GFC is 10.2 Gbps. Physical interface variants fall into two categories:

- Serial variants which assume 64B/66B encoding of the data and therefore have a serial baud rate of 10.51875 Gbaud/s
- Four-lane parallel variants (either transmitting on four separate fibers or transmitting four wavelengths using a CWDM PMD) which assume byte striping of data across four lanes and 8B/10B encoding of each lane, and operate at 3.1875 Gbaud/s on each lane.

Note that Table 5.20 additionally provides a column for reference to an Ethernet variant. Several of the 10GFC physical interface variants are essentially the same device as can be used for the referenced Ethernet PMD layer. In addition, 10GFC specifies a few additional physical interface variants which do not have an equivalent optics specification in Ethernet. The framing and signaling specification for 4-lane 10GFC FC-0 (PMD) layer devices is similar to the PCS/PMA layer defined by Clause 48 of the IEEE 802.3 Ethernet standard. In addition, 10GFC defines the equivalent XGMII function to transport data across XAUI interfaces within the enclosure. The framing and signaling specification for serial 10GFC FC-0 (PMD) layer devices is similar to the PCS/PMA layer defined by Clauses 49 and 51 of the IEEE 802.3 Ethernet standard.

Table 5.20 10GFC physical interface variants

Variant	Specification	Baud rate	Laser type	Ethernet reference
1200-Mx-SN4P-I (x = 5,5E,6)	10GFC	4 lane by 3.1875 Gbps	4 lane parallel	Not applicable
1200-Mx-SN-I (x = 5,5E,6)	10GFC	10.51875 Gbps	Serial	10GBASE-SR
1200-Mx-SN4-I (x = 5,5E,6)	10GFC	4 lane by 3.1875 Gbps	4 lane CWDM	Not applicable
1200-SM-LL-L	10GFC	10.51875 Gbps	serial	10GBASE-LR
1200-Mx-LC4-L (x = 5,6)	10GFC	4 lane by 3.1875 Gbps	4 lane CWDM	10GBASE-LX4
1200-SM-LC4-L	10GFC	4 lane by 3.1875 Gbps	4 lane CWDM	10GBASE-LX4

5.5 PCI Express

PCI Express is a high-performance I/O bus architecture for interconnecting peripheral devices in computing and communication platforms. It is the third generation that evolved from PCI and PCI-X and is specifically designed to overcome the performance limitations of those busses. In addition to the performance advantage, PCI Express also reduces system cost because it requires fewer package pins on the device and a smaller number of system board wires. Furthermore, PCI Express systems are backward compatible with software for PCI/PCI-X so that existing operating systems may be ported to PCI Express with no changes to drivers or application programs.

PCI Express is developed by the membership of the Peripheral Component Interconnect Special Interest Group (PCI-SIG). The initial PCI Express Base Specification was published in April 2003, with various errata being corrected in the Revision 1.1 document published in March 2005. This document specifies what has come to be known as PCI Express Generation 1, which utilizes 2.5-Gbps Serdes technology. Revision 2.0 of this document, commonly called PCI Express Generation 2, was published in December, 2006, and adds support for a 5.0-Gbps baud rate.

The "PCI Express Card Electromechanical 2.0 Specification" is an additional companion specification which defines the electromechanical form factors of PCI Express devices. These include:

Card electromechanical.

This form factor is for standard PC add-in cards similar to PCI cards, and supports link widths from x1 to x16. Form factor is predominantly used by graphics cards, although there are other cards available such as USB and network adaptors.

Mini-card electromechanical.

This form factor is similar to PCMCIA cards for laptops and other applications which require small size and power. Only x1 link width is supported; a reduced signal amplitude is used.

Express module.

This form factor is for hot-pluggable cards in rack-mount servers. Form factor supports link widths x1 to x8 in a single-wide module, and x1 to x16 in a double-wide module.

HSS core feature settings for implementations which comply with the above electromechanical specifications can be characterized and may be specified by applications documentation available from the vendor. However, chip-to-chip links and custom implementations still require the same signal integrity analysis to determine feature settings as is required for other applications. PCI Express cards which may be used in multiple systems should consider providing provisionability of HSS features to facilitate use in nonstandard electromechanical environments.

5.5.1 PCI Express Architecture

The topology of a PCI Express bus and the device protocol layers associated with PCI Express is described in this section.

5.5.1.1 Physical Topology

Computers have an ever increasing demand for data bandwidth. In the past, PCI and PCI-X systems delivered adequate performance using multidrop parallel busses in which several devices shared each bus. Performance boosts were achieved by increasing the effective bus clock frequency from 33 MHz, 66 MHz, 133 MHz, and beyond. However, these architectures required limitations on the number of devices per bus to control electrical loading as the frequency was increased.

The PCI Express hierarchy has a significantly different physical structure from that of PCI. Instead of a parallel bus routed to several components, the physical layer of PCI Express is a point-to-point connection similar to other standards which utilize Serdes technology. The hierarchy routes traffic to components through a switch. Each switch port is a virtual PCI to PCI bridge; this allows existing PCI software to enumerate the hierarchy of components.

Figure 5.28 illustrates the main components in a typical PCI Express system, including the CPU, memory, Root Complex, endpoint devices, switches, and bridges to legacy PCI/PCI-X busses. The *Root Complex* connects the CPU and memory subsystem to the PCI Express fabric. It generates PCI Express transaction requests for the CPU and transmits them across one of its ports to the destination device (either an endpoint or a switch). When a request arrives at an endpoint device, the device completes the transaction by reading the requested data or writing data to the target location. In addition, the endpoint can initiate its own transactions across the link to the Root Complex, another endpoint, or switch. A USB device and Ethernet NIC are two examples of endpoint devices.

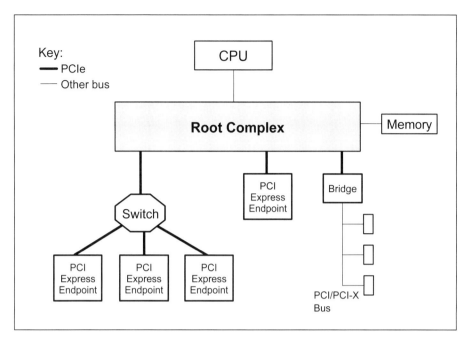

Fig. 5.28 PCI express bus physical topology

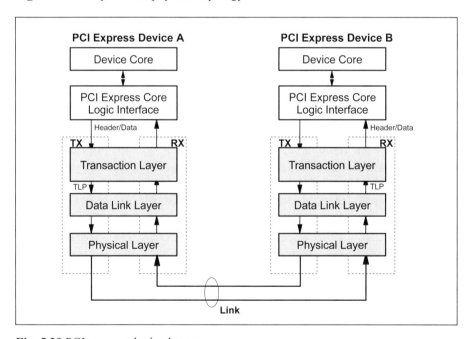

Fig. 5.29 PCI express device layers

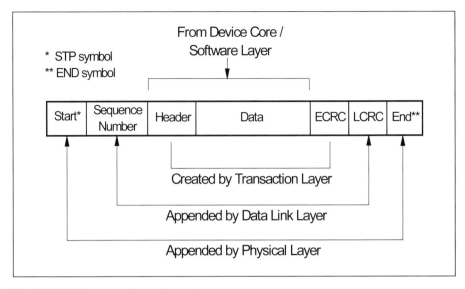

Fig. 5.30 PCI express frame format

Table 5.21 PCI express bandwidths

	x1	x2	x4	x8	x12	x16	x32
Gen 1, 2.5 GTps	0.5 GBps	1.0 GBps	2.0 GBps	4.0 GBps	6.0 GBps	8.0 GBps	16 GBps
Gen 2, 5.0 GTps	1.0 GBps	2.0 GBps	4.0 GBps	8.0 GBps	12 GBps	16 GBps	32 GBps

A *switch* attaches multiple PCI Express devices together. Each port of the switch behaves logically as a PCI-to-PCI bridge since its sole purpose is to forward packets from the incoming port (Ingress Port) to the destination (Egress Port). Each switch contains arbitration logic to determine the priority of the packets being sent.

Device Layers

Figure 5.29 illustrates the layered architecture of a PCI Express device. The top layer, *Device Core*, initiates transactions onto the PCI Express fabric and executes transactions that it receives from other devices. This layer communicates to/from the fabric through the *PCI Express Core Logic Interface*. PCI Express does not specify the Device Core and PCI Express Core Logic Interface; these layers are application specific logic and the design of these layers varies from one implementation to the next. Root Complex core logic and an endpoint Ethernet controller are two examples of Device Cores.

Every device must support the functionality of the bottom three layers of the stack: the *Transaction Layer*, the *Data Link Layer*, and the *Physical Layer*. These layers are composed of a transmit section that processes outbound traffic through the transmit (Tx) side of the link, and a receive section that handles incoming traffic from the receive (Rx) side. Each of these layers contributes to the generation and processing of fields in the PCI Express packet shown in Fig. 5.30.

The Transaction Layer assembles information from the Device Core into a Transaction Layer Packet (TLP) for transmission. This packet includes a header, up to 1024 32-bit words of data (4 kbytes), plus an optional End-to-End CRC (ECRC) field used for error detection. The TLP is forwarded to the Data Link Layer, which concatenates a sequence number and link CRC (LCRC), and stores the resulting TLP in a local *Retry Buffer*. The purpose of the LCRC is to detect errors in the transmission. Next, the TLP arrives at the Physical Layer which appends a one byte Start delimiter symbol and a one byte End delimiter symbol. Bytes of the packet are 8B/10B encoded by the Physical Layer for transmission.

On the other side of the link, the receiving device follows the reverse procedure to process fields in the packet and forward the TLP to the Transaction Layer. First, the Physical Layer deserializes the incoming bitstream, decodes the 8B/10B symbols, and removes the Start/End delimiter symbols. The Data Link Layer then processes the sequence number and LCRC fields to check for errors, and removes these fields. If there are no errors, the Data Link Layer sends an acknowledge (ACK) Data Link Layer Packet (DLLP) to the transmitting device to confirm successful delivery of the packet, allowing the transmitting device to remove the packet from its Retry Buffer. (In the event the Data Link Layer detects an LCRC error or other errors in the packet, a Negative Acknowledge (NAK) DLLP is sent to the transmitting device and the packet is resent.) Then the Data Link Layer forwards the TLP to the Transaction Layer which uses the ECRC field to check for errors in the end-to-end path, and forwards the TLP to the Device Core.

5.5.2 Physical Layer Logic

The PCI Express physical layer includes both physical layer logic and physical layer electrical specifications. The physical layer electrical specifications are generally implemented with HSS cores, and the logical specifications are implemented with companion logic.

A block diagram of the physical layer logic is shown in Fig. 5.31. An n-wide port consists of n transmit and n receive HSS lanes and associated logic to perform the physical layer function of the transmit and receive paths. State machine logic to perform training is also a required element of the port.

5.5.2.1 Physical Layer Transmit Logic

Functions of the transmit data path in Fig. 5.31 are described below.

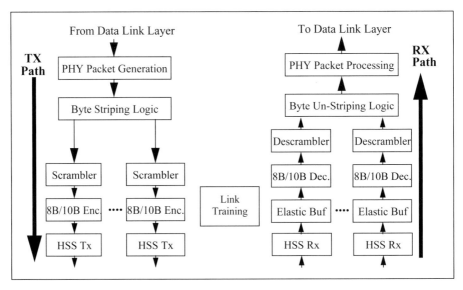

Fig. 5.31 Physical layer logic block diagram

Physical Packet Generation. The physical layer receives the TLP or DLLP from the Data Link Layer, and forms a physical packet by adding control symbols. The following control symbols are defined for PCI Express and correspond to 10-bit control symbols in the 8B/10B block code. These symbols are added in the transmission path as needed.

STP / SDP – Start Symbols (K27.7, K28.2). The physical layer adds a Start symbol to delimit the beginning of the packet. Two Start symbols are defined: Start TL Packet (STP) indicates the start of a TLP, and Start DLL Packet (SDP) indicates the start of a DLLP.

END / EDB – End Symbols (K29.7, K30.7). The physical layer also adds an End symbol to delimit the end of a packet. Two End symbols are defined: End Good (END) indicates the normal end of a packet, and End Bad (EDB) indicates the end of a packet which is to be ignored (possibly because it contains an error).

COM – Comma Symbol (K28.5). The COM symbol is the first symbol of an Ordered Set. Ordered Sets are described below.

SKP – Skip Symbol (K28.0). The SKP symbol is transmitted as part of a Skip Ordered Set. Ordered Sets are described below.

FTS – Fast Training Sequence Symbol (K28.1). The FTS symbol is transmitted as part of a Fast Training Sequence Ordered Set. Ordered Sets are described below.

EIE – Electrical Idle Exit (K28.7). The EIE symbol is transmitted as part of an Electrical Idle Exit Ordered Set used in systems operating at speeds greater than 2.5 GT/s. Ordered Sets are described below.

PAD – Pad Symbol (K23.7). PAD symbols are transmitted at the end of a packet, if needed, on unused lanes.

IDL – Idle Symbol (K28.3). IDL symbols are transmitted as part of an Electrical Idle Ordered Set. Ordered Sets are described below.

TS1 / TS2 – Training Sequence Symbols (D10.2, D5.2). The TS1 and TS2 symbols are data symbols transmitted within training sequence ordered sets to identify the type of ordered set.

The following Ordered Sets are defined for PCI Express:

Training sequences. Training sequence ordered sets TS1 and TS2 are transmitted during link training in order to negotiate port parameters. This is described in more detail later in this section.

Skip ordered set. The SKIP ordered set is scheduled to be transmitted at periodic intervals. The periodic interval must be at least once every 1,538 symbol times, and no more than once every 1,180 symbol times. This ordered set consists of the symbol sequence (COM,SKP,SKP,SKP). Receiver logic may add/drop these symbols in the Elastic Buffer stage in order to compensate for frequency differences between the recovered receive clock and the local clock reference.

Note that transmission of a packet cannot be interrupted by a SKIP ordered set. Therefore, if the SKIP is scheduled for transmission while a packet transmission is in process, the SKIP is not transmitted until after the packet transmission is complete. This might result in more than 1,538 symbol times between the occurrence of SKIPs on the interface.

Electrical idle ordered set. This ordered set is transmitted by the transmit port to inform the receive port that the transmitter wants to transition the link into the Electrical Idle power management state. The ordered set consists of the symbol sequence (COM,IDL,IDL,IDL).

Fast training sequence ordered set. This ordered set is transmitted by the transmit port to quickly train the receiver and allow the receiver to achieve symbol lock after exiting the Electrical Idle power management state. The ordered set consists of the symbol sequence (COM,FTS,FTS,FTS).

Electrical idle exit ordered set. When operating at speeds greater than 2.5 GT/s, this ordered set is transmitted by the transmit port after exiting the Electrical Idle power management state, and before transmitting the TS1/TS2 Ordered Sets. It is also transmitted periodically between TS1/TS2 Ordered Sets. This ordered set has sufficient low frequency components to ensure electrical idle detection circuits at the receiver recognize the exit condition. (Some circuit implementations would have difficulty detecting the exit condition on a higher loss channel without low frequency content.) The ordered set consists of the symbol sequence (COM, 14 EIE symbols, TS1).

The Physical Packet Generation stage either transmits TLP or DLLP data contents, control symbols, or ordered sets. If none of the above needs to be transmitted, then Logical Idle symbols are transmitted ("00"h bytes).

Byte-Striping:

PCI Express specifies ports which consist of a *link width* of 1, 2, 4, 8, 12, 16, or 32 electrical links in each direction. Of these permitted widths, only link widths of 1, 4, 8, and 16 are commonly used in the industry. Link widths, and corresponding data bandwidth, are summarized in Table 5.21. The link width for a given port is determined dynamically during training. For an *n*-link port, bytes are striped to the lanes such that the first byte is transmitted on lane 0, the second byte on lane 1, and so forth.

When the link width of the port is greater than 1, the following rules apply for alignment of packets and ordered sets across the links of the port:

- The Start Packet delimiter is always transmitted on lane 0 unless the packet is immediately following the end of another packet. In such cases, the Start packet delimiter must be transmitted on a lane with a lane number divisible by 4 (i.e., lanes 0, 4, 8, etc.)

- The End Packet delimiter is always transmitted on a lane with a lane number divisible by 4 –1 (i.e., lanes 3, 7, 11, etc.). If another packet does not start following the end of the packet, then the remainder of the line is filled with PAD symbols.

- Ordered Sets and Logical Idle sequences must be transmitted on all lanes simultaneously. Ordered Sets are not byte-striped across lanes, but rather are transmitted in parallel on all lanes.

Scrambler:

Scrambling is performed using a sidestream scrambler which implements the following polynomial:

$$G(x) = x^{16} + x^5 + x^4 + x^3 + 1. \tag{5.9}$$

The state value of the scrambler for each link of a multilink port must be synchronized such that all links have the same scrambler state value at all times. Data symbols and Logical Idle Sequences are scrambled; ordered sets and other control symbols are not scrambled.

The scrambler state is updated by 8 bits for each byte transmitted regardless of whether the byte is scrambled or not. The exception to this is SKP symbols. The scrambler state does not advance when SKP symbols are transmitted since receiver logic may add/drop SKP symbols prior to the descrambler stage.

The scrambler is reset to an all 1's state whenever a COM symbol is transmitted.

8B/10B Encoder

This stage encodes each byte into the corresponding 8B/10B encoded symbol, and drives the parallel data input of the HSS transmitter. Block 8B/10B coding was described in Sect. 4.2.2.1.

5.5.2.2 Physical Layer Receiver Logic

Functions of the receive data path in Fig. 5.31 are described below.

Elastic Buffer

This stage consists of an elastic FIFO to compensate for clock frequency differences between the receive clock from the HSS core and the local reference clock. This compensation is performed by dropping SKP symbols as needed if the receive clock is faster than the local clock, or adding additional SKP symbols after SKP symbols in the incoming data if the receive clock is slower than the local clock.

Implementations of this stage generally include a function to perform symbol alignment on the input of the buffer. This is performed by searching for COM symbols in the incoming data, and pulsing RXxDATASYNC (or alternatively by stepping the position of a barrel shifter), if needed, to adjust the alignment. PCI Express refers to this process as achieving *symbol lock*.

8B/10B Decoder

This stage decodes each 8B/10B symbol into the corresponding data or control byte. Block 8B/10B coding was described in Sect. 4.2.2.1.

Descrambler

This stage descrambles data in a similar manner to the scrambling performed on the transmit data path.

Byte Un-Striping

This stage performs the inverse of the byte-striping process used at the transmitter. A key function of this stage in a multi-lane port implementation is lane-to-lane deskew of data. Although PCI Express does not specify precisely how to do this, the common method is to add delay as needed such that COM symbols for ordered sets are aligned with each other on all lanes.

Physical Packet Processing

This stage removes the Start and End of packet framing symbols from the packet and forwards the TLP or DLLP to the Data Link Layer.

5.5.2.3 Link Training

The physical layer also implements state machines associated with link training as implied by Fig. 5.31. Detailed descriptions of the training sequence and state machine implementations are beyond the scope of this text, however, the process is summarized below.

The following port parameters are negotiated as part of the link training sequence:

Link width. The transmit and receive ports on a given link may have different link widths. The training sequence determines the width of each port on the link and picks a width that is supported by both ports.

Polarity inversion. PCI Express allows that a physical implementation may intentionally or unintentionally reverse the true/complement signals of the electrical differential signal. This results in the polarity of received data being inverted. The training sequence detects that the polarity has been inverted, and inverts the received data, if needed, to correct this.

Link data rate. PCI Express Gen 1 ports use a baud rate of 2.5 Gbaud/sec on each link, while PCI Express Gen 2 ports may additionally support a baud rate of 5.0 Gbaud/s. Future generations of the specification will define higher baud rates. During training each node advertises supported speeds, and the link is initialized to pick a speed that is supported by both ports.

Lane reversal. PCI Express allows a physical implementation to reverse the physical wiring of lanes relative to the logical lane assignment. This might be desirable to reduce the circuit board cost, as an example. Optionally, a port may support detection that the assignment of lane numbers is reversed, and support reordering of bytes accordingly in the Byte Un-Striping stage. If the port does not support lane reversal, then the negotiation may instead result in a lower link width in order to establish the communications path.

The transmit port sends TS1 or TS2 ordered sets on each link during the training sequence. These ordered sets are 16 symbols long, and advertise various port characteristics in order to negotiate the above port parameters. Each port determines the characteristics of its partner based on the advertised characteristics in the received training sequence, and then adapts accordingly.

The receiver link also trains its CDR circuit, achieves symbol lock, and performs lane-to-lane deskew during the training sequence. The receiver must be fully functional and ready to receive data before the training sequence can complete.

5.5.3 Electrical Physical Layer

This section describes the electrical parameters and features applicable to implementation of the PCI Express Physical Layer.

5.5.3.1 Differential Signal Parameters

The PCI Express specification tests transmitter and receiver compliance using eye masks for the transmitter output and for the jitter tolerance signal at the receiver input. Electrical parameters for the transmitter are listed in Table 5.22. Electrical parameters for the receiver are listed in Table 5.23. Note that some receiver parameters vary depending on the baud rate. Also note that different specifications apply at 5.0 Gbps dependent on whether or not the transmitter and receiver use the same or different reference clock sources.

Table 5.22 PCI express transmitter electrical parameters

	PCI Express Gen 1	PCI Express Gen 2
Baud rate	2.50 Gbps ± 300 ppm	5.00 Gbps ± 300 ppm
Transmitter waveform	Eye mask is specified	Eye mask is specified
Output signal voltage	800–1,200 mVppd	800–1,200 mVppd [1]
DC common mode voltage	0 – 3.6 V	0 – 3.6 V
Rise/fall time	0.125 UI	0.125 UI
Total jitter (TJ)	0.30 UI	0.30 UI

[1] Some electromechanical form factors specify a lower differential amplitude for the transmitter

Table 5.23 PCI express receiver electrical parameters

	PCI Express Gen 1	PCI Express Gen 2 (common clock architecture)	PCI Express Gen 2 (data clocked architecture)
Baud rate (± 300 ppm)	2.50 Gbps	5.00 Gbps	5.00 Gbps
Jitter tolerance waveform	Eye mask	Eye mask	Eye mask
Input signal voltage	175–1,200 mVppd	120–1,200 mVppd	100–1,200 mVppd
Deterministic jitter (DJ)	Not specified	0.30 UI	0.24 UI
Total jitter (TJ)	0.60 UI	0.40 UI	0.34 UI
AC coupling capacitor	75–200 nF	75–200 nF	75–200 nF

The baud rate for all variants of PCI Express is specified with the tolerance of ±300 ppm. This tolerance is intended not only to allow for variation in frequency between clock references at each end of the link, but also to allow for Spread Spectrum Clocking (SSC). SSC varies the reference clock frequency within a range in order to reduce EMI emissions, as was discussed in Sect. 2.3.10, for the HSS EX10 core.

The transmitter specifications allow for deemphasis on the output signal. An FFE in the transmitter can correct for intersymbol interference induced by the channel. PCI Express assumes there is no equalization in the receiver.

5.5.3.2 Special Electrical Signaling Support

HSS cores which support PCI Express are required to implement a number of unique features. Additional support is desirable for optional features of the

PCI Express specification. These features are summarized in this section. Note that several of these features are related to power management states which are discussed in Sect. 5.5.4.

Electrical Idle

When in an electrical idle state, the transmitter outputs are driven to the common mode voltage. Before transitioning into this state, the transmitter must transmit one or more Electrical Idle Ordered Sets. To return to an operating state, the transmitter must transmit FTS (for exit from the L0s Power State) or TS1/TS2 ordered sets (for exit from the L1 Power State).

The receiver enters an electrical idle state upon receiving the Electrical Idle Ordered Set. The receiver exits this state upon seeing a differential voltage on the input in excess of the value allowed during an Electrical Idle state.

An HSS transmitter which supports PCI Express must provide the capability to drive an electrical idle state on the link. The HSS receiver may continue to operate and receive data while the transmitter is sending an electrical idle, or may be partially powered down. Either way, the RXxSIGDET output indicates there is no signal amplitude. When the HSS transmitter resumes sending serial data, the receiver must be powered back on (if applicable), retrain the CDR circuit, and reacquire symbol lock. If the receiver is powered down, the signal detect circuit must remain powered on so that a wake-up event can be detected. Obviously, retraining the CDR and reacquiring symbol lock requires more time if the receiver was powered down.

Receiver Detection

The transmitter checks the link after a reset to determine whether a receiver is connected at the other end. This is done by driving an abrupt change in the DC Common Mode voltage of the link (either from ground rail to VDD power rail or vice versa), and monitoring the amount of time it takes for the common mode voltage on the wire to settle to the new value. If a receiver is present, charging the AC coupling capacitance of the receiver causes this settling time to be relatively slow. If no receiver is present, the settling time is relatively fast. An HSS transmitter which supports PCI Express must provide the capability to drive a receiver detection event and the detection circuitry to determine the results.

Beacon Signaling

Beacon signaling is optional in the PCI Express specification, and may be used in the implementation of L2 power states. The transmitter sends a low-frequency high/low waveform called a beacon signal on the link to indicate a desire to exit the L2 power state and return to a full-on state. The beacon signal has a pulse width of at least 2 ns and no more than 16 μs. The receiver is powered down in the L2 power state, but must have some circuitry active which can detect the beacon signal and notify the power management function of the port. An HSS receiver which supports PCI Express must provide the capability to detect the beacon signal.

5.5.4 Power States

PCI Express defines the L0, L0s, L1, L2, and L3 power management states, each of which is intended to use progressively less power.

L0 Power State. The link transmitter and receiver are fully operational in this power state.

L0s Power State. If the physical layer transmit logic has no TLP or DLLP traffic to transmit for some length of time, the transmitter can initiate entry into the L0s power state. The protocol for entry/exit to/from an electrical idle state was described in "Electrical Idle" under Sect. 5.5.3.2. When the transmitter once again has useful traffic to transmit, it exits the electrical idle state and retrains the receiver.

Note that detailed rules exist in the PCI Express standard for when a link should enter and exit the L0s power state. These rules vary based on the type of device in the PCI Express architecture. There are also physical layer specifications for maximum response times to enter and exit the L0s power state, as well as specifications for the minimum time the transmitter must stay in this power state once it has been initiated. These specifications are beyond the scope of this text.

L1 Power State. The physical layer may enter the L1 power state upon direction by the Data Link Layer. Both the transmit and the receive ports must negotiate and agree to enter the L1 power state. This negotiation is performed through DLLP traffic between the Data Link Layers of each node, the description of which is beyond the scope of this text.

The transmitter enters the L1 power state in a similar manner to entry into the L0s state as described in "Electrical Idle" under Sect. 5.5.3.2. The transmitter exits the L1 power state by sending TS1/TS2 ordered sets. The receiver may be powered down while in the L1 power state. To exit the L1 power state, the signal detect circuit at the receiver (which is not powered down in this state), detects the signal and the port logic restores power to the receiver. Recovery time to retrain the receiver is expected to be longer when exiting L1 power state than in the case of L0s power state.

As was the case for the L0s power state, the PCI Express specifies requirements for physical layer state transitions which apply to entry/exit to/from L1 power state. These specifications are beyond the scope of this text.

L2 Power State. The physical layer enters and exits the L2 power state upon request. In this power state, a substantial portion of the HSS core can be powered off, including portions of the PLL logic. A port which is in the L2 power state may request exit from this state by transmitting a beacon signal to its link partner. Portions of the receiver which are still powered on can detect this beacon signal and propagate this request for action by the system power management software.

5.5.5 PCI Express Implementation Example

Figure 5.32 shows an IBM implementation of the PCI Express protocol stack. This implementation is used to illustrate one example of how PCI Express layers can be partitioned into functional blocks. The layers of the PCI Express protocol are partitioned into the cores in Fig. 5.32 as follows:

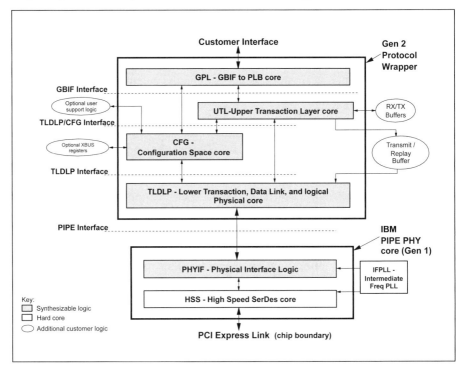

Fig. 5.32 IBM set of cores for the PCI express protocol (Gen 1)

GBIF to PLB Core (GPL). GPL implements the PCI Express Core Logic Interface layer, connecting the Transaction Layer logic to a vendor-specific processor interface. This layer is application dependent and is not specified by the PCI Express standard.

Upper Transaction Layer Core (UTL). UTL implements part of the Transaction Layer logic, including transaction generation and ordering, end-to-end CRC generation and checking, and virtual channel arbitration.

Configuration Space Core (CFG). CFG implements the configuration control registers specified by the PCI Express standard. This core supports interfaces to user support logic and to an XBUS interface.

Transaction, Data Link, and Logical Physical Core (TLDLP). TLDLP implements the remaining Transaction Layer functionality, the Data Link Layer, and part of the Physical Layer logic, including Physical Layer Packet Generation and Processing, Byte Striping and Unstriping, scrambling and descrambling, and Training Sequence generation and control.

Physical Interface Logic (PHYIF). PHYIF implements the remaining Physical Layer logic function, including 8B/10B encoding and decoding and elastic buffers for clock compensation.

The PHY Interface for PCI Express Architecture (PIPE) connects the TLDLP core of the Gen 2 Protocol Wrapper to the PHYIF core. The specification for this interface is an industry standard developed by several companies and published by Intel. This standard is intended to facilitate interoperability of the lower levels of the Physical Layer with higher levels of the protocol stack supplied by different vendors. This specification also dictates the allocation of physical layer functionality between the TLDLP and PHYIF cores. The reader is referred to the standards document for more information.

The HSS EX10 core described in Chap. 2 included features specifically intended to support the PCI Express protocol. These features were described in Sect. 2.2.9 for the transmitter slice and in Sect. 2.3.13 for the receiver slice. Additional power management features to support PCI Express were described in Sect. 2.4.8.

5.6 References and Additional Reading

The following standards documents are relevant to "SONET/SDH Networks" as described in Sect. 5.1:

1. "ANSI T1.105-2001 Synchronous Optical Network (SONET) – Basic Description including Multiplex Structure, Rates, and Formats", American National Standards Institute, Inc., 2001.

2. "ITU-T G.707 – Series G: Transmission Systems and Media, Digital Systems and Networks, Digital Terminal Equipment – General, Network Node Interfaces for the Synchronous Digital Hierarchy (SDH)", International Telecommunications Union, 1996.

3. "ITU-T G.709 – Series G: Transmission Systems and Media, Digital Systems and Networks, Digital Terminal Equipment – General, Interface for the Optical Transport Network (OTN)", International Telecommunications Union, 2001.

4. "ITU-T G.783 – Series G: Transmission Systems and Media, Digital Systems and Networks, Digital Terminal Equipment – Characteristics of SDH Equipment Functional Blocks", International Telecommunications Union, 2006.

The following reading is recommended for more information regarding "SONET/SDH Networks":

5. "SONET", Second Edition, Walter J. Goralski, McGraw-Hill, New York, 2000.

The following standards documents are relevant to "OIF Protocols" as described in Sect. 5.2:

6. "System Interface Level 5 (SxI-5): Common Electrical Characteristics for 2.488–3.125 Gbps Parallel Interfaces", OIF-SxI-5-01.0, Optical Internetworking Forum (http://www.oiforum.com), Oct. 2002.

7. "Common Electrical I/O (CEI) – Electrical and Jitter Interoperability agreements for 6G+ bps and 11G+ bps I/O", OIF-CEI-02.0, Optical Internetworking Forum (http://www.oiforum.com), Feb. 28 2005.

8. "Serdes Framer Interface Level 5 (SFI-5): Implementation Agreement for 40Gb/s Interface for Physical Layer Devices", OIF-SFI5-01.0, Optical Internetworking Forum (http://www.oiforum.com), Jan. 29 2002.

9. "Serdes Framer Interface Level 5 Phase 2 (SFI-5.2): Implementation Agreement for 40Gb/s Interface for Physical Layer Devices", OIF-SFI5-02.0, Optical Internetworking Forum (http://www.oiforum.com), Oct. 2 2006.

10. "Scalable System Packet Interface (SPI-S) Implementation Agreement: System Packet Interface Capable of Operating as an Adaption Layer for Serial Data Links", OIF-SPI-S-01.0, Optical Internetworking Forum (http://www.oiforum.com), Nov. 17 2006.

11. "Common Electrical I/O – Protocol (CEI-P) – Implementation Agreement", OIF-CEI-P-01.0, Optical Internetworking Forum (http://www.oiforum.com), Mar. 2005.

The following additional references are also relevant to "OIF Protocols":

12. Information on StatEye software: http://www.stateye.org.

The following standards documents are relevant to "5.3 Ethernet Protocols" as described in Sect. 5.3:

13. "IEEE Standard for Information Technology – Telecommunications and Information Exchange Between Systems – Local and Metropolitan Area Networks – Carrier Sense Multiple Access with Collision Detection (CSMA/CD) Access Method and Physical Layer Specifications", IEEE 802.3-2005, Institute of Electrical and Electronic Engineers, Dec. 12 2005.

14. "Amendment: Ethernet Operation over Electrical Backplanes", IEEE P802.3ap, Draft 3.3, Institute of Electrical and Electronic Engineers, Jan. 26 2007.

15. "INF-8077 10Gb Small Form Factor Pluggable Module", Revision 4.5, 10 Gigabit Small Form Factor Pluggable (XFP) Multi Source Agreement (MSA) Group (http://www.xfpmsa.org), Aug. 2005.

The following standards documents are relevant to "5.4 Fibre Channel (FC) Storage Area Networks" as described in Sect. 5.4:

16. "ANSI INCITS 404-2006 For Information Technology – Fibre Channel – Physical Interfaces-2 (FC-PI-2)", American National Standards Institute, Inc., International Committee for Information Technology Standards, Aug. 11 2006.

17. "ANSI INCITS 424-2007 For Information Technology – Fibre Channel – Framing and Signaling-2 (FC-FS-2)", American National Standards Institute, Inc., International Committee for Information Technology Standards, Aug. 9 2007.

18. "ANSI INCITS 424-2007 AM1 For Information Technology – Fibre Channel – Framing and Signaling-2 – Amendment 1 (FC-FS-2/AM1)", American National Standards Institute, Inc., International Committee for Information Technology Standards, Aug. 9 2007.

19. "INCITS Project 1647-D – Fibre Channel – Physical Interfaces-4 (FC-PI-4), Rev 8.00", INCITS Working Draft Proposed American National Standard for Information Technology, May 21, 2008.

20. "ANSI INCITS 364-2003 For Information Technology – Fibre Channel – 10 Gigabit (10GFC)", American National Standards Institute, Inc., International Committee for Information Technology Standards, Nov. 6 2003.

The following standards documents are relevant to "5.5 PCI Express" as described in Sect. 5.5:

21. "PCI Express Base Specification, Revision 2.0", Peripheral Component Interconnect Special Interest Group (PCI-SIG) (http://www.pcisig.com), Dec. 20 2006.

22. "PHY Interface for the PCI Express Architecture", Draft Version 1.90, Intel Corporation, 2007.

23. "PCI Express Card Electromechanical Specification, Revision 2.0", Peripheral Component Interconnect Special Interest Group (PCI-SIG) (http://www.pcisig.com), 2007.

The following reading is recommended for more information regarding "5.5 PCI Express":

24. "PCI Express System Architecture", Ravi Budruk, Don Anderson, and Tom Shanley, Mindshare, Inc., 2004.

Interested IBM employees and IBM ASIC customers may also wish to consult the following IBM HSS databooks and application notes for more information regarding IBM ASIC core offerings.

25. "High Speed Serdes (HSS) – PCI Express Gen 2 for Cu-08 Core Databook", SA15-5846-02, IBM.

26. "Implementing a PCI Express Device with IBM Cores", SA15-5976-00, IBM.

The following standards documents were additionally mentioned in this chapter, although not covered in detail:

27. "ANSI INCITS 376 For Information Technology – Serial Attached SCSI (SAS)", American National Standards Institute, Inc., International Committee for Information Technology Standards, Jan. 1 2003.

28. "Infiniband Architecture Specification Volume 2, Release 1.2", Infiniband Trade Association, October 2004.

29. "Serial ATA Revision 2.5", Serial ATA International Organization (http://www.sata-io.org), Oct. 27 2005.

5.7 Exercises

1. Answer each of the following questions to classify the SONET/SDH protocol characteristics.

    (a) Does the protocol use Synchronous or Plesiosynchronous clocking?

    (b) Does the protocol use Packet-based or Continuous Transmission?

    (c) Which block code and/or scrambling does the protocol use (if any)?

    (d) What type of error detection is used by the protocol (if any)? Does the protocol support error correction?

2. Answer each of the following questions to classify the SONET/SDH protocol characteristics.

    (a) Does this protocol use a serial bit stream or parallel lanes? Specify the number of lanes, if applicable.

    (b) Specify the baud rate(s) associated with each lane.

    (c) If the protocol uses parallel lanes, briefly describe how deskew is performed at the receiver.

3. Answer the questions in Exercises 1 and 2 for the OIF SFI-5.2 protocol.

4. Answer the questions in Exercises 1 and 2 for the OIF SPI-S protocol assuming the protocol carries CEI-P lane traffic.

5. Answer the questions in Exercises 1 and 2 for the IEEE 802.3 10GBASE-KX4 variant of the Ethernet protocol.

6. Answer the questions in Exercises 1 and 2 for the IEEE 802.3 10GBASE-KR variant of the Ethernet protocol.

7. Answer the questions in Exercises 1 and 2 for the serial variants of the INCITS T11 Fibre Channel protocol. Note that some answers may be different for 8.5 Gbps from that of lower baud rates.

8. Answer the questions in Exercises 1 and 2 for the 10GFC variant of the INCITS T11 Fibre Channel protocol.

9. Answer the questions in Exercises 1 and 2 for the Gen 1 variant of the PCI Express protocol.

10. Answer the questions in Exercises 1 and 2 for the Gen 2 variant of the PCI Express protocol.

11. Design logic (Verilog or VHDL) for a scrambler (or descrambler) for the SONET/SDH which has 16-bit parallel data inputs and outputs. Requirements for this scrambler are described in Sect. 5.1.2.1. Design your scrambler to assume STS-1 frames. Note that your answer for Chap. 4 Exercise 17 may be useful to get you started.

12. Modify the logic for the scrambler from Exercise 12 to scramble STS-3 frames.

13. Design logic (Verilog or VHDL) for a barrel shifter with 16-bit parallel data inputs and outputs. The implementation should include control logic to search for an "F628"h framing pattern. This framing pattern may occur on the input with any arbitrary bit alignment. The control logic should set the barrel shifter such that this pattern is aligned on the proper 16-bit boundary on the data output.

14. The SONET/SDH protocol does not scramble the framing pattern and therefore the logic in Exercise 13 can be used to detect and align on this framing pattern. However, the data stream may also contain scrambled data which matches this framing pattern. It is possible for *aliasing* to occur such that the barrel shifter incorrectly aligns based on this data. In order to avoid such aliasing, the control logic in Exercise 13 needs to be modified to build *hysteresis* into the state machine decisions. Assuming an STS-3 frame, modify the Verilog or VHDL logic from Exercise 13 such that the control logic implements the state transitions below:

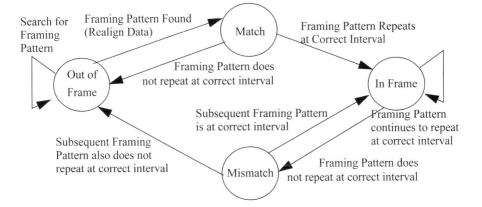

15. SONET/SDH Line Overhead H1/H2 pointer operation for the case of *positive stuffing* was illustrated in Fig. 5.3. Draw a similar illustration of the case of *negative stuffing*.

16. SONET/SDH Line Overhead pointer operation allows pointer justification (either positive or negative stuffing) to occur at most once every four SONET frames. Assuming an STS-3 frame, what frequency tolerance (in ppm) is required for the clock sources in order to avoid needing to make adjustments more often? What is the frequency tolerance assuming STS-12 frames?

17. Design logic (Verilog or VHDL) for a serial SFI-5.2 deskew channel generator which has four one-bit inputs and generates the one-bit output corresponding to Fig. 5.8.

18. Design logic (Verilog or VHDL) for a parallel SFI-5.2 deskew channel generator which has four 10-bit inputs and generates the 10-bit parallel output corresponding to Fig. 5.8. The 10-bit output will be serialized by the HSS transmitter. Your design must take this into account and ensure that the deskew channel is transmitted with the correct timing relationship and bit order.

19. Design logic (Verilog or VHDL) for a parallel SFI-5.2 deskew channel framing function. This logic should include a barrel shifter with a 10-bit input and a 10-bit output. The control function for the barrel shifter should align the data such that the bits of the sequence shown in Fig. 5.8 are aligned on the 10-bit boundary.

20. Design additional logic for the logic from Exercise 19 that aligns the DATA[0] channel to the deskew channel. This logic should include a barrel shifter with a 10-bit input and a 10-bit output. The control function for the barrel shifter should compare the applicable data bits to the data bits on the deskew channel and update alignment accordingly.

21. The skew and wander specification for the SFI-5.2 protocol is described in Table 5.5, and an example of a skew budget for a receiver at the R_i or R_e compliance points was described in Table 4.4. Develop an equivalent skew budget for a receiver at the R_{es} assuming:
    - Signal routing differences in the receiver are reduced to 1.5 UI.
    - The deskew logic in the receiver chip performs the deskew at the full baud rate of the interface with no deserialization.

22. Given the block formats defined in Fig. 5.10 for CEI-P, show the 66-bit codewords for transmitting the following data packet:
    - Address = "13FC"h
    - Flags = "04"h
    - CRC = "457"h (this is not the actual calculated value for this packet)
    - Data = "01"h, "02"h, "03"h, "04"h, "05"h, "06"h, "07"h, "08"h,

"09"h, "0A"h, "0B"h, "0C"h, "0D"h, "0E", "0F"h, "10"h, "11"h, "12"h, "13"h, "14"h, "15"h, "16"h, "17"h, "18"h, "19"h, "1A"h, "1B"h, "1C"h, "1D"h, "1E"h, "1F"h.

23. Translate the 66-bit codewords of the packet described in Exercise 22 to the equivalent T bits and payload within a CEI-P coding frame as described in Fig. 5.11.

24. The IEEE 802.3 Ethernet standard uses the following acronyms in relation to the interfaces defined in clauses 47 and 48: XGMII, XAUI, and XGXS. Specify what these acronyms stand for and define them. How are these three acronyms related?

25. The IEEE 802.3 Ethernet standard defines the PCS and PMA layers for 10GBASE-X variants in clause 48. Is the functionality described by this clause sufficient to implement the XGXS function defined by clause 47? Is the functionality described by this clause sufficient to implement the 10GBASE-KX4 Backplane Ethernet variant? Is additional functionality required in either of these cases?

26. Given the following data packet, show the corresponding bus cycles on the XGMII:
    - Destination Address = "112233"h
    - Source Address = "445566"h
    - Data = "01"h, "02"h, "03"h, "04"h, "05"h, "06"h, "07"h, "08"h, "09"h, "0A"h, "0B"h, "0C"h, "0D"h, "0E", "0F"h, "10"h, "11"h, "12"h, "13"h, "14"h, "15"h, "16"h, "17"h, "18"h, "19"h, "1A"h, "1B"h, "1C"h, "1D"h.
    - FCS = "778899"h (this is not the actual calculated value for this packet)

27. Given the XGMII packet in Exercise 26, show the corresponding bus cycles on the four lanes of the XGXS interface to the HSS EX10 core. These bus cycles will be using the equivalent 8B/10B code words.

28. Show 33 consecutive bus cycles on the four lanes of the XGXS interface to the HSS EX10 core during an idle period between two packets. At some point during this period a sequence ordered set is inserted with the 8B/10B codewords corresponding to the data bytes "21"h, "22"h, and "23"h.

29. The XGXS function defined in Sect. 5.3.3 deskews data across the lanes using the Align ordered set. This ordered set is transmitted at intervals which range from every 16 symbols to every 31 symbols. Assume that two lanes of the interface are skewed by 13 symbols (130 UI). Explain why the deskew logic may not correctly align the interface on the first attempt. Draw a timing diagram showing an example of a sequence of Align ordered sets received on two lanes of the interface. Construct this example to show the circumstances under which the interface is misaligned on the first attempt, but is correctly aligned upon receiving a subsequent Align ordered set.

30. The IEEE 802.3 Ethernet standard defines the PCS layer for 10GBASE-R variants in clause 49. This PCS layer is commonly called "XFI." This clause includes a definition of the 64B/66B block code which is summarized in Fig. 5.19 and Table 5.13. Given the XGMII packet in Exercise 26, show the corresponding 64B/66B codewords as transmitted by the HSS EX10 core. (Assume all control bits/bytes are idles.)

31. Your answer to Exercise 30 started the sequence with a control block using one of the "Start" block types. Repeat this exercise using the other "Start" block type.

32. Assume that an HSS EX10 receiver is connected to "gearbox" logic which buffers the 32-bit output of the HSS core into 66-bit blocks. When properly aligned, one 64B/66B code word is transferred on each cycle of this 66-bit bus. Design logic (Verilog or VHDL) which implements a state machine that monitors the output of the gearbox logic, and pulses RXxDATASYNC if the codeword is not properly aligned.

33. Design logic (Verilog or VHDL) to implement the "gearbox" logic described in Exercise 32. Hint: The logic design is simplified if you assume that the input and the output of the logic operate at the same clock rate, and that a control signal indicates cycles when new data is available on the output.

34. The autonegotiation protocol defined in Clause 73 for the IEEE 802.3 Ethernet standard was described briefly in the text. One function of this protocol is to advertise which Backplane Ethernet variants are supported by a node, and to negotiate which of these variants to use. Suggest a truth table which determines which variant to use based on the variants supported by this node and the variants supported by the link partner. Note that the variant to be used should adhere to the following priority order: 10GBASE-KR (highest priority), 10GBASE-KX4, 1000BASE-KX (lowest priority).

35. The training algorithm for the 10GBASE-KR variant of the IEEE 802.3 Ethernet standard was described briefly in the text. As noted, the standard does not define the algorithms which should be employed to determine optimal equalizer settings. Create a flow chart suggesting a possible algorithm for optimizing the FFE coefficients. (The algorithm can use a brute force approach.)

36. Draw a complete end-to-end block diagram of an INCITS T11 Fibre Channel link, including: a host chip connected across a backplane to an optical PMD, one PMD connected across an optical cable to another PMD in a second enclosure, and the PMD in the second enclosure connected across a backplane to another host chip. Label the interoperability points on this diagram.

37. Assume that the optical PMDs in Exercise 36 use linear optical receivers. How does this affect your labeling of the interoperability points?

38. Equations for a 32-bit parallel self-synchronizing scrambler using the following polynomial were described in Sect. 4.2.2.3:

$$G(x) = x^{58} + x^{39} + 1.$$

Design the logic (Verilog or VHDL) for this scrambler. Add a *K_bit* control signal. When K_bit = "1," data is not scrambled and the scrambler state is reset to "029438798327338"h. When K_bit = "0", data is scrambled and the scrambler state is advanced.

39. Design the logic (Verilog or VHDL) for the descrambler corresponding to the scrambler in Exercise 38.

40. The data flow for a Fibre Channel 8GFC transmitter and receiver are shown in Figs. 5.25 and 5.26, respectively. The data paths shown are 32-bits wide throughout. Fibre Channel transmits and receives all Ordered Sets on this 32-bit boundary, and requires that data packet length be an integral number of 32-bit words. Discuss the merits of this approach in relation to the logic complexity of the design.

41. The data flow for a Fibre Channel 8GFC transmitter and receiver are shown in Figs. 5.25 and 5.26, respectively. Scrambling in this data flow occurs prior to the 8B/10B encoder. Discuss the merits of this hierarchy (as opposed to performing 8B/10B encoding and then scrambling the encoded symbols) in relation to the spectral characteristics of data through the optical devices.

42. The 10Gbps variants of Fibre Channel are listed in Table 5.20. Which of these variants uses the XGXS logic defined in Sect. 5.3.3, and which of these variants uses the XFI logic defined in Sect. 5.3.4?

43. Contrast the baud rates for the 10-Gbps Fibre Channel variants listed in Table 5.20 with their Ethernet counterparts.

44. The PCI and PCI-X protocols which were the predecessors for PCI Express needed to limit the number of devices on the bus to control electrical loading as the frequency was increased. Given the discussion of source synchronous busses in Chap. 1, explain why.

45. The PCI Express uses plesiosynchronous clocking and the protocol uses *Skip Ordered Sets* to perform clock justification. When a node receives a Skip Ordered Set, it may drop or add one *SKP* symbol in order to compensate for clock frequency differences. Assuming the topology diagram in Fig. 5.28, what is the maximum and minimum number of *SKP* symbols in a Skip Ordered Set received by any PCI Express endpoint from any other PCI Express endpoint?

46. Given the following packet contents, show the corresponding transmission of the TLP (including start and end delimiters) as byte-striped across 4, 8, or 16 PCI Express lanes.
    • Data = "01"h, "02"h, "03"h, "04"h, "05"h, "06"h, "07"h, "08"h, "09"h, "0A"h, "0B"h, "0C"h, "0D"h, "0E", "0F"h, "10"h, "11"h, "12"h, "13"h, "14"h, "15"h, "16"h, "17"h, "18"h, "19"h, "1A"h, "1B"h, "1C"h, "1D"h, "1E"h.

47. As noted in the text, PCI Express endpoints must schedule a Skip Ordered Set for transmission at least once every 1,538 symbol times. However, if this transmission is scheduled immediately after the endpoint has started transmitting a packet, then the Skip Ordered Set cannot be transmitted until the packet finishes. What is the maximum number of symbol times that can occur between Skip Ordered Sets as a result of this behavior? What link width does this assume?

48. Assuming the maximum time between Skip Ordered Sets determined in Exercise 47, calculate the minimum size required for the elastic buffers in the receiver logic of the PCI Express endpoint.

49. Illustrate the symbol transmission across four PCI Express lanes for each of the following Ordered Sets:
    • Electrical Idle Ordered Set
    • Skip Ordered Set
    • Fast Training Ordered Set
    • Electrical Idle Exit Ordered Set

50. Assume that the board designer for a PCI Express node determines that reversing the physical wiring of both the transmit and receive lanes will reduce the number of board layers and thereby reduce costs. The PCI Express protocol logic driving the interface supports the lane reversal option. Is it permissible for the board designer to implement this lane reversal? Are there any potential impacts to system performance?

51. A PCI Express lane uses HSS EX10 cores at both the transmitter and receiver end of the link:

    (a) Describe the sequence of events which causes a PCI Express link to transition from an L0 to an L0s power state.

    (b) Describe the sequence of events which causes a PCI Express link to transition from an L0s to an L0 power state.

52. Describe which pins on the HSS EX10 transmitter and receiver are asserted while the corresponding PCI Express lane are in an L1 or L2 power state.

53. Compare the transmitter differential amplitude specifications for the following:

    (a) OIF CEI-11G-LR and Ethernet 10GBASE-KR

    (b) OIF SxI-5, Ethernet electrical variants which use the XGXS logic defined in Sect. 5.3.3, electrical variants of Fibre Channel 4GFC, and PCI Express Gen 1

    (c) OIF CEI-11G-SR, OIF CEI-11G-MR, and OIF CEI-11G-LR

Chapter 6
Reference Clocks

A significant consideration contributing to the performance of any high-speed serial data link is the availability of a stable, low jitter reference clock. In this chapter, on-chip clock distribution architectures are discussed, as well as the electrical analysis of the clock distribution network.

6.1 Clock Distribution Network

When designing a circuit network for distributing the reference clock to HSS devices, the following considerations warrant discussion.

- Type of signals (single-ended vs. differential)
- Direct distribution vs. use of an Intermediate Frequency PLL; and
- Any special requirements:
  - Skew requirements for transmitter serial data outputs
  - Loop timing (as discussed in Sect. 5.1.4)
  - Manufacturing test considerations

6.1.1 Single-Ended vs. Differential Reference Clocks

Power distribution is a significant concern on high-density chips using submicron fabrication processes. Voltage at any given circuit on the chip depends on the current being drawn by the circuit and neighboring circuits, and the resistance of the power distribution path. Additionally, as circuits in the same vicinity switch states, localized transient noise occurs on the power supply network. Power supply decoupling capacitance is often added in the chip layout to reduce power supply noise; however, such capacitance is only effective at reducing transient noise in the immediate vicinity of the capacitor. Steady-state voltage drop and lower-frequency noise components are not generally improved by on-chip power supply decoupling.

Critical noise-sensitive analog circuits in HSS cores are often powered from a separate analog power supply to reduce the impacts of transient noise. Generally, the analog power distribution is localized. The analog power supply input for an HSS core is near the circuits which are connected to this power supply, and the core input pin is generally connected directly to a chip pin. Unlike these critical analog circuits, elements of a clock distribution network are *not* localized and are scattered throughout the chip. Clock distribution circuits are generally powered by the same noisy power supply used to supply power to the bulk of the digital logic on the chip.

The effects of power supply compression on a single-ended signal are illustrated in Fig. 6.1, which shows two signals. One signal is illustrated by the solid line which swings across the approximate full range of the power supply. The other signal is illustrated by the dotted line where the power rail has been compressed due to voltage drop in the power distribution network and/or transient noise, and the signal amplitude has been reduced accordingly. Keeping in mind that the sink device on this signal does not necessarily experience the same power supply compression, a fixed switching threshold is assumed at approximately half of the uncompressed power supply voltage. This switching threshold is illustrated by the horizontal line at 0.6 V.

When power supply voltage to the circuit is reduced, the circuit switches more slowly. Additionally, the midpoint of the signal transition shifts in relation to the switching threshold of the sink circuit. As shown in Fig. 6.1, the point at which signal transitions intersect the switching threshold is affected, thereby creating jitter. This jitter on the reference clock input of an HSS core degrades the jitter performance of the transmitter, and degrades jitter tolerance of the receiver.

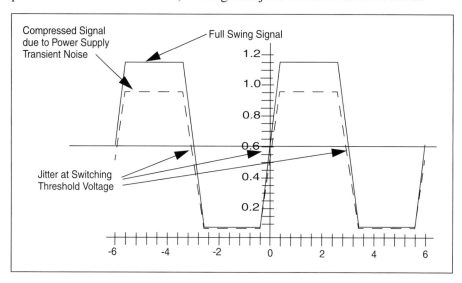

Fig. 6.1 Single-ended signals in presence of transient power supply noise

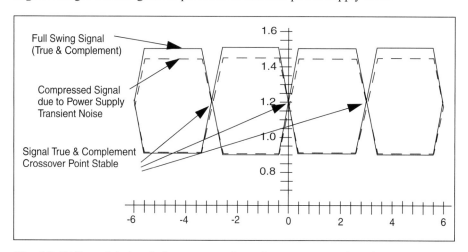

Fig. 6.2 Differential signals in presence of transient power supply noise

The corresponding effects of power supply compression on a differential signal are illustrated in Fig. 6.2. The solid lines in this figure illustrate the true and complement legs of a differential signal which is swinging across the approximate full single-ended range of each signal leg. The dotted lines illustrate the true and complement legs of a differential signal which is experiencing power supply compression and therefore has reduced amplitude.

Unlike single-ended signals, the downstream differential receiver circuit does not have a fixed signal switching threshold. Rather, the sink circuit switches based on the crossover point of the true and complement legs of the differential signal. As shown in Fig. 6.2, the amplitude reduction of the differential signal affects both the true and complement legs of the signal in a roughly equal fashion. This shifts the common mode voltage, but the crossover point of the signal legs is not affected.

The power supply noise described above is one form of *common mode noise*. Common mode noise sources affect the true and complement legs of the signal equally. Other signals in the chip routed in parallel to the differential clock can also cause coupled noise, a significant portion of which is common mode noise. Differential buffers have superior noise rejection qualities for common mode noise, and therefore coupled noise is less of a concern when differential clocks are used.

As the reader may assume from the above analysis, differential clock networks are preferred for distributing reference clocks across chips to HSS cores. All of the following reasons contribute to this preference:

1. As illustrated by the prior analysis, differential buffers have superior noise rejection qualities. Noise rejection is as much as 20 times better in the lower frequency ranges than equivalent single-ended circuits.

2. Random noise levels generated by differential buffers are very minimal. Common mode noise does not affect switching of the sink device, and differential noise is very low.

3. Differential buffers are linear circuits with a relatively constant current draw independent of switching state. Therefore, differential buffers do not contribute significantly to chip-level noise. (This is unlike single-ended clock buffers, where the combination of a high switching factor, the need for crisp rise/fall transitions, and high loading of the output pin combine to make these buffers significant noise aggressors on the chip.)

6.1.2 Reference Clock Sources

The input to the clock distribution network which drives the HSS core reference clocks may either be driven directly from chip I/O pins, or may come from an intermediate frequency (IF) phase locked loop (PLL). These two options are shown in Fig. 6.3, and are discussed below.

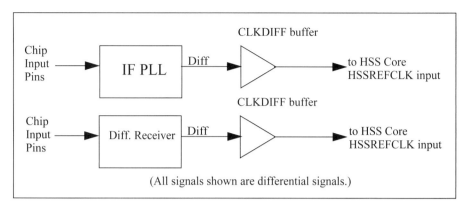

Fig. 6.3 Reference clock network options

6.1.2.1 Differential Receivers

The reference clock to the HSS core may be supplied directly from an off-chip clock source. In such cases, a differential receiver is used to receive the external differential signal as shown in Fig. 6.3, and to drive the on-chip clock distribution network. In some cases, the differential receiver and clock distribution network may be entirely contained in the HSS core with the clock inputs of the core connecting directly to chip I/O pads. If there is more than one HSS core on the chip which requires connection to the reference clock, then it is generally preferable to provide one differential input to the chip, and to distribute the clock to the various HSS cores on the chip. The on-chip clock distribution network generally includes one or more differential clock buffers to redrive the clock as needed to maintain signal integrity.

The disadvantage of directly using the external reference clock is that the on-chip reference clock frequency must be the same as the frequency of the reference clock on the circuit board. The HSS EX10 core described in Chap. 2 is fairly typical of HSS cores designed for 10 Gbps baud rates. The HSSREFDIV and HSSDIVSEL[1:0] input pins on this core (see Table 2.1) control the divide ratios for clock dividers in the PLL slice. The reference clock must therefore be at one of the selectable ratios in the range of one-eighth to one-fortieth of the baud rate. (The range for HSS cores designed for lower baud rates is typically one-fourth to one-twentieth of the baud rate.) Higher reference clock frequencies which use lower PLL divider ratios result in better jitter performance for the HSS transmitter, and better jitter tolerance performance for the HSS receiver.

Although higher frequency reference clocks may be desirable from the viewpoint of HSS jitter performance, distributing these higher frequencies on the circuit board is not desirable from the standpoint of electromagnetic Interface (EMI). Most systems are required to meet government standards for EMI generation. (In the United States, these standards are set by the Federal Communications Commission.) The EMI contribution for a signal is related to its frequency and the length over which the signal is driven. Higher-frequency clock signals distributed across the circuit board make a significant contribution to the EMI of the system, and therefore it is desirable to use lower frequencies on the circuit board.

6.1.2.2 Intermediate Frequency (IF) PLL

An alternative configuration uses an on-chip, low-jitter PLL to step up the frequency of the external reference clock source. As shown in Fig. 6.3, the chip I/O pads connect directly to the IF PLL, and the output of this PLL drives the on-chip clock distribution network. This allows the system to distribute a lower-frequency reference clock on the circuit board, while still providing a higher-frequency reference clock to the HSS cores. The overall jitter performance of the clock distribution is improved using this topology, and overall EMI is reduced.

6.1.2.3 Multiple Baud Rates

The same reference clock frequency and clock distribution network may feed HSS cores that are operating at different baud rates. Figure 6.4 shows an example where the same reference clock drives two HSS EX10 cores operating at different baud rates. An external 106.25 MHz reference clock is stepped up to 531.25 MHz by an IF PLL and then is distributed to two HSS EX10 cores.

The HSSDIVSEL inputs of the HSS EX10 cores are set differently. For the first core:

HSSDIVSEL[1:0] = 01, and
HSSREFDIV=0.

This clock divider selection causes the HSS PLL slice to step up the 531.25 MHz clock by a factor of 20, resulting in a 10.62 Gbps baud rate.

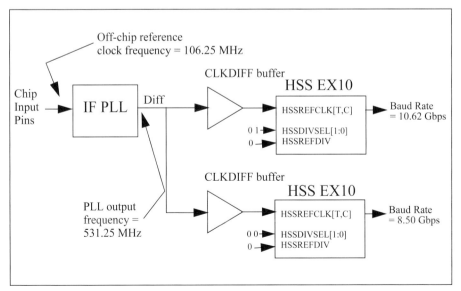

Fig. 6.4 Example of single reference clock for multiple baud rates

For the second core:

HSSDIVSEL[1:0] = 00, and
HSSREFDIV=0.

This selection causes the HSS PLL slice to step up the clock by a factor of 16, resulting in an 8.5 Gbps baud rate.

Additional flexibility exists if multiple IF PLLs are used. The clock dividers of the IF PLLs provide additional multiplier options, including fractional multiplier ratios. A single external reference clock can feed one or more IF PLLs configured for different multiplier ratios, and thereby be used to generate a wide variety of baud rates.

6.1.3 Special Timing Requirements

Some applications have requirements which drive the need for special considerations when designing the reference clock distribution network. These considerations are discussed below.

6.1.3.1 Transmit Data Skew

Skew was defined in Sect. 4.1.2.5 as the constant portion of the difference in the arrival time between the data of any two in-band signals. This can be visualized as jitter on one signal relative to the other signal (used as a reference signal for this measurement) at DC (0 Hz). Skew results from differences in the propagation delay of the reference clock to various HSS cores on the clock distribution network, differences in clock routing to various transmitter or receiver slices within the HSS core, and signal time-of-flight differences due to routing of the signals through the package and circuit board.

Skew becomes significant to the reference clock distribution network in cases where a multibit interface is implemented by more than one HSS core, and a maximum skew is specified between various signals of the interface. In such cases, the clock distribution network must be balanced to minimize clock skew such that the clock arrives at all of the HSS cores at the same time. Reference clock skew is one component contributing to skew between bits of the transmit data; the chip designer must determine the skew budget allocation for the various skew contributors such that the overall maximum skew specification is met for the transmitted signals.

For applications where there is no skew specification, or where all bits of the applicable interfaces are driven by the same HSS core using a common PLL slice, there is no need to balance the reference clock distribution network.

6.1.3.2 Loop Timing

Loop timing was discussed generally in Sect. 4.1.3.1, and in Sect. 5.1.4 as it relates to the SONET standard. Devices using loop timing are required to retransmit data at the same frequency as received data. OIF SPI-S interfaces, as described in Sect. 5.2.3, also have a requirement to transmit status channels at the same frequency as received data.

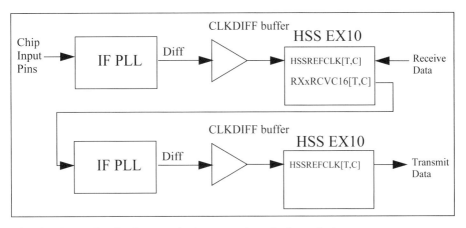

Fig. 6.5 Example of reference clock connections for loop timing

To support these requirements, the HSS core must provide a differential clock output which is recovered from the receive data and has acceptable jitter characteristics to support its use as an input to a PLL. The RXxRCVC16[T,C] outputs of the HSS EX10 core, as defined in Table 2.3, meet these requirements and support the implementation of loop timing. An example of the resulting HSS core interconnections is shown in Fig. 6.5. The reference clock for the HSS core used to recover the receive clock is driven by an external clock source through an IF PLL. The differential RXxRCVC16 clock output of this core is connected to an additional IF PLL, and the output of this PLL drives the HSSREFCLK input of the HSS cores used to transmit loop-timed data.

It is important to ensure that the HSS core's RXxRCVC16 clock output is not connected to the same core's HSSREFCLK input. Such a feedback path would not result in stable system operation.

Some applications may require a loop timing mode to be programmable. In such cases, a differential multiplexor is added to drive the HSSREFCLK of the transmitting HSS core, and selected between the recovered RXxRCVC16 clock output and the IF PLL clock source. This multiplexor may be implemented either as separate logic outside of the HSS core (as would be the case for the HSS EX10), or may be incorporated into the HSS core design.

6.1.3.3 Spread Spectrum Clocks

Another technique for EMI reduction makes use of *spread spectrum* clock sources. Such clock sources produce a reference clock where the frequency is deliberately swept over a frequency range of up to a few thousand parts-per-million around the nominal frequency value. This reduces the amplitude of the EMI energy peaks at the nominal frequency, and makes it easier for system vendors to pass the corresponding EMI tests.

If spread spectrum clocking is being used, any PLLs driven by this clock must support the use of a spread spectrum clock. PLLs which do not support such clocks will loose lock as the frequency of the reference clock input changes. The IF PLL and the PLL in the HSS core must both support the spread spectrum range being generated by the external clock source.

Some protocol standards, such as Serial ATA (SATA), require use of spread spectrum clocks [2].

6.1.4 Special Test Requirements

As a final consideration for reference clock distribution network topologies, note that the manufacturing test requirements for the HSS core may depend on a PLL to provide reference clocks at the necessary frequencies to conduct at-speed testing. The clock frequencies required to test the HSS core may be too high to be supplied directly by inexpensive manufacturing test equipment. Even if the application does not require an IF PLL, it may be necessary to add a multiplexor to allow selection of a clock from a PLL for executing the manufacturing test sequence.

An example of this is shown in Fig. 6.6. As shown, a test mode selection signal allows the reference clock to be driven by a PLL for manufacturing test purposes. For normal operation, the reference clock is driven by an off-chip source through a differential receiver.

6.2 Clock Jitter

Jitter is discussed more generally as it relates to serial link operation in Chap. 8. This section discusses jitter specifically in the context of clock signals.

Clock jitter is an important signal integrity concern which impacts the performance of the serial link. The amount of jitter present on a serial link is related to the achievable bit error rate (BER). If the jitter on a serial link is excessive, data errors result and overall link performance is degraded, possibly to unacceptable levels. Jitter generation is therefore a concern for PLLs in the system, including both IF PLLs and high-frequency PLLs embedded in the HSS core.

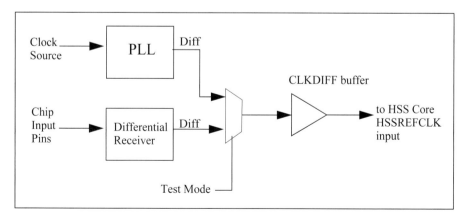

Fig. 6.6 Example of additional PLL for manufacturing test

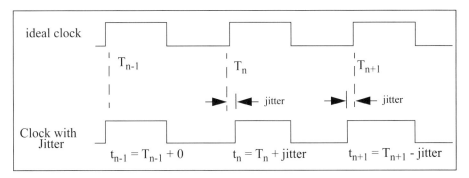

Fig. 6.7 Jitter on a clock signal

6.2.1 Jitter Definitions

The topic of *jitter* was first introduced in Sect. 1.4.3 of this text, and is defined more formally in Sect. 8.2 as the deviation in the arrival time of a signal from its ideal or expected arrival time. Figure 6.7 illustrates timing variation of the edges of a clock signal when compared to an ideal clock with no jitter. The difference between a clock edge t_n and the corresponding edge of an ideal clock is the jitter associated with the clock edge. This jitter can be quantified in terms of *phase jitter*, *period jitter*, or *cycle-to-cycle jitter*.

6.2.1.1 Phase Jitter

An ideal clock signal as shown in Fig. 6.7 has no jitter, and therefore every clock cycle has the same clock period. Using the notation T to designate this period, and assuming that edge T_0 occurred at time 0, then $T_n = n\,T$, and the difference between any two consecutive edges $T_n - T_{n-1}$ is T_{cyc}. The variation of the actual clock edge from the ideal clock edge is the *phase jitter* of the clock edge and is expressed mathematically as:

$$J_{phase}(n) = t_n - n T_{cyc} \qquad (6.1)$$

where n designates a specific clock edge, and $J_{phase}(n)$ is the absolute jitter in units of time for clock edge n.

Note that the value of phase jitter is absolute and accumulates over time. This accumulation does not occur without bound. Assume the range of phase jitter values for $J_{phase}(n)$ is defined as $\pm J_{phase}$. Phase jitter is defined relative to the ideal clock which has no jitter and therefore does not accumulate any deviation. Therefore, on any clock edge n, the maximum deviation from the ideal clock edge is still within the range $\pm J_{phase}$. Jitter on prior clock edges have a combination of positive and negative jitter values which accumulate such that the maximum deviation remains within this range.

This can be further visualized by examining Fig. 6.7. The clock edge at t_n deviates from the ideal clock by $+jitter$, resulting in a clock cycle period of:

$$t_n - t_{n-1} = T_{cyc} + jitter$$

Assume that the value of jitter associated with this deviation is $+J_{phase}$. Given the deviation associated with this clock cycle, the next clock cycle cannot have the same period because this would cause the next edge to deviate by more than $+J_{phase}$. Given that:

$$t_n - t_{n-1} = T_{cyc} + J_{phase}$$

the next clock cycle is limited such that:

$$T_{cyc} - 2 J_{phase} \leq t_{n+1} - t_n \leq T_{cyc}$$

Prior cycle times continue to accumulate such that phase jitter remains within the proscribed bounds. It should be clear that restricting the accumulation of phase jitter is necessary if the clock is to have the specified frequency (and clock period). If phase jitter were allowed to accumulate without bound, then the average cycle time of the clock would differ from T_{cyc} without bound. If this occurs, then by definition the clock frequency is not $1 / T_{cyc}$.

6.2.1.2 Period Jitter

Period jitter is the deviation of the period of a given clock cycle from that of an ideal clock cycle. A clock cycle is defined by two consecutive clock edges: t_n and t_{n-1} . The period of the corresponding clock cycle is therefore: $t_n - t_{n-1}$, and the period jitter is defined mathematically as:

$$J_{period}(n) = (t_n - t_{n-1}) - T_{cyc} \tag{6.2}$$

While phase jitter is an absolute value that accumulated mathematically over time, period jitter is a relative value determined by the jitter of two consecutive clock edges. Period jitter may also be calculated by the equation:

$$J_{period}(n) = J_{phase}(n) - J_{phase}(n-1) \tag{6.3}$$

Assume the range of phase jitter values for $J_{phase}(n)$ is defined as $\pm J_{phase}$, where J_{phase} is the maximum phase jitter for a given clock. The maximum period jitter J_{period} is bounded as follows:

$$J_{period} = 2 \times J_{phase} \tag{6.4}$$

where the range of jitter values for $J_{period}(n)$ is defined as $\pm J_{period}$. This maximum period jitter results when the phase jitter of one clock edge is $+J_{phase}$ followed by the next clock edge having a phase jitter of $- J_{phase}$ (or vice versa). This corresponds to a frequency of jitter variation (f_{jitter}) such that:

$$f_{jitter} = 0.5 f_{clock} = 0.5 / T_{cyc}$$

This is improbable in a real system. More realistically, phase jitter varies based on a periodic jitter function with frequency components significantly less than the clock frequency.

Assume a clock signal for which the phase jitter variation is represented by a sinusoidal function such that:

$$J_{phase}(n) = A \sin (2\pi \ f_{jitter} \ tn) \tag{6.5}$$

where A is the amplitude of the phase jitter.

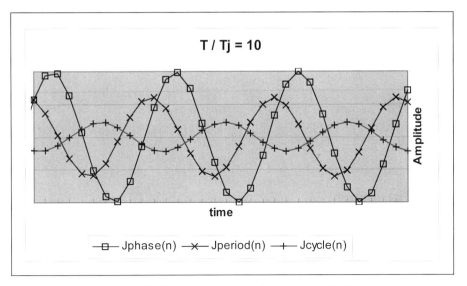

Fig. 6.8 Comparison of clock phase, period, and cycle-to-cycle jitter

Figure 6.8 compares the phase jitter to the period jitter for the case of:

$$f_{\text{jitter}} = 0.1 f_{\text{clock}} = 0.1 / T_{\text{cyc}}$$

As is shown, the amplitude of the period jitter is less than that of the phase jitter, and can be shown to have a peak amplitude of:

$$J_{\text{period}} = 2 \times \left| \sin(\pi f_{\text{jitter}} T_{\text{cyc}}) \right| \times J_{\text{phase}} \qquad (6.6)$$

For the case of $f_{\text{jitter}} = 0.5 f_{\text{clock}}$, (6.6) reduces to (6.4.)

6.2.1.3 Cycle-to-Cycle Jitter

Cycle-to-cycle jitter is the deviation of the period of a given clock cycle from that of the prior clock cycle as defined by:

$$J_{\text{cycle}}(n) = J_{\text{period}}(n) - J_{\text{period}}(n-1) \qquad (6.7)$$

Clock cycles are defined by consecutive clock edges: t_n and t_{n-1}, and t_{n-1} and t_{n-2}. The period deviation of these two consecutive clock cycles is

$$(t_n - t_{n-1}) - t(n-1 - t_{n-2})$$

and the cycle-to-cycle jitter may also be calculated using the equation:

$$J_{\text{cycle}}(n) = (t_n - 2t_{n-1} + t_{n-2}) \qquad (6.8)$$

Assuming the range of period jitter values for phase jitter and period jitter are designated using the notation in (6.4), the maximum cycle-to-cycle jitter J_{cycle} is therefore bounded as follows:

$$J_{\text{cycle}} = 2 \times J_{\text{period}} = 4 \times J_{\text{phase}} \qquad (6.9)$$

where the range of jitter values for $J_{\text{cycle}}(n)$ is defined as $\pm J_{\text{cycle}}$. Maximum period jitter results when the phase jitter of one clock edge is $+J_{\text{phase}}$, followed by the next clock edge having a phase jitter of $-J_{\text{phase}}$, followed by the next clock edge having a phase jitter of $+J_{phase}$ (or vice versa). As was the case for (6.4), this requires an improbable frequency of jitter variation. Assuming sinusoidal variation of phase jitter as was defined in (6.5), the corresponding cycle-to-cycle jitter can be shown to have a peak amplitude of:

$$J_{period} = 4 \times [\sin(\pi f_{jitter} T_{cyc})]^2 \times J_{phase} \qquad (6.10)$$

Figure 6.8 compares cycle-to-cycle jitter to the phase and period jitter of this clock. For this example, the amplitude of the cycle-to-cycle jitter is further reduced from that of the period jitter. Equation (6.10) reduces to (6.9) for:

$$f_{jitter} = 0.5 f_{clock}$$

6.2.1.4 Phase Noise

The phase jitter described by (6.1) in the time domain is the equivalent of *phase noise* in the frequency domain. Consider the time domain equation for a sinusoidal signal:

$$v(t) = V_{max} \sin(\omega t + \theta) \qquad (6.11)$$

where $v(t)$ is the instantaneous time domain voltage of the signal, V_{max} is the amplitude of the signal, ω is the frequency (in rad s^{-1}), and θ is the initial phase at time zero. For a nonideal clock source, the θ value includes a noise component which varies according to a probability distribution. This noise on the phase component tends to shift the sinusoidal waveform either left or right. The waveform in Fig. 6.9 illustrates this variation, comparing a sinusoidal signal with phase noise to an ideal signal. It should be obvious from the figure that the phase noise causes phase jitter in the time domain.

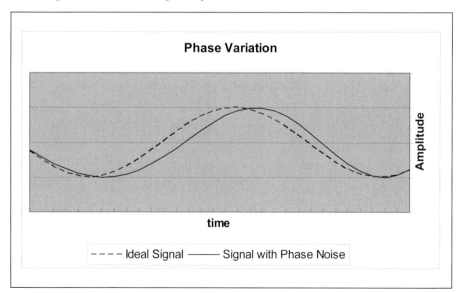

Fig. 6.9 Phase noise on a sinusoidal signal

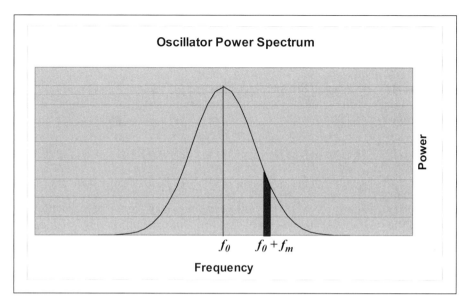

Fig. 6.10 Phase noise in the frequency domain

In the frequency domain, phase noise appears as modulation of the carrier frequency. A typical power spectrum for a sinusoidal oscillator is illustrated in Fig. 6.10. While an ideal oscillator would have a single spectral line at the carrier frequency f_0, real devices have phase noise which distributes this power to adjacent frequencies and results in sidebands.

Figure 6.10 illustrates one sideband at frequency $f_0 + f_m$. The phase noise at a given sideband offset is defined as the ratio of the power in a 1-Hz bandwidth centered at $f_0 + f_m$ to the total power of the carrier. The phase noise is generally specified in units of $dBc\,Hz^{-1}$, where dBc is the ratio in dB relative to the carrier.

The noise power in the sidebands can be converted to random jitter in the time domain. The phase jitter $J_{phase}(n)$ in (6.1) was defined as the difference between the time of the zero crossing of the actual signal to an ideal reference signal with period T_{cyc}. $J_{phase}(n)$ for signal edge n can be expressed in terms of the phase noise as:

$$J_{phase}(n) = n \cdot \left(\frac{2\pi}{T_{cyc}}\right) \theta \tag{6.12}$$

where θ_n is the phase noise, in radians, for the signal at edge n. The RMS value of the phase jitter may therefore be derived from the phase noise $\theta(n)$ by summing over the phase noise over N edges and allowing N to approach ∞:

$$(J_{pheas-rms})^2 = \left(\frac{2\pi}{T_{cyc}}\right)^2 \lim_{N \to \infty} \frac{1}{N} \sum_{j=1}^{N} \theta_n^2 \tag{6.13}$$

The summation in (6.13) may be approximated by [5]:

$$(J_{\text{phase} - \text{rms}})^2 = \left(\frac{2\pi}{T_{\text{cyc}}}\right)^2 \lim_{T \to \infty} \frac{1}{T} \int_{-T/2}^{T/2} \theta_n(t) dt \qquad (6.14)$$

where the limit represents the average power of θ_n. Parceval's theorem may be used to equate this integral to the area under the curve for the power spectrum in Fig. 6.10:

$$(J_{\text{phase} - \text{rms}})^2 = \left(\frac{2\pi}{T_{\text{cyc}}}\right)^2 \int_{-\infty}^{\infty} S_n(f) df \qquad (6.15)$$

where $S_n(f)$ denotes the noise power in the sideband centered at f.

The above description of phase noise for sinusoidal signals is sufficient to illustrate the relationship between phase noise and phase jitter. Digital clock signals are square waves, not sinusoidal waveforms, and may contain both RJ and DJ phase components. A more generalized and in-depth treatment of this subject is found in [4] and [5].

6.2.2 Jitter Effects

Clock jitter on the high-speed clocks within the HSS core degrades performance of the serial link, and additionally affects timing analysis of logic connected to the HSS core.

6.2.2.1 Serial Link Performance

Any jitter on the high-speed clock used to clock flip-flops in the transmitter driver circuit results in jitter on the transmitter serial data output. It is necessary to minimize clock jitter to minimize jitter on the serial data.

Jitter on the high-speed clock used for the clock and data recovery (CDR) circuit in the receiver degrades link performance. The function of the CDR circuit is to choose a sample point for the received serial data signal which is approximately in the center of the data eye. Jitter on the high-speed clock reference creates uncertainty in the sample point which reduces the jitter tolerance of the receiver, and therefore increases the bit error rate of the link. Obviously, it is desirable to minimize clock jitter to improve jitter tolerance and reduce bit errors.

6.2.2.2 Digital Logic Timing

The high-speed clocks used by the transmitter and receiver slices (*TXxDCLK* and *RXxDCLK* on the HSS EX10 core) are divided to produce parallel data clock outputs used by logic driving or latching data to/from the HSS core.

Jitter on the high-speed clocks accumulate and result in jitter on the parallel data clocks. Although the parallel data clocks accumulate jitter over multiple cycles of the high-speed clock, this jitter must still accumulate per the description in Sect. 6.2.1.1. Jitter on the parallel data clocks is therefore consistent with the J_{phase}, J_{period}, and J_{cycle} jitter ranges of the corresponding high-speed clocks.

Jitter on parallel data clocks must be taken into account when analyzing timing of digital logic which uses these clocks. The minimum clock period is

$$T_{min} = T_{cyc} - J_{period} \tag{6.16}$$

Any given cycle of the parallel data clock may have a period of T_{min}, and propagation delays associated with the digital logic must allow proper operation given this cycle time. Accounting for clock jitter in timing analysis is described further in Sect. 10.3.1.

6.2.3 PLL Jitter

As noted earlier in this chapter, IF PLLs are widely used to generate reference clock inputs to HSS cores. In addition, most HSS devices contain PLL slices used to step up the reference clock frequency to a baud rate (or half baud rate) clock.

A block diagram of a PLL is shown in Fig. 6.11(a). The PLL consists of a phase detector, a low-pass filter (LPF), a voltage controlled oscillator (VCO), and a divider (or multiplier) circuit used to generate the feedback clock. The divider ratio determines the ratio between the frequency of the clock output and the frequency of the reference clock input.

Phase jitter is a key parameter for evaluating the performance of the PLL circuit. Phase jitter may result from the VCO operation. Additionally, phase jitter may occur on the clock output as the result of jitter on the reference clock input which is within the tracking bandwidth of the PLL.

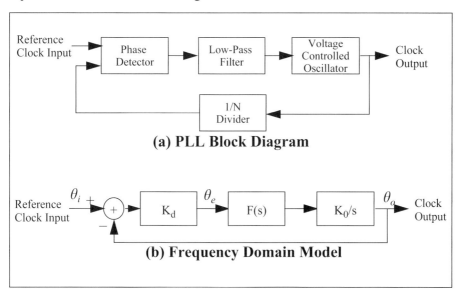

(a) PLL Block Diagram

(b) Frequency Domain Model

Fig. 6.11 PLL block diagram and frequency domain model

6.2.3.1 Jitter Transfer Function

A PLL frequency domain model is shown in Fig. 6.11(b) and corresponds to the block diagram in Fig. 6.8(a). This model is specified in the s-domain using the Laplace transforms of the equivalent time domain functions. The low- pass filter has been represented by an unspecified function $F(s)$ in this model. The corresponding PLL system transfer function is

$$H_0 = \frac{\theta_0(s)}{\theta_i(s)} = \frac{K_d K_0 F(s)}{s + K_d K_0 F(s)} \qquad (6.17)$$

The derivation of the above transfer function may be found in [4], along with analysis of the magnitude and phase characteristics for various $F(s)$ functions.

The discussion in this text is limited to some key metrics used to evaluate the PLL which are based on the plot of the magnitude of (6.17). The general form of this plot is shown in Fig. 6.12, and is primarily determined by the low-pass filter (LPF) transfer function $F(s)$. Two characteristics of this curve are of significance: the amplitude of the jitter peaking, and the frequency at which the curve crosses the -3-dB magnitude, called the *jitter transfer bandwidth* (f_{bw}). These characteristics are discussed further in the next few sections.

6.2.3.2 Jitter Tolerance Mask

HSS cores frequently use a PLL circuit as part of the CDR circuit in the receiver. (The HSS EX10 core uses a phase rotator circuit instead of a PLL.) In this application, the PLL locks to the frequency of the serial data and generates the sampling clock for the receiver.

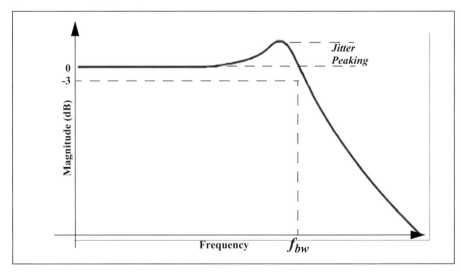

Fig. 6.12 PLL jitter transfer function

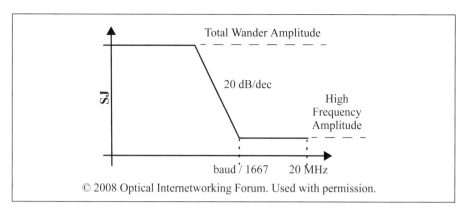

© 2008 Optical Internetworking Forum. Used with permission.

Fig. 6.13 Jitter tolerance mask example

An example of a jitter tolerance mask is shown in Fig. 6.13. Jitter tolerance testing using this mask is discussed in Sect. 7.5.2.1, and the significance of sinusoidal jitter to jitter tolerance testing is generally discussed in Sect. 8.2.6. The current discussion concentrates on the correspondence between characteristics of the curve in Fig. 6.13 and the LPF function in Fig. 6.12.

The jitter tolerance mask in Fig. 6.13 has a corner frequency at "baud/1667"; above this frequency the jitter is filtered by the CDR circuit, and below this frequency the jitter is tracked by the CDR. This corner frequency corresponds to the f_{bw} corner frequency of the LPF in Fig. 6.12.

Below the corner frequency in Fig. 6.13, the slope of the curve is specified as "20db per decade." This corresponds to the slope of the curve in Fig. 6.12 above the f_{bw} corner frequency. The slope of this curve is determined by the order of the LPF implementation. A second-order filter is commonly used to produce a slope of 20 dB per decade.

Jitter peaking is an important design parameter, but is not specifically addressed by Fig. 6.13. Excessive jitter peaking causes jitter amplification, causing overcorrection of the CDR sampling point and increasing bit errors in the system. However, overdamping to reduce or eliminate jitter peaking results in a longer time required for the PLL to lock (i.e., data acquisition time).

6.2.3.3 Jitter Transparent Applications

Some system applications contain elements which use loop timing to retransmit data without any significant filtering of jitter on the input. Consider the block diagram shown in Fig. 6.14. The CEI-11G-SR variant of the OIF CEI standard[1] is used in the reference model shown in Fig. 5.7. The compliance points in Fig. 6.14 are consistent with the reference model in Fig. 5.7.

The *egress data conditioner* and the *ingress data conditioner* devices in this block diagram are typically optical transmitter and receiver devices which perform an electrical-to-optical and optical-to-electrical conversion, and retransmit data using a loop timing architecture. If the sample clock of the CDR is used directly to retransmit data without using an additional PLL for jitter clean-up (as suggested by the loop timing description in Sect. 6.1.3.2), then any jitter on the input to the conditioner device is transferred to the output. This is called a *jitter transparent* application.

The jitter in jitter transparent applications accumulates as the data passes through the devices between the T_E and R_I compliance points. Interface standards for jitter transparent applications must allocate the jitter that may be contributed by each element, and must specify characteristics of the jitter transfer function for jitter transparent conditioner elements.

Although the data conditioner elements in Fig. 6.14 do not filter the recovered clock, there is still an inherent bandwidth associated with the CDR in the receiver. This bandwidth must comply with appropriate standards to ensure interoperability. Also, jitter peaking is specified to ensure stability of the overall system.

An example of these specifications is shown in Table 6.1. This table applies to the ingress signal conditioner for telecom applications which must comply with [3], and is driving a CEI-11G-SR link as specified in [1]. Other tables in [1] specify requirements for the egress signal conditioner for telecom applications, and specify requirements for signal conditioners that must comply with various datacom applications. The jitter peaking requirements in this table are intended to specify both the maximum amplitude of the jitter peaking, and the minimum frequency at which the jitter peaking can start to increase the magnitude of the transfer function.

Table 6.1 Telecom signal conditioner, ingress direction

Characteristic	Symbol	Condition	Min.	Typ.	Max.	Unit
Jitter transfer bandwidth	BW	Data,[a]			8	MHz
Jitter peaking		Frequency <120 kHz			0.03	dB
		Frequency >120 kHz			1	dB

[a]PRBS-31, OC-192/SDH-64 sinusoidal jitter tolerance mask

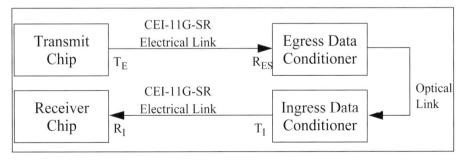

Fig. 6.14 Block diagram of a jitter transparent application

6.2.3.4 Intermediate Frequency PLLs

The jitter trade-offs associated with an IF PLL are different from PLLs used as part of CDR circuits. In the latter case, the PLL must be designed to respond relatively quickly to changes in the frequency and phase of the serial data input. Conversely, IF PLLs typically have a continuous clock input at a stable frequency. Therefore, IF PLLs are generally designed with a much lower f_{bw} corner. This has the effect of filtering any jitter which does occur on the reference clock above this corner frequency, and thereby results in a stable, low-jitter clock source.

Spice simulations are used to determine various analog characteristics of the clock signal as it arrives at the HSS cores.

6.3 Clock Floorplanning

The chip floorplan must consider the relative placement of IF PLL and HSS devices on the chip. Once this is determined, the physical design of the clock tree must be determined, and the clock must generally be prewired. A poorly designed clock tree will introduce signal integrity impairments. Likewise, clock wiring with excessive bends and vias will also impair signal integrity.

6.3.1 Clock Tree Architecture

The relative placement of IF PLL and HSS devices on the chip determines the distance over which the clock must be distributed. Differential clock buffers are typically used to redrive clock signals between the clock source and the HSS cores. However, there are practical limits on the capacitive load which can be driven by these buffers. Excessive load degrades signal amplitude and slew rate. In addition, driving the clock through an excessive number of stages results in excessive duty cycle distortion as will be described in Sect. 6.4.2. As the clock frequency increases, the impact of these factors on signal integrity of the clock increases. Chip floorplanning must therefore determine:

- The placement of the IF PLL and HSS devices on the chip
- The number of buffer levels in the clock tree that are needed to drive the clock from the IF PLL to the HSS cores
- Fanout of each clock buffer stage

Table 6.2 Example of clock buffer max. load/levels vs. clock frequency

Frequency (MHz)	Max load (fF)	Max. levels	Frequency (MHz)	Max load (fF)	Max. levels
751–800	1080	10	451–500	1450	16
701–750	1130	12	401–450	1540	16
651–700	1180	14	351–400	1650	16
601–650	1230	14	301–350	1800	16
551–600	1300	16	0–300	1950	16
501–550	1370	16			

The silicon vendor may provide characterization information regarding the drive capabilities of the clock buffer which may be used as a first-order approximation in designing the clock tree. Table 6.2 specifies the maximum load and maximum number of stages for an example of a differential clock buffer. This specification is provided as a function of the clock frequency.

The maximum load specification limits the length of the wire which may be driven by the clock buffer. For example, using the specification in Table 6.2, assume a clock buffer output is driving a wire which is connected to a single clock input. Furthermore, assume:

- The clock frequency is 375 MHz
- The capacitive load due to the clock buffer output pin is 300 fF
- The capacitive load due to the clock buffer input pin is 300 fF
- The capacitive load of a wire is 292 fF mm[1].

Given this clock frequency, the maximum load that can be driven by the clock output is 1,650 fF. Since the output pin of the driving buffer contributes 300 fF, and the input pin of the downstream buffer contributes another 300 fF, only 1,050 fF of this can be due to the wire. The maximum length of this wire is therefore: $1,050 \text{fF}/(292 \text{fF mm}^{-1}) = 3.6 \text{mm}$.

The above example assumed a clock buffer driving a fanout of one. If the clock source is supplying a clock to more than one HSS core, the clock tree must often be constructed such that the clock buffer drives more than one buffer (or HSS) input. As the fanout is increased, the length of the wire from the buffer output to any of the loads is reduced. Consider that the above example is modified such that the clock buffer is driving the clock inputs of two buffers. Given the buffer can drive 1650 fF, after allowing for load associated with the two inputs, only 750 fF of load can be due to the wire. This corresponds to a wire length of: $750 \text{fF} / (292 \text{fF mm}^{-1}) = 2.57 \text{mm}$.

Note that this wire length indicates the *total* wire length that can be driven. Assume that the clock signal wiring from the above example is split at the output of the clock buffer, with one wire going to one buffer input in one direction, and another wire going to another buffer input in the opposite direction. The *sum* of these wire lengths cannot exceed 2.57 mm.

As can be surmised from this example, the number of clock buffers in the clock tree and the placement of these buffers must be carefully planned to ensure the clock can be driven over the required distance to all of the HSS cores. Once elements of the clock tree have been placed on the chip and wired, signal integrity is analyzed for the actual layout as described in Sect. 6.4.

6.3.2 Clock Tree Wiring

Once PLLs, HSS cores, and clock buffers have been placed on the physical layout of the chip, the clock network is prewired. The true and complement legs of the signal must always be routed parallel to each other, otherwise the common mode noise rejection property of differential signals is negated. Also, any significant mismatch in length or load results in substantial DCD.

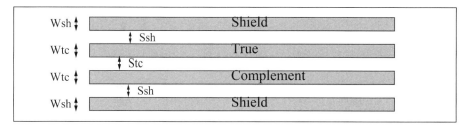

Fig. 6.15 Parallel wiring of differential clocks

True and complement signal wiring is illustrated in Fig. 6.15. In addition to the clock signal wires, shield wires connected to an AC ground are routed in adjacent tracks. Wire widths and spacing is carefully selected to minimize signal impairments and control the transmission line impedance.

6.4 Signal Integrity of the Clock Network

The run length of clock traces in the reference clock distribution network, the number of clock buffers in the network, and the number of HSS cores which can be driven from a single IF PLL source impact the signal integrity of the reference clock. Spice, an industry standard circuit simulation tool, is typically used to check the analog characteristics of the differential clock tree and the propagation of the clock signal through the differential clock buffers and wire segments.

6.4.1 Analog Signal Levels and Slew Rates

Each leg of a differential signal can be viewed individually as a single-ended signal. The rate of change of voltage of this signal is related to the current that can be sourced by the clock buffer, and the capacitance being driven. This relationship is defined by:

$$\mathrm{d}V = \frac{I}{C}\mathrm{d}t \qquad (6.18)$$

where $\mathrm{d}V$ is the change in voltage, I is the current sourced by the clock driver, C is the lump sum capacitance of the wires and input circuits connected to the driver, and $\mathrm{d}t$ is the change in time.

The value of I is primarily determined by the size of the transistors used in the clock driver, and is a fixed value for a given clock buffer. According to (6.18), as C increases, the slew rate of the signal increases. This in turn can limit the differential amplitude (V_{diff}) of the signal as shown in Fig. 6.16. The signals of the "1×Clock Waveform" in this figure has sufficient time to transition to the maximum signal amplitude. However, at four times the frequency ("4×Clock Waveform"), the clock signals only have time to transition across a fraction of the dynamic range. Increasing C (thus increasing the slew rate) for a fixed frequency has the same affect. In addition to reducing the differential amplitude, large values of C can also impact the V_{hi} and V_{lo} rail voltages of the signal, and can shift the common mode voltage (V_{cm}).

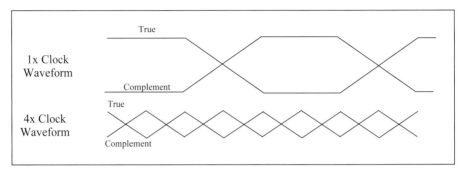

Fig. 6.16 Effect of slew rate on maximum amplitude

Clock signals must meet specifications for analog signal levels and slew rates at the input of each clock buffer in the clock distribution network, as well as at the reference clock inputs to the HSS cores. Rail voltages beyond their maximum/minimum limits and excessive V_{diff} levels can cause distortion of the signal. Insufficient V_{diff} levels, shifts in V_{cm}, and excessive slew rates can introduce duty cycle distortion and random jitter. The following subsections provide more detailed definitions for each of these signal parameters, and provide examples of measurements from Spice simulations.

6.4.1.1 Input Signal Levels

Each leg of a differential signal swings between a high and a low rail voltage (V_{hi} and V_{lo} respectively). For the first of the measured differential signals shown in Fig. 6.17, these levels correspond to:

V_{hi} = approximately 1,300 mV
V_{lo} = approximately 620 mV

resulting in a dynamic range of 679 mV. The second signal shown in this figure has a dynamic range of 667 mV.

6.4.1.2 Differential Amplitude

The *differential amplitude* (V_{diff}) is defined as:

$$V_{\text{diff}} = 2 (V_{\text{hi}} - V_{\text{lo}}) \tag{6.19}$$

where V_{hi} and V_{lo} are high and low rail voltages of the signal, respectively, as defined previously. This amplitude is generally expressed in units of millivolts peak-to-peak differential (mVppd), or alternatively in volts peak-to-peak differential (Vppd). For the differential signals in Fig. 6.17, V_{diff} equals 1,358, and 1,334 mVppd, respectively.

Differential receivers require V_{diff} be constrained within specified limits order to ensure circuits have sufficient input signal amplitude to provide an adequate signal to noise ratio, and to ensure the receiver input is not overdriven (which may result in nonlinear distortion). V_{diff} must be checked to ensure it is within specified ranges at the input to every circuit in the clock distribution network, including the inputs of both clock buffers and HSS cores.

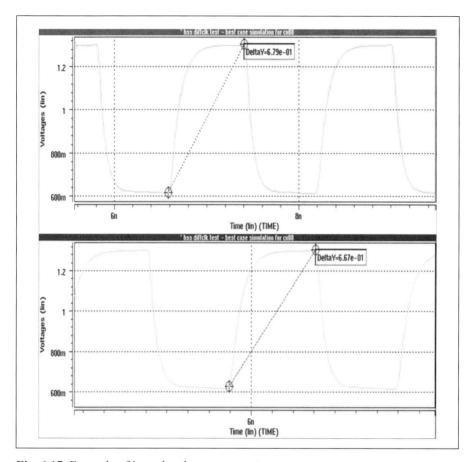

Fig. 6.17 Example of input level measurements

6.4.1.3 Common Mode Voltage

Common mode voltage (V_{cm}) is the average voltage on any one leg of a differential signal pair. By definition:

$$V_{cm} = (V_{hi} + V_{lo}) / 2 \qquad\qquad (6.20)$$

where V_{hi} and V_{lo} are high and low rail voltages of the signal, respectively, as defined previously.

Since the true and complement legs of the signal are always driven to opposite rails, V_{cm} may also be defined as the average voltage of the two legs of the differential signal. Figure 6.18 shows an example of a differential signal and its corresponding common mode voltage. In this example, the common mode voltage is 952.92 mV.

Differential receivers require V_{cm} be constrained within specified limits to ensure circuits are within their linear operating ranges. A V_{cm} value outside of the specified limits may cause transistors to saturate and reduce the dynamic range of the signal swing. This would reduce the resulting differential amplitude of the signal output. V_{cm} must be checked to ensure it is within specified ranges at the input to every circuit in the reference clock distribution network, including the inputs of both clock buffers and HSS cores.

6.4.1.4 Signal Rise/Fall Times

Rise time (t_{rise}) is the time required for a signal to transition from a specified low value to a specified high value. *Fall time* (t_{fall}) is the time required for a signal to transition from a specified high value to a specified low value. The term *slew rate* is also used to refer to the rise and fall times of signals.

Rise and fall times are generally measured from some percentage of the low value to some percentage of the high value (or vice versa). For differential signal shown in Fig. 6.19, the rise time transition is 0.12 ns as measured from the 10% to 90% points of the signal swing. The corresponding fall time is also 0.12 ns.

Slow slew rates create a larger window of uncertainty as to exactly when the receiver circuit detects the signal crossover and switches states. This degrades the jitter performance of the circuit. On the other hand, excessively fast slew rates generate noise in surrounding circuits. Limits may be specified at various circuit inputs for either maximum or minimum slew rates. Slew rates must be checked to ensure they are within specified ranges at the input to every circuit in the reference clock distribution network, including the inputs of both clock buffers and of HSS cores.

6.4.2 Duty Cycle Distortion

The *duty cycle* of a signal is defined as:

$$\text{Duty cycle} = T_{pwh} / T_{cyc}$$

where T_{pwh} is the pulse width high duration, and T_{cyc} is the clock cycle time or clock period. Duty cycle is generally expressed as a fraction or percentage. A perfect square wave has a duty cycle of 50%.

Duty cycle distortion (DCD) is a type of jitter which may result from either unequal rise/fall times of signals, and/or from a DC offset between the two legs of the differential signal pair. This type of jitter is described further Sect. 8.2.2.1. DCD can accumulate in the clock distribution network as illustrated in Fig. 6.20. This figure shows an IF PLL driving a clock distribution network consisting of two stages of clock buffers.

For the clock buffers in this example:

$$t_{rise} > t_{fall}$$

and the buffer output begins to transition when the input crosses the signal midpoint.

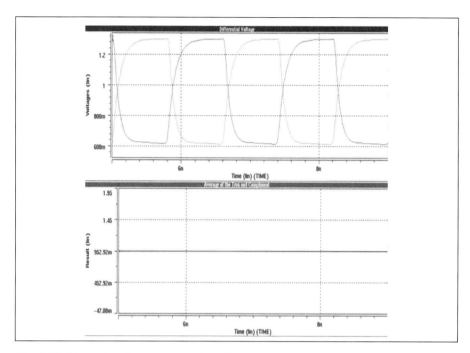

Fig. 6.18 Example of common mode voltage measurements

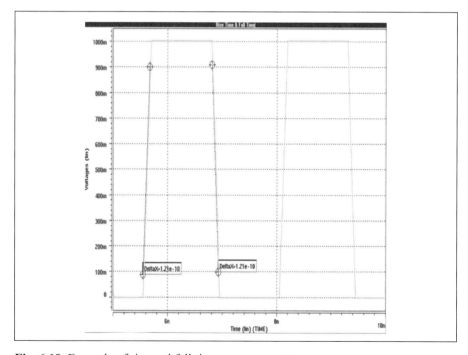

Fig. 6.19 Example of rise and fall time measurements

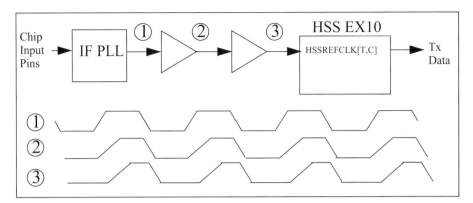

Fig. 6.20 Cumulative effect of DCD in a clock distribution network

At point (1), the output of the IF PLL has a 50% duty cycle. However, the slower t_{rise} transition of the first clock buffer reduces the duty cycle of the output at point (2). The slower t_{rise} transition at point (2) delays the start of the transition of the output of the second clock buffer at point (3), and the duty cycle at point (3) is thus reduced over that of point (2). If additional stages are added to this clock distribution network, the duty cycle of the clock would continue to be reduced with each successive stage. At some point the signal output of the clock buffer would not have time to swing across the full dynamic range, and may not switch at all.

The clock signal must meet specifications for duty cycle at the input to each clock buffer in the clock distribution network, as well as the reference clock inputs to the HSS cores driven by the network.

6.4.3 Differential Clock Analysis Methodology

The clock analysis methodology must ensure the reference clock distribution network provides clock signals to all of the HSS cores which meet the required analog characteristics necessary to guarantee proper low-jitter operation of the cores.

The basic flowchart for performing analysis of the clock network is shown in Fig. 6.21. The input to this flow is the chip layout, including a fully wired reference clock distribution network. First, circuit parasitics are extracted from the chip design for both best case and worst case process conditions using the appropriate parasitic extraction software (assumed to be IBM ChipEdit in the figure). These clock net parasitic values are then used to build Spice decks and used to run Spice simulations. Results must be verified to ensure conformance to HSS core requirements. If analog signal characteristics are not within specified ranges, the clock distribution network design must be modified and the analysis repeated.

This flow is discussed in more detail in the following sections. Examples of Spice decks and report formats are drawn from the software tools used to analyze differential clock trees for IBM ASIC chips.

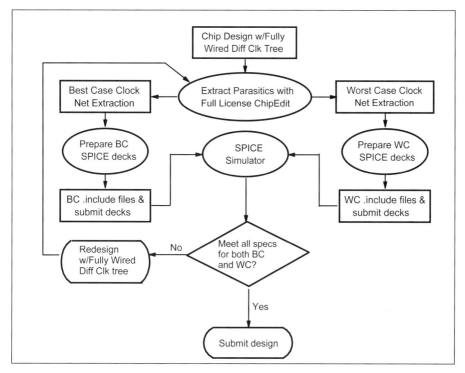

Fig. 6.21 Analysis flow chart

6.4.3.1 Extraction of Clock Tree Parasitic Values

Figure 6.22 shows a typical clock tree routed from the IF PLL to an HSS core. Parasitics for this clock tree are extracted from the chip design using IBM ChipEdit or other equivalent parasitic extraction software. When the extraction is completed there are two separate extracted netlists for each differential clock tree in the design: one for best case and one for worst case conditions.

The next step is to create appropriate differentially wired Spice decks for each differential clock tree in the design based on the parasitic information extracted from the composite differential clock nets. These decks are then simulated using Spice.

6.4.3.2 Spice Deck Creation and Simulation

Spice decks used to simulate the differential clock tree must include the following elements: (1) Instantiations of models for all clock buffers in the clock distribution network; (2) Modeling of the parasitics for clock tree nets for both the true and complement legs of the differential signals; (3) Instantiations of models for all of the HSS cores connected to the clock tree; (4) Connections for any unused inputs of cells in the clock tree (for example, unused inputs to differential multiplexors that are included in the clock tree); and (5) Spice measurement statements that measure all of the signal characteristics of interest at each cell in the differential clock tree.

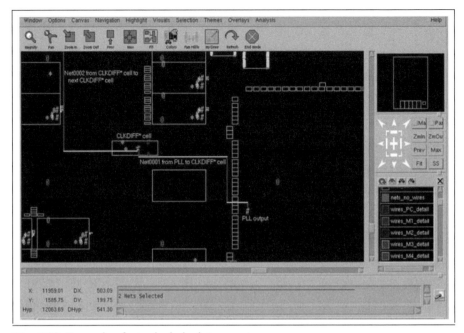

Fig. 6.22 Example of a typical clock tree route

For designers of IBM ASIC chips, IBM provides utilities which assist in the creation of Spice decks for analyzing differential clock trees. These utilities create individual Spice files that contain the above elements. Spice "submit deck" template files are also supplied which can be used to submit the required simulations after customization of the following parameters:

Clock Frequency: The reference clock frequency must be customized for the particular application that is being analyzed.

Temperature: The minimum and maximum junction temperature (T_j) limits specified for the chip must be customized for the particular chip containing the clock tree being analyzed. Note that these values should match the temperature conditions at which the netlist extraction was performed.

VDD: The minimum and maximum power supply voltage (V_{dd}) limits specified for the chip must be customized for the particular chip containing the clock tree being analyzed. Note that these values should match the power supply voltages at which the netlist extraction was performed.

An example of a Spice submit deck template file for an IBM ASIC chip is shown in Fig. 6.23.

Once the decks have been modified with application-specific parameters, they are simulated using Spice. Simulations need to be run for both best case and worst case environmental conditions for every clock tree. The next step is to analyze the simulation output to see whether the clock signals have met the required analog characteristics to guarantee the low-jitter operation.

```
* HSS DIFFCLK test - Best Case Simulation
* Filename: skel_bc.sp
* .default is to probe every node, might want to change that.
*
* ==> IBM ASIC process settings <==
.param sigma   = xyz    $ (bc nc wc) => (bc_sigma nc_sigma wc_sigma )
.temp          = 0      $ (bc nc wc) => (bc_temp   nc_temp   wc_temp )
.param vdd     = 1.7    $ (bc nc wc) => (bc_vdd    nc_vdd    wc_vdd )
.param vss     = 0
.param vavdd2  = 2.5
.param vswing  = 0.25

*.options brief
.options search='./models'
.option parhier=local


*********************************************************************
*
*.options acout=0 captab
.options post

.inc 'blk_netlist.inc'          $include extracted listing of path blocks
.inc 'blk_netlist_comp.inc'     $include extracted listing of complement path parasitics
.inc 'blk_netlist_true.inc'     $include extracted listing of true path parasitics
.inc 'blk_netlist_load.inc'     $include file with pseudo-HSS load models
.inc 'blk_netlist_meas.inc'     $include file with all pre-built .measure statements

.global vdd vss gnd lt mc
vdd    vdd   0 vdd
vss    vss   0 vss
vlt    lt    0 vss
ven    en    0 vdd
vmc    mc    0 vss
vavdd2 avdd2 0 vavdd2

.param tper=1.11ns              $ 900 MHz default (1/1.11ns)
.param trise=100ps tfall=trise tup='(tper-trise-tfall)/2'
.param tstop=100ns linedel='tper/2' t2='2*tper'

vdrv_t pad  vss pulse('(vavdd2/2)+vswing' '(vavdd2/2)-vswing' 1n trise tfall tup tper)
vdrv_c padn vss pulse('(vavdd2/2)-vswing' '(vavdd2/2)+vswing' 1n trise tfall tup tper)

.TRAN 10ps 10ns

.end
```

Fig. 6.23 Spice submit deck example

```
$DATA1 SOURCE='SPICE' VERSION='Y-2006.09-SP1 '
.TITLE '* hss diffclk test - best case simulation'

      hss01t_pp =    0.679      PASS  (minimum < Vpp)
    hss01t_trise =  2.20e-10    PASS  (trise < maximum)
    hss01t_tfall =  1.55e-10    PASS  (tfall < maximum)
hss01t_dutycycle =   49.11      PASS  (minimum < DC < maximum)
      hss01c_pp =    0.667      PASS  (mimimum < Vpp)
    hss01c_trise =  2.261e-10   PASS  (trise < maximum)
    hss01c_tfall =  1.572e-10   PASS  (tfall < maximum)
hss01c_dutycycle =   50.3147  PASS  (minimum < DC < maximum)
    hss01_vcma =    0.944      PASS  (minimum < Vcm < maximum)
    hss01_vcmb =    0.856      PASS  (minimum < Vcm < maximum)
     hss01_vcm  =    0.930      PASS  (minimum < Vcm < maximum)


*******   End of HSS core input measurements. *******

******* Intermediate CLKDIFF book measurements. *******
- - - - - - - - - - - - - - - - - - - - - - - - - - - - - - - - - - - - - - -
HSS_PCIE_CLKDIFF_U01t_pp =   0.9184  PASS  (minimum < Vpp)
HSS_PCIE_CLKDIFF_U01c_pp =   0.9182  PASS  (minimum < Vpp)
HSS_PCIE_CLKDIFF_U01_vcm = 0.8436  PASS (minimum < Vcm <maximun

******* End of HSS differential clock report.
```

Fig. 6.24 Report file example

6.4.3.3 Analysis to Determine the Integrity of the Clock Signal

After each of the Spice simulations has been successfully complctcd, the differential clock signals must be analyzed at the input to each differential clock buffer and each HSS core in the clock tree to be certain that the HSS cores will perform adequately. Figure 6.24 contains an example of a report produced for an IBM ASIC chip based on the measurement statements in the Spice include file produced for this chip. Pass/fail status for each measurement is reported in this example.

If the analysis does not meet all of the criteria for both best and worst case conditions, the differential clock must to be redesigned and the analysis must be repeated, as was indicated in Fig. 6.21. Depending on the situation, any of the following corrective actions may be necessary: select clock buffers with higher/lower drive strengths, reroute signal traces to reduce the number of vias or directional changes, reduce the distance between clock buffers, etc.

6.5 References and Additional Reading

The following interface standards documents are referenced in this chapter:

1. "Common Electrical I/O (CEI) - Electrical and Jitter Interoperability agreements for 6G+ bps and 11G+ bps I/O", OIF-CEI-02.0, Optical Internetworking Forum, Feb. 28 2005.

2. "Serial ATA Revision 2.5", Serial ATA International Organization (http:\\www.sata-io.org), Oct. 27 2005.

3. "ITU-T G.783 - Series G: Transmission Systems and Media, Digital Systems and Networks, Digital Terminal Equipment - Characteristics of SDH Equipment Functional Blocks", International Telecommunications Union, 2006.

The following reading is recommended for more information regarding clock jitter and PLL jitter transfer functions:

4. "Jitter, Noise, and Signal Integrity at High Speed", Mike Peng Li, Prentice Hall, 2007.

5. "Design of Integrated Circuits for Optical Communications", Behzad Razavi, McGraw-Hill, 2003.

6.6 Exercises

1. Figure 6.1 illustrates the effects of power supply noise on the output of a single-ended clock buffer. A novice engineer suggests that compression of the signal does not matter because the power supply compression would also reduce the switching threshold voltage of the input to the next clock buffer in the distribution network. Explain why this is a fallacy.

2. Explain why it is more important to avoid high-frequency reference clocks on the circuit board than it is to avoid them on the chip.

3. The OIF SFI-5 (version 1) protocol uses the SxI-5 electrical layer discussed in Sect. 5.2. This protocol specifies that the frequency of the reference clock into the chip is 1/4 of the link baud rate.

   a. Assuming a 2.488 Gbps baud rate on each link, what is the frequency of this reference clock?

   b. Given that the available on-chip PLLs for the target technology are limited to a reference clock input frequency of 400 MHz or less, which of the topologies in Fig. 6.3 must be used to support this configuration?

   c. Specify the logic levels on the HSS EX10 PLL slice pins and the value that must be programmed for the Tx/Rx slice Rate Select to select the correct baud rate and reference clock frequency for this interface.

d. Draw a block diagram showing the connections to the HSS EX10 HSSREFCLKT/C input pins. Assume test requirements dictate that a on-chip PLL must be used to source the reference clock during manufacturing test.

4. Two HSS EX10 cores are used in an OIF SPI-S application (see Fig. 5.5) requiring four data lanes and one status lane. The baud rate of these lanes is 10 Gbps. In the sink chip for this interface, the transmitter must transmit status at exactly the same baud rate as the received data.

   (a) Draw the clock connections between the two HSS EX10 cores that are necessary to implement this interface on the sink chip.

   (b) For the HSS EX10 being used to receive the SPI-S data, specify the reference clock frequency, PLL slice HSSDIVSEL and HSSREFDIV pin values, the *Receive Configuration Mode Register* setting, and the *SONET Clock Mode Register* setting.

   (c) For the HSS EX10 being used to transmit the SPI-S status, specify the PLL slice HSSDIVSEL and HSSREFDIV pin values, and the *Transmit Configuration Mode Register* setting.

5. Draw block diagrams of the reference clock distribution for chips which use HSS EX10 cores to implement interfaces for each pair of baud rates listed below. The interfaces for these baud rates are implemented using separate HSS EX10 cores and are operational at the same time. Use as few on-chip PLLs as possible in each case.

   (a) 8.50 Gbps and 4.25 Gbps

   (b) 10.3125 Gbps and 1.25 Gbps

   (c) 10.3125 Gbps and 8.50 Gbps

6. Draw block diagrams of the reference clock distribution for chips which use HSS EX10 cores to implement interfaces for each pair of baud rates listed below. These interfaces share pins on the chip and therefore must be implemented with the same HSS EX10 core. Software programs select which interface is being used by changing the configuration of the HSS EX10 core and/or the on-chip PLL divider settings. (The external reference clock frequency cannot be changed.) The available on-chip IF PLL supports frequency multiplication factors in the range of 2.0 – 16.0 in increments of 0.25. Use as few IF PLLs as possible in each case, and use differential multiplexors to select between different IF PLL outputs only if different reference clock frequencies are required.

   (a) 8.50 Gbps and 4.25 Gbps

   (b) 10.3125 Gbps and 1.25 Gbps

   (c) 10.3125 Gbps and 8.50 Gbps

7. Draw the block diagram of the reference clock distribution for a chip which implements the Backplane Ethernet baud rates of 1.25, 3.125, and 10.3125 Gbps. Assume the same HSS EX10 core must be used for all three cases and is provisioned by software. Also assume on-chip PLLs are available with restrictions as described in Exercise 6.

8. Draw the block diagram of the reference clock distribution for a chip which implements the Fibre Channel baud rates of 3.18750, 8.50000, and 10.51875 Gbps. Assume the same HSS EX10 core must be used for all three cases and is provisioned by software. Also assume on-chip PLLs are available with the restrictions in Exercise 6.

9. Assume the phase jitter variation of a clock signal is modeled with a uniform triangular waveform with an amplitude of 10 ps and a frequency $f_{jitter} = 0.1 f_{clock}$. Calculate the phase jitter, period jitter, and cycle-to-cycle jitter of this clock for 15 consecutive clock cycles.

10. Assume the phase jitter variation of a clock signal is modeled as a summation of two sinusoidal components, each of which can be calculated using (6.5). The first component has a frequency $f_{jitter} = 0.05 f_{clock}$ and an amplitude of 10 ps. The second component has a frequency $f_{jitter} = 0.1667 f_{clock}$ and an amplitude of 5 ps. Calculate the phase jitter, period jitter, and cycle-to-cycle jitter of this clock for 25 consecutive clock cycles.

11. Prove that when $f_{jitter} = 0.5 f_{clock}$, (6.6) reduces to (6.4), and (6.10) reduces to (6.9).

12. What is the limit of J_{period} and J_{cycle} as determined by (6.6) and (6.10) respectively, if $f_{jitter} \ll f_{clock}$.

13. Given the jitter tolerance mask in Fig. 6.5, specify the bandwidth of the PLL used in the CDR circuit given the following baud rates:

    (a). 10.3125 Gbps (b). 8.50 Gbps (c). 3.125 Gbps

14. Assume that a PLL used in a CDR circuit is designed such that: $f_{bw} < f_{baud} / 2$. Is this compliant with the mask in Fig. 6.5? Explain.

15. Assuming clock buffers with characteristics described in Sect. 6.3.1 and Table 6.2, calculate maximum wire lengths for the following cases:

    (a) $f_{clock} = 775\,\text{MHz}$, fanout $= 2$
    (b) $f_{clock} = 250\,\text{MHz}$, fanout $= 3$
    (c) $f_{clock} = 500\,\text{MHz}$, fanout $= 1$

16. Given the wire lengths calculated in Exercise 15, assume the buffer inputs for the indicated fanouts are equidistant from the driving buffer, and are in opposite directions on the chip. What is the maximum distance between the driving buffer and each buffer input?

17. Describe the effects which may be encountered if the following analog parameters are beyond their specified ranges on the input to a differential clock buffer:

    (a) V_{cm} too high or too low
    (b) V_{diff} too low
    (c) V_{diff} too high
    (d) Slew rate too fast
    (e) Slew rate too slow

18. Describe how mismatches in the t_{rise} and t_{fall} parameters of the single-ended signals can contribute to a mismatch in propagation delay resulting in deterministic jitter on the output waveform.

19. The high and low voltages of a differential signal are provided below for various systems. For each pair of voltages, calculate the corresponding V_{cm} and V_{diff} values.

    (a) 1.05 V, 0.35 V
    (b) 1100 mV, 600 mV
    (c) 675 mV, 155 mV
    (d) 920 mV, 330 mV

20. Given the following V_{cm} and V_{diff} values, calculate the V_{hi} and V_{lo} voltages:

    (a) $V_{cm} = 550$ mV, $V_{diff} = 300$ mVppd
    (b) $V_{cm} = 600$ mV, $V_{diff} = 400$ mVppd
    (c) $V_{cm} = 750$ mV, $V_{diff} = 120$ mVppd
    (d) $V_{cm} = 700$ mV, $V_{diff} = 1200$ mVppd

21. A clock distribution network for a 400-MHz reference clock consists of a string of four differential clock buffers. The design of these buffers is such that there is a mismatch between the propagation delay for the rising and falling edges of the clock signal given the net capacitance for the current chip layout. For each buffer:
    - t_{pd} (rise) = 300 ps
    - t_{pd} (fall) = 500 ps

    Draw a timing diagram showing the resulting differential amplitude waveform and duty cycle on the output of each clock buffer stage.

22. Assume that the clock buffers in a clock distribution network require a 500 ps minimum pulse width to guarantee the buffer output switches. Given the clock distribution network in Exercise 21, draw conclusions as to whether the output of each stage is usable.

Chapter 7
Test and Diagnostics

HSS devices incorporate features which support various levels of testing by the chip manufacturer, the system manufacturer, and by the end user (as part of a diagnostic test suite). These levels of testing include:

JTAG 1149.1/1149.6 Test. The JTAG 1149.1 and 1149.6 standards [1,2] define a method of performing chip-to-chip stuck-fault testing during circuit board manufacture. This requires test structures which must be incorporated in all chips to support this testing.

Pseudo-Random Bit Sequence (PRBS) Test. HSS cores generally provide a means of transmitting a PRBS pattern and checking it at the receiver. Such test sequences are used for manufacturing test and characterization of HSS cores, as well as characterization of serial data links in systems.

Logic Built-In-Self-Test (LBIST). The chip designer often includes LBIST capabilities on the chip which support in-system diagnostics testing. LBIST implementations are not standardized; each system design team develops their own methodology.

Manufacturing Test. The chip manufacturer runs a series of tests on each chip after wafer fabrication, and again after module assembly. Manufacturing tests used to test HSS devices are developed by the HSS design team. The details of such tests are usually not of importance to the chip designer, however the chip designer must sometimes provide controllability or observability of certain pins on the HSS core to facilitate such tests.

Characterization Test. As part of the design development process, the HSS core design team generally builds test chips containing the HSS core for the purposes of laboratory test and measurement. Certain features of the HSS core are not tested on each chip as part of the manufacturing process, but rather are guaranteed through laboratory measurement on test chips and through analysis of parameter variation based on the tolerances of the manufacturing process. The results of design characterization test are documented in a characterization report. HSS core designs often include features which are intended to support characterization testing and facilitate the measurement of various design characteristics. In some cases these design features may also be used by the system designer to characterize the chip usage within the system design.

Although portions of the above topics may have been covered to some extent in prior chapters, this chapter covers these test topics in depth. This chapter primarily concentrates on standards, typical implementations, and general approaches that are relevant to chip designers using HSS devices. Other topics, such as detailed manufacturing test flows or detailed descriptions of characterization features, vary significantly from one HSS implementation

D. R. Stauffer et al., *High Speed Serdes Devices and Applications,*
© Springer 2008

to another and are therefore discussed at a more general level in this text. This chapter extends the description of the HSS EX10 core that was used as a tutorial example in Chap. 2 to add appropriate test features.

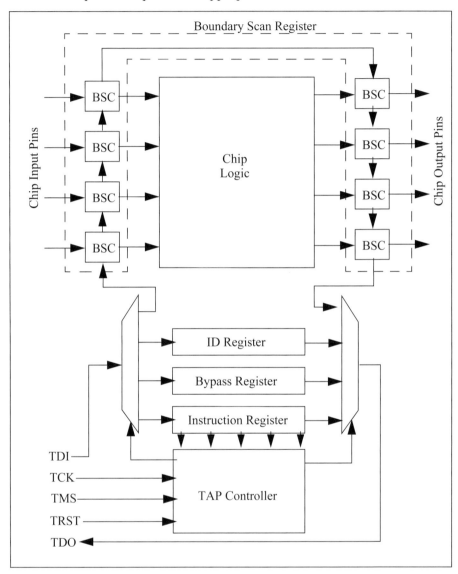

Fig. 7.1 JTAG boundary scan interface architecture

7.1 IEEE JTAG 1149.1 and 1149.6

The Joint Test Action Group (JTAG) was formed in 1985 to develop better methods of performing manufacturing test on circuit boards. The increasing use of multilayer circuit boards and nonlead-frame ICs was making test points inaccessible to test probes. To provide sufficient test of the chip interconnect, it was necessary to develop the ability for each chip in the system to drive outputs with arbitrary logic levels and capture logic levels on inputs. A standard interface, common to all chips and with a minimal number of pins, was needed to access the control and observe test points. This interface could then be used during circuit board manufacturing test to perform stuck-fault testing of the entire interconnect (including the connection of the chips to the circuit board) and achieve high test coverage. The result of this work was published in 1990 as IEEE Std. 1149.1-1990, and is commonly called JTAG 1149.1.

7.1.1 JTAG 1149.1 Overview

Fig. 7.1 illustrates a conceptual block diagram for the implementation of JTAG 1149.1 on a chip. Using the 5-pin standard interface bus and the test access port (TAP) control circuitry, data can be launched through the outputs of one chip and captured by another chip on the card, thus effectively testing the chip solder bumps or wirebonds, the solder connections on the package and card, the wiring on the card, plus any cables and connectors.

The JTAG 5-pin standard interface bus consists of the following pins, as shown in Fig. 7.1:

Test Clock (TCK). This pin is the JTAG clock input which clocks all JTAG registers and the TAP Controller.

Test Mode Select (TMS). This pin is a control input used to select the state of the TAP Controller.

Test Data In (TDI). This pin is the JTAG data input used to serially scan data into JTAG registers.

Test Data Out (TDO). This pin is the JTAG data output used to serially scan data out of JTAG registers.

Test Reset (TRST). This pin is the JTAG input used to force a reset of JTAG registers and the TAP Controller state. This input is optional; a reset state may also be forced through assertion of TMS for five clock cycles.

The key components of the JTAG 1149.1 implementation shown in Fig. 7.1 are described in the subsections which follow.

7.1.1.1 TAP Controller

The TAP Controller is a finite state machine which implements the state diagram defined in the IEEE 1149.1 standard. This state diagram is shown in Fig. 7.2. State transitions within this state diagram occur based on logic level on the TMS pin during rising edges of the TCK pin. The TMS logic values corresponding to these transitions are shown on the arcs in Fig. 7.2.

Note that it is possible to uniquely return to the *Test-Logic-Reset* state from any state in this state diagram within five TCK cycles by asserting TMS = 1. Once in this initial state, additional TMS values and TCK cycles transition the TAP controller to the appropriate desired state.

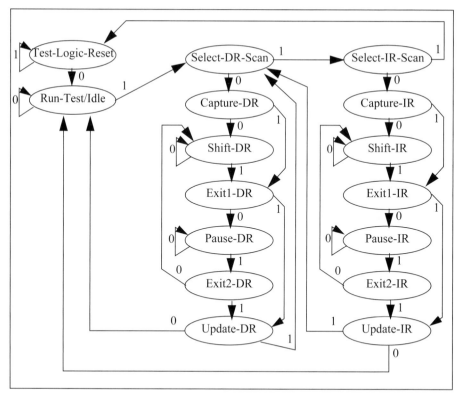

Fig. 7.2 Tap controller state diagram

JTAG registers may be loaded and read by scanning data through the TDI/TDO pins when the TAP Controller is in the appropriate state. The *Instruction Register* is selected for scanning when the TAP Controller is in the *Shift-IR* state. The *ID Register*, *Bypass Register*, *Boundary Scan Register*, or other implementation-specific registers may be scanned when the TAP Controller is in the *Shift-DR* state. The contents of the *Instruction Register* determine which of these data registers is selected.

When registers are serially scanned, data is actually scanned through a shadow register so that the contents of the actual register are not corrupted during the scan process. Data is transferred in parallel from all bits of the shadow register to the actual register when the TAP Controller is in either the *Update-IR* or *Update-DR* state. Likewise, the *Capture-IR* and *Capture-DR* states are used to capture status data in the *Instruction Register*, or capture logic values from chip inputs in the *Boundary Scan Register* prior to serially scanning this data out of the chip.

7.1.1.2 Instruction Register

The contents of the *Instruction Register* determines the address and control information that selects which of various data registers is to be accessed. Values for this register are called *test instructions*. The JTAG 1149.1 standard defines three required test instructions, six optional instructions, and allows for other implementation-specific instructions. The instructions defined by the JTAG 1149.1 standard are described briefly as follows:

Required Instructions:

BYPASS Instruction. When the Instruction Register contains this instruction, serial data is scanned through the *Bypass Register* without affecting operation of the chip.

SAMPLE/PRELOAD Instruction. When the Instruction Register contains this instruction, serial data is scanned through the *Boundary Scan Register* while the chip remains in a functional mode.

EXTEST Instruction. When the Instruction Register contains this instruction, serial data is scanned through the *Boundary Scan Register* and the chip is placed in a mode where the Boundary Scan Register drives chip outputs and receives chip inputs.

Optional Instructions:

INTEST Instruction. When the Instruction Register contains this instruction, serial data is scanned through the *Boundary Scan Register* and the chip is placed in a mode where the Boundary Scan Register controls internal inputs to the chip logic and captures internal outputs of the chip logic.

RUNBIST Instruction. When the Instruction Register contains this instruction, serial data is scanned through a user-specific data register and runs a Built-In-Self-Test sequence.

CLAMP Instruction. When the Instruction Register contains this instruction, serial data is scanned through the *Bypass Register* and the chip is placed in a mode where the current contents of the *Boundary Scan Register* drive chip outputs.

HIGHZ Instruction. When the Instruction Register contains this instruction, serial data is scanned through the *Bypass Register* and the chip is placed in a mode where chip outputs are driven to a high impedance state.

IDCODE Instruction. When the Instruction Register contains this instruction, serial data is scanned through the *ID Register* without affecting operation of the chip.

USERCODE Instruction. When the Instruction Register contains this instruction, serial data is scanned through the *ID Register* without affecting operation of the chip. Additional user-defined data is captured in the *ID Register* as part of this instruction.

7.1.1.3 Bypass Register

The *Bypass Register* is a required one-bit register which allows data to be scanned through an abbreviated path from the TDI pin to the TDO pin. The BYPASS instruction is loaded into chips which are not involved in a test to shorten the serial scan path on the circuit board and thereby speed-up the test sequence.

7.1.1.4 ID Register

The *ID Register* is an optional register which contains a unique identification code for the chip. This allows the test program to automatically identify chips on the circuit board and the order in which these chips are connected into the scan chain. The test program can then reference information about the chip to determine the test sequence.

7.1.1.5 Boundary Scan Register

The *Boundary Scan Register* is a required register which consists of one *Boundary Scan Cell (BSC)* corresponding to each chip input pin, chip output pin, and chip output enable control signal. Depending on the instruction loaded in the Instruction Register, the BSC may control the logic level driven on output pins and may capture logic levels on chip input pins. Optionally, the BSC may also override chip input logic levels being driven to internal chip logic and capture the logic levels being driven by internal chip logic to outputs.

The frequency of operation of serial data signals for HSS cores precludes connecting additional logic between the Serdes device and the chip pins. HSS cores must therefore provide some means of bypassing the normal Serdes functions, driving BSC logic levels to output pins, and capturing logic levels on input pins. Without such support, the serial data interconnect would not be testable, and chips using these cores would not comply with the JTAG 1149.1 standard. This is discussed further in the next section.

7.1.2 HSS Core Support for JTAG 1149.1

When the EXTEST instruction is loaded in the Instruction Register, the chip outputs are driven by the logic levels loaded into the corresponding BSCs, and the logic levels on chip inputs are captured in BSCs associated with the input pins. In addition, BSCs associated with output enable control signals determine whether outputs are driven or are in a high impedance state. The launch and capture elements for a chip-to-chip interconnect path are shown in Fig. 7.3. Stuck-fault testing of the interconnect path is performed by launching data from the BSC associated with the output of the driver chip, and capturing the corresponding data in the BSC at the receiving chip.

Signal integrity requirements for an HSS core dictate that the Serdes driver and receiver circuits connect directly to chip pins. Intervening logic would not be able to operate at the necessary data rates, and even the connection of additional gate inputs to the signal would create impedance mismatches and degrade signal quality. To support JTAG 1149.1, the HSS core must provide a transmit bypass path so that a BSC may be used to drive serial data outputs, and a receive bypass path so that the logic level on the serial data input may be captured in a BSC. The logic levels being driven and captured are essentially DC levels, and therefore the normal serializer and deserializer functions of the HSS core must be bypassed to be able to drive and receive these signals.

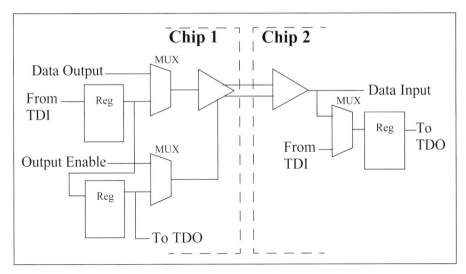

Fig. 7.3 Chip to *Chip Interface with JTAG 1149.1 Test Features*

The HSS EX10 core discussed in Chap. 2 includes bypass functions on both the transmitter and receiver slices which were primarily intended to support JTAG 1149.1. The transmitter bypass function is described in Sect. 2.2.6 and the receiver bypass function is described in Sect. 2.3.6. Connection of the various signals to the JTAG Boundary Scan Cells is shown in Fig. 2.6.

7.1.3 HSS Core Support for JTAG 1149.6

When applied to high-speed serial data interconnect, there are several drawbacks to testing within the constraints of the JTAG 1149.1 standard.

First, JTAG 1149.1 is limited to DC stuck-fault testing. Changing the logic levels being driven on the interconnect requires successively loading data through the JTAG scan chain into the JTAG Boundary Scan Registers on the chips of the circuit board, an inherently slow process. If the interconnect is DC coupled, then DC stuck-fault testing may be adequate. However, if the high-speed serial data link includes decoupling capacitors, then a DC logic level cannot be driven across the interconnect. In such cases, JTAG 1149.1 would be unable to test the serial data interconnect.

Second, JTAG 1149.1 tests the serial data link as if it were a single wire with a DC logic level driven on it. In fact, the serial data consists of a differential signal. When a logic level of 0 is driven, then one leg of the differential pair is driven low and the other is driven high (and vice-versa for a logic 1 level). This was described in Sect. 1.3.4 and was illustrated in Fig. 1.16 and Fig. 1.18.

It is possible for one leg of the differential pair to be unconnected, and yet have JTAG 1149.1 perform a successful test of the link. To illustrate this, Fig. 1.16 has been modified as shown in Fig. 7.4. Although the JTAG 1149.1 logic levels would be DC levels, signal transitions are shown in the figure for the purposes of covering both the logic 0 and logic 1 cases. The first logic 1 and logic 0 level shown in Fig. 7.4 is normal with both the true (T) and complement (C) signals at proper logic levels.

However, for the second logic 1 and logic 0 level shown in the figure, the complement leg of the signal (the dashed line) has failed and is biased to the common mode voltage. The corresponding received differential signal is shown in Fig. 7.5. As shown, the failure of the complement leg of the signal reduces the amplitude of the received signal, but the correct logic levels are still received. Because JTAG 1149.1 is limited to treating the differential serial data link as a single interconnect path, the test cannot distinguish between the normal and failure cases in Fig. 7.5, and the test passes even in the presence of the failure.

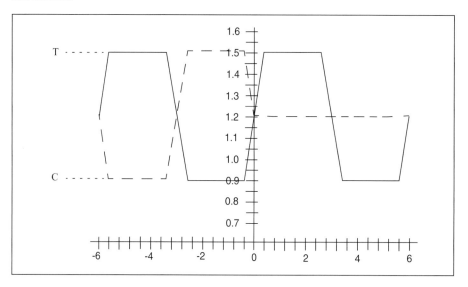

Fig. 7.4 Single-ended signals showing failure of complement leg

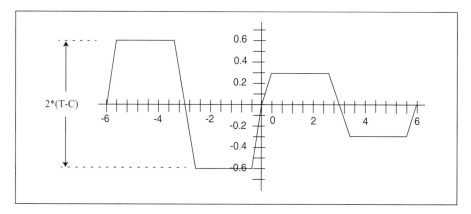

Fig. 7.5 Differential peak-to-peak signal with failure of complement leg

Consequently, in 2003 the IEEE Joint Test Action Group defined the IEEE 1149.6 standard which addresses deficiencies in the 1149.1 standard related to differential I/O testing and AC coupled I/O testing. The IEEE 1149.6 test standard, also known as "AC JTAG," requires independent observability of both the true and complement side of a differential I/O. Two additional TAP functions are defined which support generation of pulses on AC coupled links. The EXTEST_PULSE instruction generates a single pulse, while the EXTEST_TRAIN instruction generates a sequence of pulses. The standard also requires a test receiver cell, implemented as a hysteretic comparator, which is used to observe pulses on AC coupled links. In AC mode, the test receiver cell is an edge detector which can detect either a positive-going or negative-going edge, indicating the presence or absence of a pulse propagating through a capacitor. In DC mode, the test receiver acts like a buffer.

Fig. 7.6 illustrates the modifications to Fig. 7.3 required for AC coupled serial data links for compliance with the 1149.6 "AC JTAG" standard.

The HSS EX10 core described in Chap. 2 includes edge detector functions in the receiver to support the JTAG 1149.6 standard. These functions are described in Sect. 2.3.6, and connection of the various signals to JTAG Boundary Scan Cells is shown in Fig. 2.6. The additional functions associated with the JTAG Boundary Scan Cell on the transmitter do not require any additional logic in the HSS transmitter design.

Note that while the signals generated and received by JTAG 1149.6 are sufficiently high frequency to pass through any decoupling capacitors included in the interconnect, they are still typically well below the intended baud rate for the core. Pulses are generated by TCK, the maximum frequency of which is limited by the slowest interconnect path on the circuit board. Clock frequencies well below 1 MHz are typical. JTAG 1149.6 is not intended to perform at-speed test.

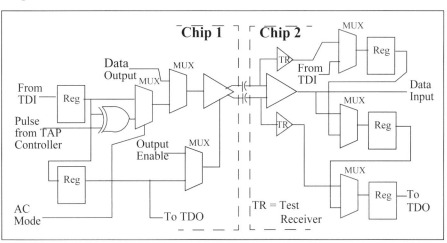

Fig. 7.6 AC coupled serdes with 1149.6 test structures

7.2 PRBS Testing and Loopback Paths

As was discussed in prior chapters, HSS cores generally provide pseudo-random bit sequence (PRBS) generators and checkers for diagnostic purposes. These circuits are used for manufacturing test, characterization test, and in-system diagnostic tests to determine the integrity of Serdes circuits and the link interconnect. The capabilities of the PRBS generator in the HSS EX10 core are described in Sect. 2.2.7, and the capabilities of the PRBS checker are described in Sect. 2.3.8.

7.2.1 Loopback Paths

The ability to loop data at various points within the protocol stack is an important diagnostic capability of serial protocols and interfaces. Some protocols require certain loopback paths be present in hardware to facilitate system diagnostic functions. Even when diagnostic requirements are not specified by the protocol standard, chip development teams require certain diagnostic paths to facilitate chip and system characterization. Often, PRBS pattern generator and checker logic is included in lower layers of the protocol logic and/or the HSS cores. This logic allows basic checking of the hardware datapath and interconnections without requiring a fully functional software/hardware protocol stack.

Fig. 7.7 illustrates an egress data path consisting of protocol transmit logic and an HSS transmitter, and an ingress data path consisting of an HSS receiver and protocol receiver logic. Possible loopback paths are shown in the figure. (Note that the names of these paths as shown in the figure and used in this text are the author's names. Names of diagnostic loopback paths used in other sources may vary. There is no standard set of names for loopback paths.)

In should be noted that not all of these diagnostic loopback paths will exist in a given chip. Also, Fig. 7.7 oversimplifies the implementation of these loopback paths. For example, the parallel data output of the HSS receiver cannot be directly connected to the parallel data input of the HSS transmitter without an elastic buffer to perform clock compensation. These loopback paths are discussed in somewhat more detail in the following subsection.

7.2.1.1 Serial Diagnostic Loopback

The serial diagnostic loopback path wraps data from the output of the HSS transmitter to the input of the HSS receiver, and can only exist for Full Duplex HSS cores which contain both transmitters and receivers in the same core. This path provides a means for performing diagnostics on the ASIC chip through the HSS transmitter and receiver. If data can be transmitted and received through this diagnostic path, then most of the chip's functional path has been tested. (For some HSS core implementations there may be a small amount of analog circuitry that is not exercised by this loopback path. Also, the connection through the chip package is not tested.)

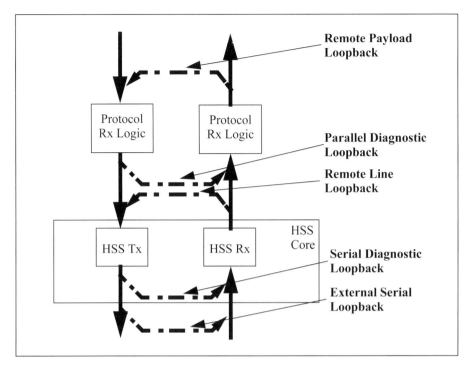

Fig. 7.7 Possible loopback paths

The Serial Diagnostic Loopback is also useful for HSS core test and characterization. The PRBS generator in the HSS transmitter sends data, and this data is looped to the HSS receiver and checked by the PRBS checker in the HSS receiver. This PRBS test does not require higher layers of the protocol stack (either implemented in hardware or software) to be operational. By running the PRBS test repeatedly and for sustained periods of time, and measuring the number of PRBS errors that occur, the bit error rate (BER) of the HSS hardware can be estimated.

Full Duplex HSS cores which implement these loopback paths always pair transmitters and receivers in the core such that specific transmitter slices are wrapped to specific receiver slices. Chip designers need to be cognizant of the chip and application requirements associated with the Serial Diagnostic Loopback path, and reconcile these requirements with the pairing implemented by the HSS core. If there is a requirement for traffic to be generated by the protocol transmit logic and looped through the Serial Diagnostic Loopback path to the receiver, then the interconnection of HSS transmitters and receivers to the protocol logic is constrained by the pairing of these transmitters and receivers in the HSS core.

7.2.1.2 External Serial Loopback

The External Serial Loopback path is similar in function to the Serial Diagnostic Loopback path, except that it is implemented external to the chip by the engineer in the lab. How this loopback connection is made depends on the design of the printed circuit board. It may consist of jumper wires, a test jig plugged into a connector, or some other means. Because it requires physical reconnection of the link external to the chip, this loopback path typically only exists in a lab environment.

The diagnostic purpose of this path is similar to the Serial Diagnostic Loopback path, albeit with better coverage. (The chip package interconnect and all stages of the HSS transmitter and receiver are included in the path being tested.) Also, since any arbitrary connection can be made outside of the chip, the limitations of transmitter/receiver pairing that applied to the internal Serial Diagnostic Loopback path do not apply for this path.

7.2.1.3 Remote Line Loopback

The Remote Line Loopback path takes data received through the HSS receiver, and retransmits this data on the HSS transmitter. This path supports system diagnostics by allowing an interface port to send test data to the remote device, and have that device retransmit the data back without requiring processing by higher layers of the protocol. Using this function, system diagnostics can test the signal integrity of the electrical interconnect. Some protocols, such as PCI Express, require implementation of this path.

As noted previously, implementation of this path is not as simple as connecting the parallel data output of the HSS receiver to the HSS trans-mitter. If plesiosynchronous reference clocks are used for each end of the link, then the receive data clock is at a slightly different frequency from that of the transmitter. Even if the two ends of the link are synchronously clocked, phase differences will still exist between the parallel data clocks for the receiver and the transmitter. Therefore an elastic buffer is required in this path. In the case of plesiosynchronous clocking, this elastic buffer may need to add or drop bytes (or symbols) as defined by the protocol to perform clock compensation.

This elastic buffer is already typically part of the Protocol Receive Logic. Therefore, the existing elastic buffers in this logic are generally included in the Remote Line Loopback path; there would be no benefit in implementing separate elastic buffers for diagnostic operations. The result is that the Remote Line Loopback path, when implemented, is an integral part of the function of the Protocol Logic. It is generally implemented by wrapping the output of the elastic buffers in the Protocol Receive Logic to the transmit data path of the Protocol Transmit Logic.

An alternative implementation of the Remote Line Loopback path retransmits data using a loop timing mode in which the receive data clock is used as the reference clock for the transmitter. However, this assumes loop timing mode is supported by the selected HSS cores, and is implemented in the chip design. The logic requirements for implementing loop timing mode were discussed in Sect. 6.1.3.2.

7.2.1.4 Parallel Diagnostic Loopback

The Parallel Diagnostic Loopback path wraps data from the output of the Protocol Transmit Logic to the input of the Protocol Receive Logic. The purpose of this diagnostic path is similar to the function of the Serial Diagnostic Loopback, except that the HSS cores are excluded from the data path. This is useful for testing higher layers of the protocol in cases where the HSS hardware is not yet operational (either because the configuration has not been programmed correctly, because of hardware problems, or because of interconnect problems). It may also be used in conjunction with the Serial Diagnostic Loopback path to isolate problems to the HSS cores. If data is successfully wrapped through the Parallel Diagnostic Loopback path, but cannot be wrapped through the Serial Diagnostic Loopback path, then the problem may be assumed to be in the HSS cores.

Implementation of this path requires both data and clocks to be switched at the inputs to the Protocol Receive Logic. Normally, the parallel data clock for the parallel data input of the Protocol Receive Logic is driven by the HSS Receiver. When the Parallel Diagnostic Loopback path is active, this clock must be connected to the same clock source that is driving the Protocol Transmit Logic (usually the parallel data clock output of the HSS transmitter).

7.2.1.5 Remote Payload Loopback

The Remote Payload Loopback path wraps data from the output of the Protocol Receive Logic to the input of the Protocol Transmit Logic. This diagnostic path differs from the Remote Line Loopback path in that a substantial amount of the protocol processing is included in both the receive and transmit data paths. While the Remote Line Loopback path typically retransmits data as received (possibly with some adjustment for clock compensation), the Remote Payload Loopback path retransmits the same payload data but with different management information.

When this loopback path is executed in a SONET system, for example, the SONET section and line overhead are processed in the receive logic and SPEs are wrapped to the transmit logic. The transmit logic maps SPEs into SONET frames with locally generated overhead, complete with pointer management as needed. The section and line overhead functions of both ends of each section continue to operate and exchange information normally, even while the SPE is wrapped. System diagnostics in the SONET network can implement fault isolation by wrapping the SPE at various LTE and PTE elements.

7.2.2 PRBS Circuits and Data Patterns

A string of binary digits ("1" and "0" bits) is called a pseudo-random sequence if it meets two conditions:

- Local randomness (i.e. the next output of a sequence generator is roughly equally likely to be a "1" as a "0"); and
- Reproducibility (i.e., when the sequence generator is reset to an initial state, the exact random sequence of "1" and "0" bits is replicated).

7.2.2.1 PRBS Generator Circuits

Linear Feedback Shift Registers (LFSRs) are the most common method of generating pseudo-random sequences. A shift register is a collection of flip-flops or other storage elements connected such that the state of each element is shifted to the next element in response to a clock signal. Feedback in an LFSR occurs when the outputs of selected stages of the shift register are summed (XOR'd) and connected back to the shift register input. The length of the sequence that is generated and the likelihood of producing a certain maximum run length are mathematically dependent on the number of stages in the shift register and the details of the feedback connections. If the sequence generated by an n-stage LFSR has a period of $2^n - 1$, it is a *maximum-length sequence*.

The LFSR implementation of a PRBS-7 generator is shown in Fig. 7.8. The PRBS-7 pattern consists of seven flip-flops with the appropriate feedback term implemented using an XOR gate (i.e., a binary summation function). In Fig. 7.8, the flip-flops have been assigned indices "1" through "7"; the feedback term is generated by XOR'ing the outputs of flip-flops "6" and "7." Mathematical polynomials are often used as a short-hand method of specifying the PRBS pattern being implemented. The polynomial for PRBS-7 is

$$PRBS7 = x^7 + x^6 + 1 \tag{7.1}$$

The powers of the "x" terms indicate time-shift delay and directly correlate to the positions of the feedback taps in the LFSR implementation.

The LFSR implementation shown in Fig. 7.8 is a serial implementation which produces one bit of the PRBS sequence for each clock cycle. Parallel implementations of PRBS generators which produce any number of bits of the sequence per clock cycle are also possible. Such circuits are designed by mathematically calculating the necessary logic equations using the PRBS polynomial.

For example, assume an implementation of (7.1) using seven flip-flops with the initial states defined below, where the number in the parenthesis indicates the advance of $n = 0$ bits of the sequence:

$$x^7(0) = x^7 \qquad\qquad x^3(0) = x^3$$
$$x^6(0) = x^6 \qquad\qquad x^2(0) = x^2$$
$$x^5(0) = x^5 \qquad\qquad x^1(0) = x^1$$
$$x^4(0) = x^4$$

After the sequence is advanced by one bit, the contents of a serial implementation LFSR would shift by one bit, and the feedback term would be loaded into the first bit:

$$x^7(1) = x^6 \qquad\qquad x^3(1) = x^2$$
$$x^6(1) = x^5 \qquad\qquad x^2(1) = x^1$$
$$x^5(1) = x^4 \qquad\qquad x^1(1) = x^7 + x^6$$
$$x^4(1) = x^3$$

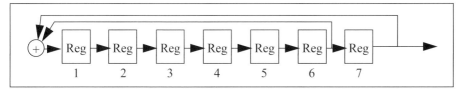

Fig. 7.8 PRBS7 generator implementation using a LFSR

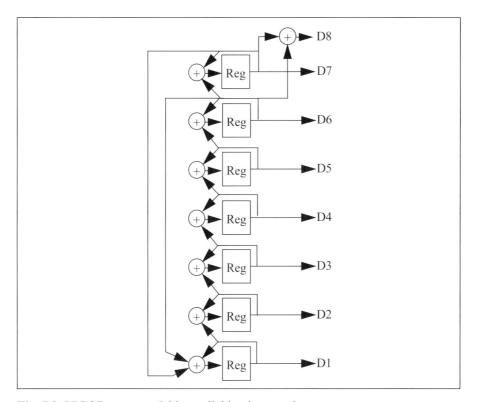

Fig. 7.9 PRBS7 generator 8-bit parallel implementation

As the sequence continues to advance, the contents of the LFSR after 8 bits of the sequence would be:

$$x^7(8) = x^7 + x^6 \qquad x^3(8) = x^3 + x^2$$
$$x^6(8) = x^6 + x^5 \qquad x^2(8) = x^2 + x^1$$
$$x^5(8) = x^5 + x^4 \qquad x^1(8) = x^7 + x^6 + x^1$$
$$x^4(8) = x^4 + x^3$$

These equations are implemented in the circuit shown in Fig. 7.9. Each clock cycle advances this PRBS generator by eight bit positions. The current eight bits of the PRBS sequence are provided as a parallel output of this circuit. Seven of these bits are determined directly by the current state of the PRBS generator, while the eighth bit is derived from a combination of state bits for the current state.

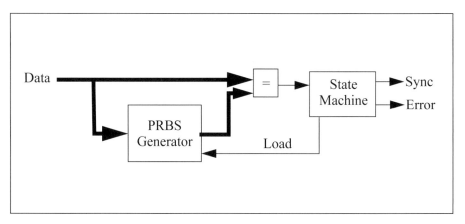

Fig. 7.10 PRBS7 checker block diagram

HSS cores commonly include PRBS generators and checkers, and generally support a selection of PRBS test patterns. However, PRBS generators and checkers are also frequently implemented in higher layers of the protocol stack. Any such circuits generally use an n-bit parallel implementation to match the width of the data path at this layer.

7.2.2.2 PRBS Checker Circuits

A typical implementation of a PRBS checker consists of a PRBS generator circuit, a comparator, and a state machine which controls the initialization sequence of the checker. A block diagram of this is shown in Fig. 7.10.

Before checking can begin, the PRBS checker must initialize its LFSR state based on the data being received so that it can successfully predict the expected data. This is called *synchronizing* to the PRBS pattern. For a PRBS-n pattern, n bits must be received to initialize the LFSR. After initializing the LFSR most designs then check some number of additional bits to make sure the PRBS checker is in fact predicting the expected data correctly. Miscompares during this period may indicate a bit error occurred while the LFSR state was being initialized. If the data compares correctly, then the PRBS checker asserts the *Sync* status indication and begins to check data.

Once the State Machine of the PRBS checker asserts the *Sync* status, any subsequent miscompares between the LFSR state and the incoming data are assumed to represent bits that have been corrupted by the serial data path, and result in the State Machine asserting the *Error* status output. Depending on the implementation of the PRBS checker, it may be possible to count the exact number of bits which were received incorrectly. Note that with this architecture, any discontinuity in the PRBS pattern inserted by the generation source (for example, skipping bits, freezing the pattern for a few bits, reinitializing the LFSR, etc.) causes the PRBS checker to be out of sync and report continuous errors in the pattern. If such a discontinuity occurs, the PRBS checker must be reinitialized.

An alternative self-synchronizing implementation of a PRBS checker is sometimes used which reloads the LFSR state from the datastream continuously. The next n-bits of the PRBS sequence are predicted based on the previous n-bits received; these bits are compared to the next n-bits received and any errors are reported. This architecture may require less logic to implement and adapts to any discontinuity in the PRBS pattern inserted by the generation source. The drawback of this architecture is that any bit errors in the received pattern not only cause miscompares on the bits which are actually in error, but also cause some number of future bits to be incorrectly predicted. A single bit error may therefore be reported as several bits being in error.

7.2.2.3 Data Patterns

Many PRBS polynomials exist and several of these are commonly implemented in high-speed Serdes devices. The order of the polynomial determines the number of flops required to implement the LFSR. This also determines the maximum run length of all 1's or all 0's that can exist in the PRBS pattern. Higher-order polynomials can generate longer run lengths without transitions, and therefore the data patterns contain more low frequency content. This stresses the CDR in the HSS receiver more than a data pattern which has a higher transition density. A PRBS-7 pattern (which is generated using an order 7 polynomial) may be sufficient to characterize a system which uses 8B/10B encoding, however this pattern would not contain a sufficient transition density to stress a system which uses 64B/66B encoding or scrambling. A PRBS pattern which can produce longer run lengths, such as a PRBS-31 pattern, is more desirable to test such systems.

Another consideration involves the limitations of test equipment. Assume a test setup in which the data pattern must be checked by test equipment which has a limited data buffer. Usually it is desirable that the test equipment buffer be able to capture and analyze the entire test pattern. If analysis is only based on a randomized segment of the pattern, results may depend on which portion of the pattern is captured. Although testing using PRBS-31 may be desirable, the size of the data buffer in the test equipment may preclude use of a pattern of this length. A PRBS-23 pattern may be more practical in this case.

As described in Sect. 2.2.7, the HSS EX10 Core supports the test patterns shown in Table 7.1. The various PRBS polynomials referenced above are supported. Options include both the PRBS pattern, and the corresponding inverted pattern. Unmodified PRBS patterns mathematically have slightly higher probability of generating a 1 bit than a 0 bit. Over time this accumulates a DC charge on the capacitor of an AC coupled system. Correspondingly, an inverted PRBS pattern has a slightly higher probability of generating a 0 bit than a 1 bit. Providing both patterns allows the user to decide which pattern is appropriate.

Table 7.1 Test patterns supported by the test pattern generator

Pattern generated	PRBS polynomial
PRBS7+	$x^7 + x^6 + 1$
PRBS7- (inverted)	
PRBS23+	$x^{23} + x^{18} + 1$
PRBS23- (inverted)	
PRBS31+	$x^{31} + x^{28} + 1$
PRBS31- (inverted)	
1010101....	Not applicable
(Repeating pattern of 64 "1"s then 64 "0"s)	Not applicable

7.2.3 PRBS Test Sequence

Fig. 7.11 illustrates a PRBS generator connected to a PRBS checker, along with typical control and status signals associated with this logic. The corresponding control and status signals are defined in Table 7.2. Depending on the implementation of the core, these signals may be pins on the core, internal registers, or both. The HSS EX10 PRBS generator was described in Sect. 2.2.7, and the corresponding PRBS checker was described in Sect. 2.3.8.

Execution of a PRBS test sequence involves the proper sequencing of these control signals, and monitoring of the corresponding status signals. A typical sequence proceeds as follows:

Step One: Set appropriate control signals to select the desired loopback mode (if applicable). If loopback is selected, then the transmitter and receiver used in this test sequence must be in the same chip. Otherwise, the remainder of this sequence may require programming the PRBS generator and PRBS checker circuits located in different chips.

For the HSS EX10 core, the *Full Duplex Wrap Enable* bit in the *Receive Test Control Register* defined in Table 2.7 enables a Serial Diagnostic Loopback path. This loopback path is also selected by the RXxPRBSEN pin defined in Table 2.3. The register bit drives the FDWRAP signal shown in the Receiver Concept Diagram in Fig. 2.7.

When FDWRAP is asserted, the Pseudo-Random Code Generator shown in the Transmitter Concept Diagram in Fig. 2.4 drives data through the various stages of the transmitter serializer to the Wrapback signal. The FDWRAP signal in Fig. 2.7 selects the Wrapback data as the source for data into the AGC Mux. This data is subsequently processed by the DFE and deserialization stages, and checked by the Pattern Recognition Logic at the output of the deserializer stage.

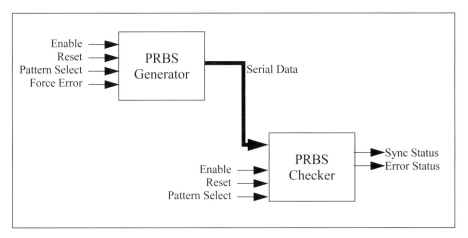

Fig. 7.11 PRBS control signals

Table 7.2 PRBS control/status ports

Block	Port	Type	Description
PRBS generator	Enable	Control Input	PRBS generator circuit enable; "0" = normal operation, "1" = enable generation of the selected PRBS pattern.
	Reset	Control Input	Reset the PRBS generator LFSR to a known state. Also reset any implementation specific status states.
	Pattern Select	Control Input	One or more bit input which selects the PRBS test pattern.
	Force Error	Control Input	Insert an error into the PRBS sequence which should trigger an error at the PRBS checker.
PRBS checker	Enable	Control Input	PRBS checker circuit enable; "0" = normal operation, "1" = enable checking of the selected PRBS pattern.
	Reset	Control Input	Reset the PRBS checker LFSR to a known state. Also reset status states.
	Pattern Select	Control Input	One or more bit input which selects the PRBS test pattern.
	Sync Status	Status Output	PRBS checker pattern sync monitor; "0" = not sync'd, "1" = pattern checker sync'd to data
	Error Status	Status Output	PRBS checker error monitor; "0" = no error; "1" = error.

Step Two: Program the PRBS generator in the HSS transmitter as follows:

1. Set control inputs to select the desired PRBS pattern.

   For the HSS EX10 core, this is selected by setting the *Test Pattern Select* bits of the *Transmit Test Control Register* in Table 2.6. When the PRBS generator is enabled using the TXxPRBSEN pin, the value of this register is overridden and the PRBS-7 test pattern is selected.

2. Reset the PRBS generator using the appropriate control input.

   For the HSS EX10 core, this is accomplished by asserting the *PRBS Reset* bit in the *Transmit Test Control Register* defined in Table 2.6, or the TXxPRBSRST pin defined in Table 2.2.

3. Enable the PRBS generator using the appropriate control input.

   For the HSS EX10 core, this is accomplished by asserting the *Test Pattern Generator Enable* bit in the *Transmit Test Control Register* defined in Table 2.6, or the TXxPRBSEN pin defined in Table 2.2.

   The PRBS generator should now be transmitting the selected data pattern.

Step Three: Program the PRBS checker in the HSS receiver as follows:

1. Set control inputs to select the desired PRBS pattern.

   For the HSS EX10 core, this is selected by setting the *Test Pattern Select* bits of the *Receive Test Control Register* in Table 2.7. When the PRBS checker is enabled using the RXxPRBSEN pin, the value of this register is overridden and the PRBS-7 test pattern is selected.

2. Reset the PRBS checker using the appropriate control input.

   For the HSS EX10 core, this is accomplished by asserting the *PRBS Reset* bit in the *Receive Test Control Register* defined in Table 2.7, or the RXxPRBSRST pin defined in Table 2.3.

3. Enable the PRBS checker using the appropriate control input.

   For the HSS EX10 core, this is accomplished by asserting the *PRBS Check Enable* bit in the *Receive Test Control Register* defined in Table 2.7, or the RXxPRBSEN pin defined in Table 2.3.

   The PRBS checker should now be monitoring the incoming serial data and attempting to synchronize to the selected data pattern.

Step Four: The PRBS checker initializes its LFSR as described previously to *synchronize* to the data pattern being received. Once this process is complete, the PRBS checker asserts the *PRBS Sync* status and begins to check data.

If *PRBS Sync* does not go active within a reasonable period of time, it is indicative that either the serial data is not reaching the receiver, or that the serial data contains too many bit errors for the PRBS checker to synchronize to the pattern. It may also indicate the PRBS generator and PRBS checker are not set for the same data pattern or baud rate. (Although if the baud rates are multiples of each other, some PRBS patterns match anyway!)

For the HSS EX10 core, *PRBS Sync* status is reported in the *RXxPRBSSYNC* bit of the *Receive Test Control Register* defined in Table 2.7, and on the RXxPRBSSYNC output pin defined in Table 2.3.

Step Five: Once *PRBS Sync* is asserted, the PRBS checker is checking the serial data. Any errors cause *PRBS Error* to be asserted. This signal may be a level or a pulse, depending on the implementation.

For the HSS EX10 core, PRBS Error status is reported in the *RXxPRBSERR* bit of the *Receive Test Control Register* defined in Table 2.7, and on the RXxPRBSERR output pin defined in Table 2.3.

Step Six: To terminate the PRBS sequence, both the generator and checker must be disabled using the corresponding *Enable* signals. Note that if the generator is disabled first, the PRBS pattern is interrupted and the PRBS checker will flag the error.

7.3 Logic Built-In-Self-Test (LBIST)

Another test feature which is frequently desirable on chips is the ability to perform an in-circuit structural self-test of the chip.

Functional test sequences exercise the chip through a sequence which is similar to the intended operation of the chip, and checks that the chip behaves consistently with the expected functional description. While it is possible to design functional BIST into a chip, this may require extensive design effort and may exercise only a fraction of the logic in the chip.

Structural test sequences differ in that they exercise the logic gates and flip-flops in the chip, and ensure that gates behave with their intended logic function. No attempt is made to construct the sequence such that it mimics actual operational function of the chip. Unlike functional BIST, structural BIST can be implemented using a more generalized approach requiring less effort, and usually results in more of the chip logic being exercised.

Self-testing of the high-speed analog circuits in HSS cores is generally performed using the PRBS generator and checker functions described previously. However, HSS cores also contain lower-speed digital logic, and it is generally desirable to include those portions of the HSS cores within the logic domains being tested by a chip-level Logic BIST (LBIST) controller.

7.3.1 LBIST Architecture

Structural testing of chips require that all flip-flops on a chip be connected into *scan chains*. Each flip-flop on the chip has two data inputs: one input pin which is the functional input data to the flip-flop, and one scan data input pin which loads data into the flip-flop when the chip is in a *scan mode*. The functional input pins of the flip-flops are connected as dictated by the chip design. The scan data input pins are connected such that the flip-flops of the chip form an arbitrary number of shift registers which can be serially scanned to load or read the current state of the chip.

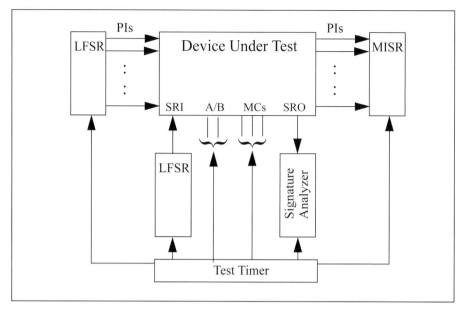

Fig. 7.12 Logic Built-In-Self-Test (LBIST) general architecture

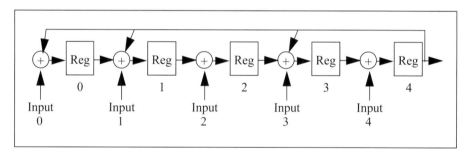

Fig. 7.13 Multiple Input Shift Register (MISR)

An at-speed LBIST controller repeatedly executes the following sequence of steps:

1. Set the chip logic to *scan mode*, and load all of the scan chains on the chip with pseudo-random data.

2. Override chip inputs and assert pseudo-random values for chip inputs to internal chip logic.

3. Deassert *scan mode*, and generate one functional clock cycle to the chip logic. The state of all flip-flops on the chip is updated based on the logic function of the combinatorial logic feeding each flip-flop.

4. Capture the logic levels of chip outputs and compare to expected values.

5. Set the chip logic to *scan mode*, and read all of the scan chains on the chip, comparing the scan chain contents to expected values.

The above sequence can be optimized by overlapping some steps. For example, reading the scan chains in step 5 can overlap loading the scan chains for step 1 of the next test iteration.

The general architecture shown in Fig. 7.12 is typically used for most LBIST implementations. The *Test Timer* in this figure is a state machine sequencer which asserts control signals to execute the various steps of the above sequence. This block directly controls all clocks and test mode control signals on the chip while LBIST is active.

Generating pseudo-random stimulus for LBIST is readily accomplished using LFSR circuits similar to those used for designing PRBS generators and checkers. As shown in Fig. 7.12, LFSRs are typically used both for generating pseudo-random values for chip inputs, as well as for loading scan chains through the *Scan Register Input* (SRI).

Checking chip outputs and analyzing the *Scan Register Output* (SRO) are typically *not* performed by directly checking the data. The expected results are dependent on the chip design, and direct checking would require an arbitrarily large amount of expected data to be coded in the chip.

Rather than do this, data are checked using *Multiple-Input-Shift-Register* (MISR) circuits. A MISR performs parallel capture of signature test data from a circuit having multiple output pins. Fig. 7.13 shows an example of a MISR. Similar to an LFSR, the MISR implements feedback taps according to a polynomial. In addition, various signals to be checked are XOR'd with the current state of the MISR at each stage. In an LBIST implementation, the various chip output and Scan Register Output signals that are to be checked are connected as inputs to one or more MISR circuits.

The LBIST controller executes an arbitrarily large number of test cycles, collecting results using MISR circuits as part of each test cycle. After a preset number of test cycles, the state of each MISR corresponds to a unique *signature* value. Any logic which is not behaving correctly will corrupt this signature such that the MISR does not end up in the correct state. (It is possible for multiple logic errors to cancel each other such that the MISR signature is correct, but this is extremely unlikely.) Therefore, the LBIST controller only needs to check that the MSIR signatures match their expected values, and does not need to check results after every test cycle.

7.3.2 LBIST Considerations for HSS Cores

During the LBIST test sequence, the LBIST controller must load and read the scan chains of the HSS core, and must be able to generate single clock cycles to update the state of flip-flops within the core. However, HSS cores generally contain PLL circuits which generate clocks internal to the core. To support LBIST, the core design must permit these internal clock sources to be disabled, and permit external control of these clocks for test purposes.

Another consideration is that analog circuits within the HSS core are generally not in a known state during LBIST. If the logic value of outputs of these analog circuits cannot be predicted, and if these unknown states affect the values captured in flip-flops which are included in the logic being tested by LBIST, then the resulting MISR signatures cannot be uniquely predicted. To support LBIST, it is necessary for the core design to *fence* analog circuit outputs such that these signals are forced to known values where they are used in the digital logic domain. This fencing logic must force these values throughout the execution of the LBIST sequence.

Assuming the HSS core design includes the appropriate test control of internal clocks and the appropriate fencing logic, then the digital logic of the HSS core can be included as part of the chip logic being tested by LBIST. However, the LBIST designer still needs to consider the speed at which this logic can be exercised. The test clock inputs to the HSS core may not support the same clock frequencies at which the high-speed internally generated clocks would operate. It is desirable for LBIST to be executed at the highest possible frequency; however, this highest possible frequency is generally gated by the slowest path within the clock domain being tested. Skew and loading levels on the clock distribution for the test clocks may further affect the frequency of LBIST operation. The LBIST designer needs to account for any frequency limitations relevant to LBIST execution that are an inherent part of the HSS core implementation.

7.4 Manufacturing Test

HSS cores are tested during manufacturing test using a combination of stuck-fault and at-speed digital test patterns, LBIST, and parametric measuring techniques. This combination of test techniques is needed to adequately test both the digital logic and analog logic present in the HSS cores. Testing falls in two categories: tests that are applied globally to the chip logic, including any HSS cores, and tests that are HSS-specific. Fig. 7.14 describes a typical manufacturing test flow.

7.4.1 Chip Level Test

Chip level manufacturing tests are performed globally on the chip. Since logic in the HSS cores is part of the logic on the chip, chip level tests also test the HSS logic. HSS cores must be designed to support these tests. As shown in Fig. 7.14, the test sequence starts by performing a short, basic test to ensure scan chains are intact. If the scan chains are intact, then subsequent tests can be executed.

7.4.1.1 Scan Test

Scan test, also called *DC Stuck-Fault Testing*, is a structural test of chip logic to verify that logic gates perform their intended functions and no stuck-at-0 or stuck-at-1 faults exist. Many types of manufacturing defects can be detected by this type of testing.

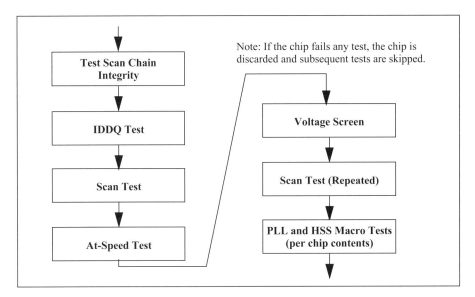

Fig. 7.14 Typical manufacturing test flow

To support scan test, all flip-flops on the chip are connected into *scan chains*. Scan chains also provide support for LBIST, and were discussed previously in this chapter. The testing concepts which apply to scan test and LBIST are similar; the difference is that scan test is applied using external tester equipment during the chip manufacturing process (either when testing chips on the wafer or when testing assembled modules), while LBIST is controlled by on-chip logic.

Scan test repeatedly executes the following sequence of steps.

1. Set the chip logic to *scan mode*, and load all of the scan chains on the chip with the desired data pattern.
2. Drive chip inputs with the desired data pattern.
3. Deassert *scan mode*, and generate one clock cycle to the chip logic. This causes the state of all flip-flops on the chip to be updated based on the logic function of the combinatorial logic feeding each flip-flop.
4. Capture the logic levels of chip outputs and compare to expected values.
5. Set the chip logic to *scan mode*, and read all of the scan chains on the chip, comparing the scan chain contents to expected values. As was the case for LBIST, this step can be overlapped with loading the scan chains for the next test iteration.

Note that in the prior discussion the data patterns for LBIST were generated using LFSRs since such generation needed to be performed on-chip. Data patterns for general scan test do not have the same restrictions. Arbitrary patterns may be used to target testing of specific circuits and minimize the number of required test patterns.

The clock frequency used in step 3 is very slow compared to normal operating frequencies of the chip. Some scan test architectures use functional clocks to execute step 3, while some scan test architectures use separate test clocks to launch and capture the functional data. In either case, traditional scan testing generally does not execute the launch/capture step at operational clock frequencies. Such testing is intended to globally target chip logic which may be designed for operation at many different clock frequencies, may contain multicycle paths, etc. Software which generates traditional scan test patterns does not consider timing. To ensure the data patterns are valid, scan test is usually applied using clock frequencies much lower than the expected operational frequencies of the chip.

7.4.1.2 At-Speed Structural Test

While DC stuck-fault testing has historically proven extremely useful for capturing manufacturing defects, *transition faults* are a prevalent failure mechanism for deep submicron chip technologies, and therefore DC stuck-fault testing is insufficient. Transition faults are manufacturing defects which result in abnormally large circuit propagation delays. DC stuck-fault testing is not a sufficient test method because the launch/capture cycle is too slow to capture the improper behavior of the circuit.

Various schemes exist for executing scan test with at-speed launch/capture cycles to test for transition faults. Such schemes rely on software that is aware of the expected circuit timing when generating the at-speed test patterns. At-speed test pattern generation software is significantly more complex than software used for the generation of DC stuck-fault test patterns. Each test pattern can only target a single clock domain (or clock domains running at the same clock frequency), and multicycle logic paths must be excluded from the test. Clock gating logic can complicate test generation. As a result, the at-speed test pattern set generally requires more test patterns and has lower coverage than a DC stuck-fault test pattern set for a given chip.

For the above reasons, both sets of patterns are used during the manufacturing test process. Traditional scan test is applied first using the shorter and more complete DC stuck-fault test pattern set. If failures occur, no further testing is required and test execution time is thereby optimized. If no defects are found with DC stuck-fault testing, then the at-speed test pattern set is applied.

7.4.1.3 Voltage Screen

Assuming a chip is initially operational and free of defects, circuits are still subject to deterioration over time which eventually causes the chip to cease to operate. Chips have the highest probability of failing either during the first few hours/days/weeks of operation (*early lifetime* failures) or beyond their expected lifetime after years of operation (*end of life* failures). It is desirable for chip manufacturing tests to incorporate methods of ensuring chips subject to early lifetime failures are not used to build systems.

One method of catching early lifetime failures would be to power on and operate the chip for several weeks prior to shipping the chip. It should be obvious this is not a practical approach since it would stall the manufacturing process and utilize expensive test equipment for long periods of time. A better approach is to apply stresses that accelerate the deterioration of the chip so that chips subject to early lifetime failures fail within seconds rather than weeks.

Voltage Screen testing is one method of doing this. Voltage Screen testing is a reliability stress test performed on digital logic of the chip. The test sequence raises the power supply voltage well beyond normal operating conditions (typically to a voltage two times higher than the nominal power supply voltage) to cause latent early lifetime failure defects to become hard failures at the tester. The voltage of the chip is elevated for some period of time, after which scan test is repeated to determine if new stuck-at faults exist.

7.4.1.4 IDDQ or Leakage Test

Leakage testing, also called *IDDQ testing*, has historically proven to be a useful screening test to find various classes of manufacturing defects. The IDDQ acronym derives from the historical association of such testing with quiescent current. However, the test actually measures leakage current.

As is discussed in Chap. 9, circuit power dissipation includes AC (active) power, DC (leakage) power, and DC quiescent power components. Support for leakage testing generally requires a control signal or other method of preconditioning the chip to disable any DC current paths, and thereby eliminate (or at least minimize) any DC quiescent power components. Many analog circuits, including those in HSS cores, have DC current paths which must be disabled, and a test control signal is usually provided to disable these circuits. Leakage testing is performed with the chip properly preconditioned, and with no signal switching occurring. In the absence of signal switching (zero activity factor), the AC active power component is also insignificant. The result is that the DC leakage power is the only significant contributor to current being drawn from the power supply.

Given these conditions, the leakage current being drawn from the power supply is measured and compared to an expected limit. Excessive leakage current is an indicator that the chip contains significant manufacturing defects. Because leakage testing does not take much time to perform, it is an efficient screening method. It can be used early in the test sequence as a quick screen test; if the test fails then no further testing is required and test execution time is thereby optimized.

7.4.1.5 PLL Macro Tests

Although PLL tests are not really chip level tests, HSS cores contain PLLs, and basic PLL tests are often conducted on all PLLs on the chip in parallel. In addition, as is discussed in Chap. 6, clock distribution networks for reference clocks to HSS cores are often sourced by intermediate frequency PLLs. It is desirable to test the interaction between the PLL driving the reference clock and the PLL in the HSS core. For these reasons the PLL macro tests have been included as part of this discussion of chip level tests.

The following PLL tests are examples of typical PLL tests which are applied in common to all PLLs on the chip. These tests are executed prior to any other HSS-specific tests, since subsequent tests are dependent on defect-free PLLs.

PLL lock/voltage regulator test. This is a test to check the functionality of the intermediate frequency PLL together with the embedded HSS PLL and verify that they have achieved lock. If the PLL slice of the HSS core contains a voltage regulator, the output voltage of this regulator is measured.

Filter Capacitor Leakage Test. This test measures the quiescent leakage current through the PLL filter capacitors to ensure they are defect-free. All PLLs on the chip are usually tested at the same time.

7.4.2 HSS Macro Test

Once chip level tests have completed, including any PLL macro tests, specialized macro tests are applied to other cores on the chip as needed. HSS cores generally require a suite of specialized tests which utilize the PRBS functionality described previously in this chapter. In this section, several typical tests are described which may be used to test an HSS core similar to the HSS EX10 core as described in Chap. 2. An actual HSS design may require some or all of these tests, and may require other tests depending on the functionality.

The simplest possible approach for HSS macro tests would be based on the assumption that the test equipment used to test the chip and module has the ability to contact any pin on the chip, and has the ability to drive and receive the high-speed serial data signals at the maximum supported baud rate. However, this would utilize state-of-the-art high-speed test equipment which would correspondingly be very expensive. Reduced pin count test approaches used by some ASIC chip manufacturers do not contact the high-speed pins on the chip, and would not support driving and receiving signals at the full baud rate even if they did. By not having as many tester channels and not requiring high-speed testers, manufacturing test costs are reduced. Such test approaches rely on loopback tests and other indirect test techniques to verify the operation of the HSS core. Reduced pin count test methods are assumed for the HSS macro tests described below.

7.4.2.1 Transmitter to Receiver Wrapback Test

The HSS Tx/Rx Wrap test is an "at speed" self-test which checks the functionality of the transmit to receive wrap path by generating pseudo-random data and applying it to the first stage of the receive logic. On the HSS EX10, this test uses the built-in PRBS capabilities and the wrap path. The test sequence generates a PRBS data pattern, and wraps the output of the transmitter to the serial input of the receiver. The PRBS checker in the receiver checks the PRBS sequence for errors. The state of the PRBS sync and error status signals upon completion of the test determines whether the test passed or failed.

HSS simplex cores may be designed to support a manufacturing wrap test. For example, a receiver core can additionally contain a PRBS generator and a test driver that injects PRBS data onto the serial data path of the receiver. Likewise, a transmitter can additionally contain a test receiver which receives the transmitter output signal and connects to a PRBS checker. Simple designs are possible for these test circuits since they only need to operate in a wrap mode where the wrap path can be assumed to be low-loss. Macro test methods for simplex cores with such test circuits are therefore similar to the corresponding full-duplex cores.

7.4.2.2 Receiver Sensitivity Test

The purpose of this test is to ensure the receiver input stage is correctly differentiating between a "0" and a "1" level. This test is not an at-speed test; the signal levels are static for each iteration of the test.

Differential voltage levels are applied to the HSS receiver serial inputs and the output of the slicer logic of the HSS receiver is monitored to determine whether the correct logic state is being received. (If the receiver contains a DFE, then the slicer logic is part of the DFE circuit.) The test sequence generally drives a nominal differential voltage level, and repeats the test for both a "0" and "1" logic level.

The output of the slicer logic (or corresponding circuit in the DFE) is an internal point within the HSS receiver. This point may be accessed through a test port on the core, but is more likely accessed by scanning out state values of the core to determine the state values of flip-flops which latch the slicer output. Most test pattern generation software has this capability.

7.4.2.3 Receiver Gain Control Test

Most HSS receivers include an Automatic Gain Control (AGC) circuit which is used to amplify the signal as needed so that the signal to subsequent circuits falls into a more limited range. The purpose of the receiver gain control test is to test that the AGC is properly amplifying the signal. To avoid requiring test equipment capability to drive precision signals, built-in circuits are utilized in the case of the HSS EX10 core. The AGC offset current digital to analog converter (IDAC) in the receiver is programmed to apply a differential input voltage level at the *RXxIP* and *RXxIN* chip pads.

In the HSS EX10 receiver design, the resulting differential level passes through gain stages in both the AGC and DFE and arrives at a summing node in the DFE. At this summing node a nulling voltage (of opposite polarity to the AGC source signal) is added to the AGC sourced voltage. In one test, the nulling voltage is of a lesser magnitude than the AGC produced level, resulting in an overall positive summed voltage which is detected as a logical "1" downstream from the DFE. In the second test, the summed voltage has a negative polarity and produces a logical "0" result. Functional clocks shift the result through the analog DFE logic to the digital CDR logic where a scan operation can observe the test results.

7.4.2.4 Receiver Offset IDAC Test

In this test sequence, the AGC Offset IDAC is itself tested. The IDAC is programmed to generate precise differential output voltage levels at the *RXxIP* and *RXxIN* differential inputs to the HSS receiver. The termination resistor that attaches to these nets remains engaged, while most other functions that interact with the *RXxIP/N* nets are disabled. The test is run with the Receiver set for DC mode and with any common mode voltage bias circuits powered down.

When the AGC Offset IDAC sinks current it draws it through the *RXxIP/N* differential inputs from the tester voltage source. The tester performs a precise analog current measurement at the *RXxIP* and *RXxIN* pins, and compares these measurements to expected ranges for each value of the AGC Offset IDAC.

7.4.2.5 Signal Detect Test

Voltage levels are applied simultaneously at all HSS receiver differential serial inputs and the signal detect logic state is scanned out and compared to expected values. Multiple test passes are performed with various signal amplitude levels. The common mode voltage (V_{cm}) of the incoming signal is also varied.

Table 2.18 described the signal detect thresholds for "good" and "bad" signals for various settings of the *Signal Detect Control Register* for the HSS EX10 core. Signal amplitudes above the "good" threshold should always be detected; signal amplitudes below the "bad" threshold should never be detected.

7.4.2.6 Receiver DC Terminating Resistance

This is a parametric test that ensures that the terminating resistor is within the specification and associated registers are wired and working correctly in the HSS receiver.

The test is performed by using manufacturing test equipment to measure the termination impedance as seen between the *RXxIP* and *RXxIN* pins.

7.4.2.7 Receiver Common Mode Voltage Bias Test

This is a test for the receiver Common Mode Bias circuits to insure defect-free operation in both DC and AC coupled modes.

The test is performed by setting the receiver for each supported coupling mode, and measuring the bias voltage on the *RXxIP* and *RXxIN* pins when these pins are not being driven by the test equipment.

7.4.2.8 JTAG Receiver Test

The purpose of the JTAG receiver test is to verify the input from the *RXxIP* and *RXxIN* pins in DC and AC modes for both a "0 to 1" and a "1 to 0" pattern. Although this test sequence requires data transitions to be driven onto the receiver inputs, these data transitions are at data rates typical for JTAG 1149.6 signals (typically no more than 1 MHz). Test equipment capable of driving higher baud rates is not required for this test.

For this test sequence, the input differential voltage applied on the *RXxIP* and *RXxIN* pins starts at the maximum allowed input voltage and then is reduced to the minimum level that will sustain the output value. Test results for each signal amplitude are determined through observation of the *RXxBSOUT* and *RXxACJZTP/N* outputs as described in Sect. 2.3.6. The test is repeated for both "0 to 1" and "1 to 0" transitions.

7.4.2.9 Transmitter Drive Strength Test

The driver used in the HSS EX10 has an FFE circuit with three taps and seven segments. Each tap contains one, two, or four segments. The drive current of each segment is controlled by values programmed into the corresponding FFE segment weighting coefficient values programmed in the *Transmit TapX Coefficient Registers,* and the value programmed into the *Transmit Power Register*. These registers are described in Table 2.6 for the HSS EX10 core. The values in these registers are converted to an analog control voltage which scales the output drive current of each driver segment.

The voltage levels for the serial driver differential output, *TXxOP* and *TXxON*, are measured for various combinations of FFE tap coefficient values and transmit power values. Too many combinations exist to expect exhaustive testing; the test engineer for the HSS core must optimize the cases to be tested with the goal of maximizing circuit coverage of the test with as few test iterations as possible. For the HSS EX10 core, a reasonable approach is to test each combination of FFE tap coefficient values using the maximum setting for the transmit power, and then to test various additional transmit power levels with arbitrary FFE tap coefficient values.

For each test iteration, the test sequence must scan the chip to set the appropriate values for tap coefficients, transmitter power, and logic state. The voltage levels on the *TXxOP* and *TXxON* pins are then measured and compared to expected values. Note that these are static voltage measurements; data does not transition during this test.

7.4.2.10 Transmitter Terminating Resistance Test

This is a parametric test that ensures that the terminating resistor is within the specification and associated registers are wired and working correctly in the HSS driver.

The test is performed by using manufacturing test equipment to measure the termination impedance as seen between the *TXxOP* and *TXxON* pins.

7.5 Characterization and Qualification Testing

Characterization testing is performed to ensure that the HSS transmitter and receiver perform both logically and electrically to specific requirements for performance, amplitude, and jitter across the full spectrum of process, voltage, and temperature (PVT) conditions. Qualification typically occurs after the HSS design has been fabricated, tested, characterized, and reliability-stressed using multiple lots of hardware including fast, nominal, and slow chips. Characterization testing is typically accomplished using an HSS test chip which contains multiple copies of transmitter and receiver links, driven by an IF PLL and/or differential clock receiver, and including FIFOs between the receivers and transmitters. The test chips are fabricated with intentional process variation for NFET and PFET threshold voltage (V_t) and effective transistor channel length (L_{eff}). For HSS characterization, these HSS test chips are typically mounted on an Evaluation Board which can be connected to clock and pattern generation and recognition equipment, cables and/or backplane channels, and whose control registers can be accessed through a parallel data port and programmed via software on a PC. An example of a characterization setup is shown in Fig. 7.15.

As discussed in Sect. 4.1.2, serial data interface standards specify that the serial data must meet requirements for amplitude, eye width, and jitter to be in compliance and to ensure interoperability between HSS transmitters and receivers designed by different vendors. In addition, the characterization lab testing of an HSS test chip supports model to hardware correlation between the Spice, behavioral, and S-parameter simulation models used for signal integrity analysis.

The following subsections discuss various topics regarding characterization testing for HSS transmitters and receivers, including examples of the types of testing that is typically performed.

7.5.1 Transmitter Tests

7.5.1.1 Test Conditions

Fig. 7.16 illustrates two possible measurement points for characterization of transmitter devices. The transmitter devices to be tested are identified as the *device under test* (DUT) blocks.

One method of performing such measurements would be to connect test equipment directly to the output of a transmitter. The test equipment provides an ideal load for the transmitter, with no intervening channel to distort the signal. Some protocol standards do specify the transmitter output pin as a compliance point, and in such cases compliance measurements are performed directly at the transmitter output.

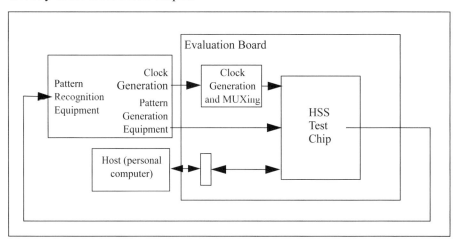

Fig. 7.15 HSS characterization and qualification lab setup

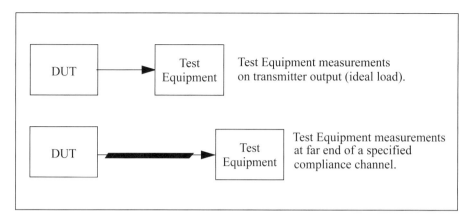

Fig. 7.16 Lab characterization conditions for the transmitter

Note, however, that in a real system the channel does distort the signal. This distortion, and associated signal integrity topics, are discussed in Chap. 8. HSS transmitters often contain an FFE circuit which is capable of altering the transmitted signal. The FFE emphasizes the transmitted signal to cancel the distortion effects of the channel, and thereby provide good signal quality at the receiver input. Measurements performed at the transmitter output must be performed using defined FFE settings, and provide no guidance as to the signal quality at the receiver.

An alternative used by many protocol standards is to define the characteristics of a *compliance channel*, and to perform transmitter characterization by measuring the transmitted signal at the far-end of this channel. This is also shown in Fig. 7.16, with the DUT driving a channel, and test equipment connected to the far-end of the channel. In this case, the FFE is expected to be set based on the channel characteristics to obtain the best possible signal quality at the test point.

HSS design teams perform characterization testing of their transmitter devices using both methods. Specifications for jitter and amplitude of the transmitter output require measurements directly at this point. Characterization of the FFE and model to hardware correlation of hardware to various signal integrity models requires measurements at the far-end of known channels. Furthermore, compliance testing for various protocol standards requires measurements be taken consistent with the test methods and conditions defined by the standards document.

7.5.1.2 Driver Transmit Jitter Output

This test quantifies the transmitter jitter output while transmitting several different serial data patterns. Link protocols generally specify the data patterns to be used for jitter compliance measurements. For example, the HSS EX10 transmitter described in Chap. 2 would be characterized with the PRBS-7 and PRBS-10 patterns, as well as the "0101" clock pattern. Clock patterns do not have pattern dependent jitter, and therefore this pattern is useful for determining the *random jitter* (RJ); the PRBS patterns are used to characterize *total jitter* (TJ). *Deterministic jitter* (DJ) can be determined mathematically once TJ and RJ are known. Jitter is discussed further in Sect. 8.2.

Jitter output testing uses a test setup similar to Fig. 7.17. Measurements are made with and without power supply noise to ensure performance in an integrated chip environment. Deterministic jitter (DJ), RJ, duty cycle distortion (DCD) and TJ at a 10^{-12} bit error rate are compared to the link compliance specifications. Software techniques are used to decompose jitter into deterministic and random components, however these techniques sometimes have trouble distinguishing between low levels of periodic and random jitter.

7.5.1.3 Driver FFE Characterization

The transmitter FFE is used to reduce intersymbol interference. Fig. 7.18 and 7.19 show the effect of the various FFE taps and their polarity. The figures use a repeating 00001111 pattern to clearly show how the tap points affect the bits before and after the transition. The FFE capabilities must be correlated to signal integrity analysis software, such as the HSSCDR simulation tool discussed in Sect. 8.4.2. HSSCDR is used to analyze the S-parameters for backplane examples, and determine proper FFE tap weights needed to equalize these backplanes. These values are then used to perform hardware measurements, and the simulated eye diagrams are correlated with the measured eye diagrams.

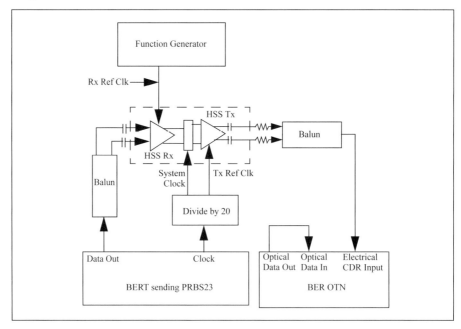

Fig. 7.17 Lab setup example for jitter generation

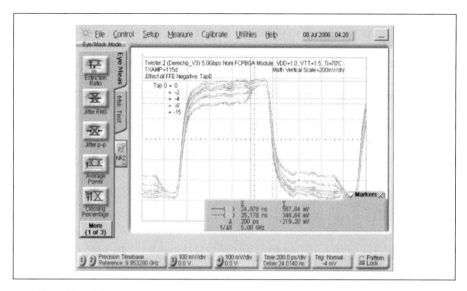

Fig. 7.18 Tap 0 (precursor) negative tap 0

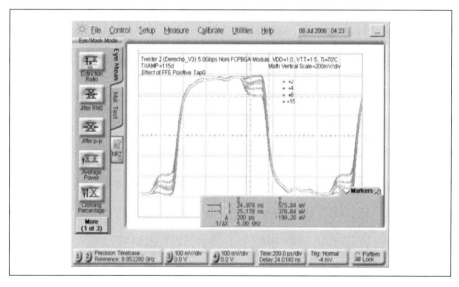

Fig. 7.19 Tap 0 (precursor) positive tap 0

7.5.1.4 Differential Amplitude

The differential amplitude of the HSS EX10 transmitter is measured with various test patterns and under various test conditions. The minimum differential amplitude of the measured data eye envelope that is observed defines the amplitude limits of the *inner eye*; the maximum differential amplitude of the measured data eye envelope defines the amplitude limits of the *outer eye.*

PRBS test patterns are used to determine overall data eye envelope limits. The frequency at which the transmitted signal switches, as well as the rise and fall times of the signal, affects the amplitude of the signal swing. If measurements are being performed through a compliance channel, then signal amplitude is also affected by the frequency response and the loss characteristics of the channel. Such effects are pattern dependent, and randomized data with appropriate spectral content is necessary to determine the amplitude limits of the data eye envelope.

Low frequency tone patterns also produce useful measurements. Such patterns are used to determine the low frequency amplitude of the driver as well as rise and fall times. For the HSS EX10 transmitter, which supports 10 Gbps baud rates, a 531.25 MHz tone is appropriate for performing these measurements.

Using a PRBS test pattern, and a test setup where measurements are performed at the end of a compliance channel, the resulting measurements do give some insight into the signal amplitudes and jitter which must be tolerated by the receiver device. However, these measurements can be misleading. Often these measurements are performed with test equipment providing an ideal impedance match; specifications in protocol standards may even require an ideal impedance match (with minimal tolerance) as part of the test conditions for measurements. In a real system, the impedance tolerance for the receiver device is generally less stringent than the specified test conditions. This creates reflections in the real system which can result in the receiver device seeing a signal amplitude that is greater than the amplitude range specified in the protocol standard. Such conditions are not a violation of the standard, and the receiver device must tolerate these amplitudes.

7.5.1.5 Eye Mask Measurements

Many protocols use eye masks to specify the amplitude and total jitter limits of the transmitted signal. Eye masks may be specified at either the transmitter output, or at the far end of a specified compliance channel.

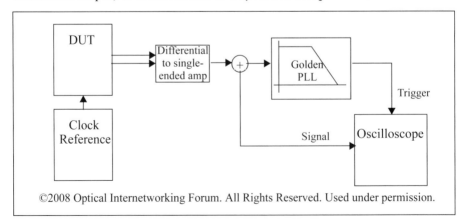

Fig. 7.20 Eye mask measurement with a Golden PLL

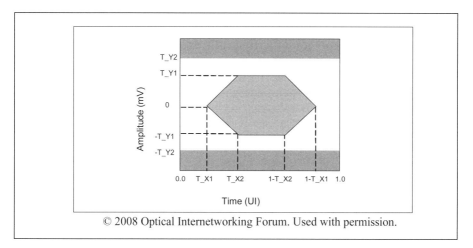

© 2008 Optical Internetworking Forum. Used with permission.

Fig. 7.21 Example of a transmitter eye mask

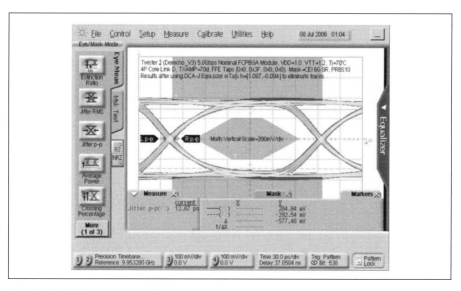

Fig. 7.22 Differential amplitude measurements

Fig. 7.20 illustrates a lab setup using an oscilloscope which may be used to perform eye mask measurements [3]. The transmitter to be tested is connected to a conversion amplifier which converts the differential signal to the corresponding single-ended signal. (In some cases this conversion amplifier may be integrated with the oscilloscope.) The output of the conversion amplifier is split and connected to the oscilloscope data input, and through a Golden PLL to the trigger input of the oscilloscope. The Golden PLL must have defined bandwidth characteristics, and is used to obtain a stable low-jitter reference clock for triggering the oscilloscope.

An example of an eye mask specification is shown in Fig. 7.21. This mask specifies both maximum and minimum limits for differential amplitude and total jitter of the transmitter output. The shape of the inner eye mask also results in an implied specification for maximum rise/fall time of the signal. When performing compliance testing for a given protocol, the measured signal eye must remain within the specified eye mask envelope. Fig. 7.22 illustrates a measured eye waveform with an eye mask also displayed. The measured eye is well within the eye mask boundaries shown on the scope trace.

Signal excursions beyond the boundaries of the eye mask specification are not necessarily an indication of noncompliance. Such excursions may occur as long as the frequency of such violations does not exceed the specified BER. The limits and shape of the eye mask waveform is a function of the target BER, and assumes continuous data collection by the test equipment. If a sampling oscilloscope is used, the eye mask must be adjusted using mathematical analysis to obtain correct results [3].

The HSS EX10 core described in Chap. 2 included a *Digital Eye* circuit which allowed certain signal quality measurements to be performed in an operational environment without test equipment. This function was described in Sect. 2.3.11. The minimum differential signal amplitude and eye width after equalization can be measured using this function.

When using test equipment, measurements are made only on the top surface of the circuit board with microstrip links to eliminate the effects of board vias (which vary significantly depending on the board design). Measurements typically include the affect of an SMP (subminiature push-on) connector, 1.6-in. board traces, 0201 decoupling capacitors, as well as the package and IC circuitry.

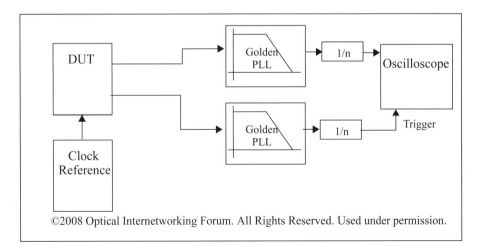

Fig. 7.23 Relative wander measurement

7.5.1.6 Multilane Synchronization

Protocols which bit-interleave or byte-stripe data across multiple serial data lanes generally specify limitations on the maximum phase difference between the various lanes to constrain the deskew range which must be handled at the receiver. In Sect. 4.1.2.5 the terms *wander* and *skew* were defined to describe variation in the arrival time of the signals of a multilane interface. *Skew* is the constant component of the phase difference between two lanes, while *wander* may vary over time due to changes in temperature, voltage, or noise. Since timing variations for both skew and wander can be tracked by the CDR in the receiver, these parameters are only of significance for multilane interfaces.

Fig. 7.23 illustrates a lab setup for measuring skew and wander [3]. Two transmit lanes of the device under test are connected to Golden PLLs which filter out the high-frequency jitter on the respective lanes. One lane is connected to the data input of the oscilloscope, while the other lane is used as a trigger. The constant portion of the phase difference between the two channels is the skew, while the time varying portion is the wander.

7.5.2 Receiver Tests

7.5.2.1 Jitter Tolerance

Jitter tolerance testing quantifies the ability of the receiver to receive incoming data in the presence of jitter. Receiver jitter tolerance is measured for both synchronous and plesiosynchronous reference clocks. Receiver jitter tolerance compliance can be tested for both optical and backplane applications, as determined by the applicable standard.

Optical mode testing of the HSS EX10 core is accomplished using a bit error rate tester (BERT) as a data source. The BERT is capable of injecting controlled amounts of jitter. Optical testing is performed using jitter that is predominantly nonequalizable. This jitter is generated by the BERT and includes random jitter (RJ), bounded uncorrelated jitter (BUJ), and sinusoidal jitter (SJ). The testing also includes a compliance channel to create intersymbol interference which is partially equalized using the DFE in the receiver. Jitter is applied over a range of frequencies and amplitudes to ascertain the maximum jitter tolerance of the HSS receiver. The jitter tolerance limit is determined by the point where one or more errors is detected in the transmitted pattern within a specified time period.

Backplane mode testing of the HSS EX10 is accomplished by inserting various backplanes between the HSS EX10 transmitter and receiver. This testing demonstrates the types of links that can be equalized when the HSS EX10 core is used at both the transmit and receive ends of the channel. Since the data source is the HSS EX10 transmitter, other types of jitter (RJ, BUJ, SJ) are determined by the characteristics of the transmitter and cannot be adjusted during this test.

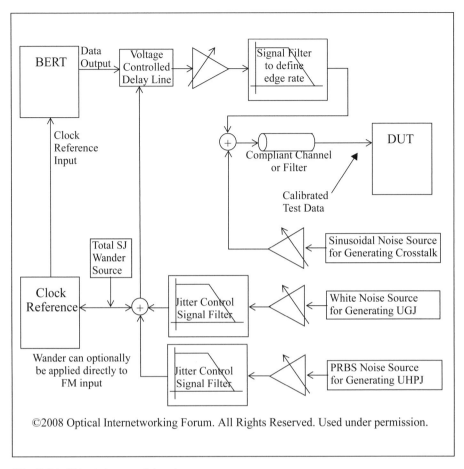

Fig. 7.24 Jitter tolerance lab setup

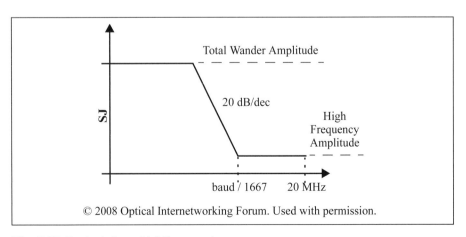

Fig. 7.25 Typical sinusoidal jitter mask

Fig. 7.24 illustrates a lab setup for jitter tolerance testing. The Noise Source and PRBS Generator are sources of RJ and BUJ, and modulate the voltage controlled delay line to inject this jitter onto the BERT output. Sinusoidal jitter is then added by a sinusoidal noise generator and intersymbol interference is added by the compliance channel.

Levels of RJ and BUJ used in jitter tolerance testing correspond to jitter specifications for the transmitter device as defined in the protocol standard. The purpose of SJ is to provide margin in the receiver design; the amount of SJ to be injected is defined as part of the test specification. SJ may be defined in terms of a mask as a function of frequency, an example of which is shown in Fig. 7.25. The sinusoidal jitter frequency is varied as part of the test sequence, with the amplitude adjusted according to the mask.

7.5.2.2 Receiver Signal Detection Test

The *signal detection test* determines the receiver input signal amplitudes (in terms of differential voltage) that cause the signal detect status indicator to go active or inactive. The signal detect specification has two limits:

- The upper limit is the minimum input voltage level where signal detect status is guaranteed to be asserted, indicating presence of a signal.

- The lower limit is the maximum input voltage level where signal detect status is guaranteed to *not* be asserted, indicating loss of signal.

These two limits are shown in Fig. 7.26. In the range between these limits, the signal indication status is not guaranteed to have any particular value. Although it would be ideal to have the upper and lower limits be the same value, tolerances seen in real hardware circuits make this impossible. Some tolerance is required to allow for variation in the actual threshold point where the circuit switches state. The difference between the upper and the lower limits reflects this variation.

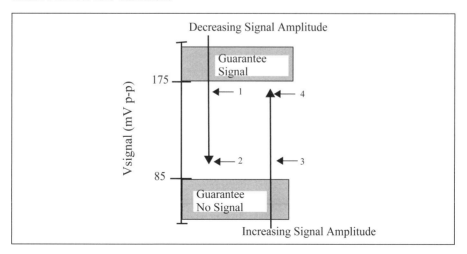

Fig. 7.26 Receiver signal detect test sequence

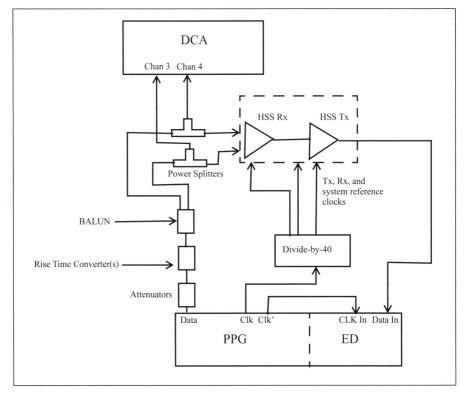

Fig. 7.27 Receiver signal detection test lab setup

Fig. 7.26 also illustrates the sequence for the signal detection test. First, the input level is set to the minimum level supported by the test setup and the input amplitude is increased until signal detect starts to toggle. This voltage is recorded as the maximum logic 0 level (point 3). The signal is incremented further until signal detect is static at logic 1. This value is recorded as the minimum logic 1 level (point 4). If the HSS signal detect circuitry has no hysteresis, then points 1 and 2 in the figure are equivalent to points 4 and 3. Otherwise, the test is repeated by setting the input level to the maximum level supported by the test setup and incrementally decreasing the signal amplitude.

Fig. 7.27 illustrates the lab setup for this test. The BERT outputs, shown at the bottom of the diagram, are attenuated and then passed through two 50-ohm power splitters. These split the signal so that the half the receiver signal is going to the DUT and the other half is observed on the scope so that amplitude can be measured.

7.5.3 General Tests

The following tests are general and apply to both transmitters and receivers.

7.5.3.1 Return Loss

As discussed in Sect. 1.4.1, the channel is an electrical transmission line composed of circuit board traces and vias, connectors and/or cables. Reflected energy due to impedance mismatches reduce the signal amplitude.

Return loss measures the returned energy from a signal launched into the package ball at either the transmitter or the receiver. Return loss is specified as the difference, in dB, between the forward and reflected logarithmic power. Return loss measurements are typically characterized for the both the HSS transmitter and receiver, and for both differential and common modes. An example of return loss measurements under various conditions is shown in Fig. 7.28.

7.5.3.2 Power Dissipation

Power dissipation is measured by turning on groups of receiver and transmitter links on an HSS test chip and monitoring the change in the supply current for various power supplies used by the HSS cores. Measurements are repeated for slow, nominal, and fast chips, across various DFE and FFE settings, and at various temperature and power supply voltage settings. The measured values are compared to the simulated power values. Power calculations are discussed in Chap. 9.

HSS implementations are often compared using a "power per link" metric. This metric typically includes the total power dissipation requirements drawn from all power supplies for one transmitter device plus one receiver device. This metric also includes the power dissipation of the PLL slice; this contribution is assumed to be amortized over all links in the core, and the per link contribution is adjusted accordingly.

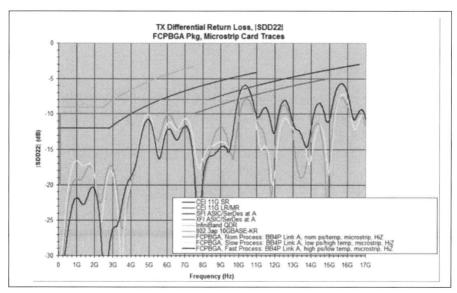

Fig. 7.28 Common mode return loss

7.6 References and Additional Reading

The following standards documents are applicable to topics in this chapter:

1. "IEEE Std 1149.1-2001 IEEE Standard Test Access Port and Boundary-Scan Architecture," Institute for Electrical and Electronic Engineers, 2001.

2. "IEEE Std 1149.6-2003 IEEE Standard for Boundary-Scan Testing of Advanced Digital Networks," Institute for Electrical and Electronic Engineers, 2003.

The following standards documents contain descriptions of jitter terminology and compliance test methods applicable to various interface standards described in Chap. 5:

3. "Common Electrical I/O (CEI) - Electrical and Jitter Interoperability agreements for 6G+ bps and 11G+ bps I/O," OIF-CEI-02.0, Optical Internetworking Forum (http:\\www.oiforum.com), Feb. 28 2005.

4. "ANSI INCITS TR-34-2004: INCITS Technical Report for Information Technology - Fibre Channel - Methodologies for Jitter and Signal Quality Specification (MJSQ)," American National Standards Institute, Inc., International Committee for Information Technology Standards, Jan. 1 2004.

The following reading is recommended for more information regarding structural testing methods:

5. "Built-in Test for VLSI, Pseudorandom Techniques," Paul H. Bardell, William H. McAnney, Jacob Savir, Wiley, 1987.

6. "An Automated, Complete, Structural Test Solution for SERDES," Stephen Sunter, Aubin Roy, J-F Cote, International Test Conference (ITC), 2004.

7.7 Exercises

1. A circuit board manufacturer requires that all chips being assembled on a circuit board must support JTAG 1149.1 or JTAG 1149.6. Explain the motivation for this.

2. Does either JTAG 1149.1 or JTAG 1149.6 perform at-speed test of signals on the circuit board? Why or why not?

3. Draw the timing diagram for a sequence of the JTAG 1149.1 TCK and TMS signals which places the JTAG tap controller in the *Shift-IR* state regardless of the initial state.

4. Draw the timing diagram for a sequence of the JTAG 1149.1 TCK and TMS signals which places the JTAG tap controller in the *Capture-DR* state assuming the tap controller is initially in the *Shift-IR* state.

5. The I/O for a given chip consists exclusively of eight HSS EX10 cores. Assuming all of the channels on these cores are used, how many JTAG 1149.1 Boundary Scan Cells must be included in the Boundary Scan Register to control and observe these I/O?

6. Assume the chip in Exercise 5 must support JTAG 1149.6. How many Boundary Scan Cells must be included in the Boundary Scan Register to comply with JTAG 1149.6?

7. What additional test coverage does JTAG 1149.6 provide which is not provided by JTAG 1149.1? Does this affect the design of the HSS transmitter and/or the receiver?

8. Figures 7.4 and 7.5 illustrate how the "complement" leg of the signal may be open and the differential signal still passes JTAG 1149.1 stuck-fault testing. Draw similar waveforms to illustrate behavior when the 'true' leg of the signal is open.

9. Explain why the *Serial Diagnostic Loopback* path in Fig. 7.7 is generally not implemented for simplex core configurations. Discuss the signal integrity implications.

10. Explain why the *Remote Line Loopback* path in Fig. 7.7 requires an elastic buffer between the HSS Rx and the HSS Tx. Discuss the implications of implementing this path for a plesiosynchronous system.

11. Design logic (Verilog or VHDL) for a 32-bit parallel PRBS generator which uses the following polynomial: $G(x) = x^7 + x^6 + 1$.

12. Design logic (Verilog or VHDL) for a 32-bit parallel PRBS checker which checks the following polynomial: $G(x) = x^7 + x^3 + 1$.

13. Design logic (Verilog or VHDL) for a 32-bit parallel PRBS generator which uses the following polynomial: $G(x) = x^{23} + x^{18} + 1$.

14. Modify the logic in exercise 13 to support generation of both PRBS23+ and PRBS23– patterns.

15. Design logic (Verilog or VHDL) for a 32-bit parallel PRBS generator which uses the following polynomial: $G(x) = x^{31} + x^{28} + 1$.

16. Write a program to generate the complete PRBS data pattern for each of the PRBS polynomials listed Table 7.1, and collect the following statistics for each of the PRBS patterns in this table:
    - Maximum run length
    - Average run length
    - Ratio of number of 1's in the pattern relative to the total number of bits
    (Run length is the number of consecutive 0's or 1's.)

17. An HSS EX10 transmitter and receiver are to be externally connected in a wrap configuration and tested using a PRBS23+ sequence. Specify a series of register write cycles (specifying register address and data) that executes this sequence.

18. In addition to various PRBS patterns, the HSS EX10 transmitter also supports data patterns which generate alternating 0's and 1's.

    (a) How are these patterns generally used?

    (b) Why is there no support on the receiver for checking these patterns?

19. An MISR circuit is connected to 32 scan outputs of a device under test, and calculates the corresponding test signature using the polynomial:

    $$C(x) = x^{31} + x^{30} + x^{26} + x^{25} + x^{24} + x^{18} + x^{15} + x^{14} + $$
    $$x^{12} + x^{11} + x^{10} + x^8 + x^6 + x^5 + x^4 + x^3 + x + 1$$

    Design logic (Verilog or VHDL) for this MISR circuit. (Note that your answer to Chap. 4 Exercise 20 or 21 may be a useful starting point.)

20. An analog block is instantiated in a digital design and must be fenced off such that LBIST may be used to test the digital design.

    (a) Draw an example of the fencing logic associated with the output of the analog block.

    (b) What fencing logic is required for inputs to the analog block?

    (c) If there is a requirement that inputs to the analog block be observable during the LBIST sequence, what logic should be added for inputs to the analog block?

21. Explain the advantage (from a test cost standpoint) of performing "Leakage Test" early in the test sequence shown in Fig. 7.14.

22. Assume an HSS core contained a significant number of analog circuits which contained DC current paths that could not be disabled for leakage test.

    (a) How would this impact the ability to perform a Leakage test for the rest of the chip?

    (b) If all of the DC current paths in the HSS core are powered from a separate analog Vdd power supply input to the chip, can you suggest a workaround that allows leakage testing of the rest of the chip?

23. The scan test and at-speed test shown in Fig. 7.14 are both structural tests while the HSS Macro Test "Transmitter to Receiver Wrapback Test" described in Sect. 7.4.2.1 is a functional test. All of these tests provide test coverage for logic in the HSS core. Contrast the differences between these tests.

24. Explain the purpose of the "Voltage Screen" test in Fig. 7.14. Is it possible for the chip to pass the first "Scan Test" and "At-Speed Test", and then fail the "Scan Test" executed after the Voltage Screen step?

25. Two test configurations are shown in Fig. 7.16 for performing characterization tests on a transmitter device. Assume an eye mask is used to specify signal characteristics of the transmitted signal.

    (a) Explain why measurements for one of these configurations must be performed without transmit equalization.

    (b) Explain why Tx equalization can be used in the other configuration.

26. If a traditional eye mask as described in Fig. 7.21 is used to specify signal characteristics, then any measurements taken at the transmitter must be performed with transmit equalization turned off. Suggest an alternative method of specifying a transmitter waveform at the output of the transmitter (not using a compliance channel) that includes the affects of transmit equalization. (Hint: IEEE 802.3 Backplane Ethernet variant 10GBASE-KR came up with one approach.)

27. Assume various test patterns are used to characterize the jitter generated by a transmitter device using the test configuration shown in Fig. 7.17.

    (a) Using a PRBS pattern, will the corresponding measured jitter be RJ, DJ, or TJ?

    (b) Do you expect significant differences in the measurement for a PRBS-7 pattern as opposed to a PRBS-31 pattern? Why or why not?

    (c) Using an alternating '00110011...'' pattern, will the corresponding measured jitter be RJ, DJ, or TJ? Why?

28. Does a test pattern exist which can be used to characterize the jitter generated by the transmitter device in Fig. 7.17 that only exhibits deterministic jitter? Why or why not?

29. The discussion of transmit equalization in Sect. 1.3.2 defined *preemphasis* and *deemphasis* classifications for FFE architectures. Which of these applies to the waveforms in Fig. 7.18, 7.19? Why?

30. Draw waveforms similar to Fig. 7.18, Fig. 7.19 for an FFE which uses an architecture that is the opposite of that in exercise 29.

31. What is the purpose of the "Golden PLL" in Fig. 7.20?

32. Assume an eye mask as described in Fig. 7.21 is used to specify signal characteristics for a transmitter device. The T_Y2 parameter corresponds to the maximum differential amplitude allowed on the transmitter output under the specified test conditions. In an actual system using a compliant transmitter device, is it possible for a receiver device to see a differential amplitude that exceeds this specification? Why or why not?

33. Fig. 7.23 illustrates a test configuration used to measure the wander and skew between two transmit data lanes of an interface. Explain why the "Golden PLL" devices are required. How would omitting the "Golden PLL" devices affect the measurement?

34. For the test configuration in Fig. 7.23:

    (a) If a single measurement is taken, does this measurement indicate the *skew* or the *wander* between the two data lanes (or some mixture)?

    (b) Suggest a test procedure to perform a series of measurements such that the value of both the *skew* and *wander* can be determined.

35. A jitter tolerance test is performed on a receiver device using the test configuration in Fig. 7.24. During the 24 hour test period, some errors are received. How many errors are allowed if the receiver device is to meet the following bit error rates?

    (a) BER $= 10^{-12}$ (b) BER $= 10^{-15}$ (c) BER $= 10^{-18}$

36. Sinusoidal Jitter is injected on the signal as part of a jitter tolerance test using the test configuration shown in Fig. 7.24.

    (a) The test configuration also injects both RJ and DJ components onto the signal. What is the purpose of injecting SJ as well?

    (b) The amount of SJ injected depends on the frequency of this jitter component, and is defined by the mask in Fig. 7.25. Why does SJ have a larger amplitude at lower frequencies?

37. A signal detection test is performed on a receiver device using the test sequence described in Fig. 7.26, and with the thresholds shown in that figure. What is the expectation for whether RXxSIGDET is asserted for each of the following signal amplitudes:

    (a) 70 mVppd   (b) 180 mVppd   (c) 100 mVppd
    (d) 175 mVppd   (e) 85 mVppd   (f) 173 mVppd

38. Given the HSS EX10 core described in Chap. 2:

    (a) Suggest a test sequence for characterizing power dissipation of the core for various equalization modes of the transmitter and receiver.

    (b) How would you convert these power dissipation numbers to a "power per link" metric?

    (c) How is the power dissipation contribution of the PLL slice handled in this metric?

Chapter 8
Signal Integrity

The data rates at which serial links typically operate are very often greater than the inherent bandwidth capabilities of the channel which connects the transmitter device to the receiver device. Signal integrity analysis is therefore required to determine whether the signal being transmitted, as distorted by the channel, is recoverable by the receiver.

This chapter focuses on the types and sources of jitter in detail. Signal integrity analysis techniques are described using both circuit simulation and statistical analysis techniques.

8.1 Probability Density Functions

A probability density function (PDF) defines the probability of a sample x having a certain value for the universe of possible values of x. In the context of jitter, x is defined as the timing deviation of a given edge of the signal, and $p(x)$ is the probability of the signal edge having this amount of deviation. Before discussing the various contributors to jitter, it is useful to review the mathematics behind two common PDFs: the PDF for a Gaussian distribution, and the PDF for a dual-Dirac distribution.

8.1.1 Gaussian Distribution

The Gaussian distribution is well studied in the field of probability and statistics. It can be used to characterize many naturally occurring physical phenomena and it does a good job of representing the random jitter component of serial data transitions. A Gaussian distribution is defined by two values: the mean, represented by the "μ" symbol, and the standard deviation, represented by the "σ" symbol. The mean is the central position of the curve on the x-axis, and the standard deviation is a measure of the width of the distribution. By definition, the Gaussian distribution is unbounded, meaning that no matter how far a value is from the mean value, there is a nonzero probability of encountering an occurrence at that value. Fig. 8.1 illustrates a PDF for a Gaussian distribution for which $\mu = 0$, $\sigma = 1$.

The PDF corresponding to Fig. 8.1 is defined by the following equation:

$$p(\tau, \sigma) = \frac{1}{\sqrt{2\pi}} \cdot \frac{1}{\sigma} \cdot e^{-\frac{\tau^2}{2\sigma^2}} \tag{8.1}$$

where $\tau = (x - \mu)$, representing the difference between the amount of deviation of a given edge transition (x) and the mean of the distribution (μ). The $p(\tau, \sigma)$ function specifies the probability distribution and can be integrated between two limits to determine the probability of x (or τ) having a value within those limits. Integration of the (8.1) is easier if τ is normalized in terms of σ, suggesting the substitution: $z = (x - \mu) / \sigma$.

D. R. Stauffer et al., *High Speed Serdes Devices and Applications,*
© Springer 2008

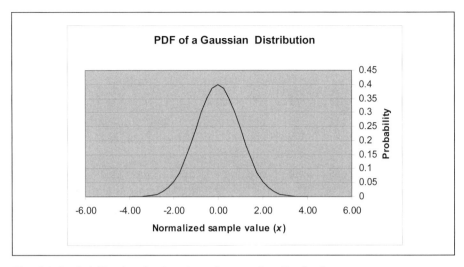

Fig. 8.1 Probability density function of a gaussian distribution

Making this substitution into (8.1), and then integrating (8.1) over the range of $-\infty$ to z with respect to t results in:

$$P(z) = \int_{-\infty}^{z} \frac{1}{\sqrt{2\pi}} \cdot e^{-\frac{t^2}{2}} dt = 0.5\left[1 + \mathrm{erf}\left(\frac{z}{\sqrt{2}}\right)\right] \qquad (8.2)$$

where error function $erf(z)$ is defined as:

$$\mathrm{erf}(z) \quad \frac{2}{\sqrt{\pi}} \cdot = \int_{0}^{z} e^{-t^2} dt \qquad (8.3)$$

Equation (8.3) is called a cumulative distribution function (CDF), and represents the cumulative probability that the deviation of a given edge transition (x) is in the range: $-\infty \leq x \leq z$. The CDF in (8.2) may be used to determine the probability of z being between two limits:

$$P\ Z_1 \leq z \leq Z_2) = P(Z_2) - P(Z_1) \qquad (8.4)$$

Using this formula, it can be demonstrated that approximately 68.26% of the events are within one σ of the mean (μ), 95.45% of the events are within two σ of the mean, and 99.73% of the events are within three σ of the mean.

A Gaussian distribution is inherently unbounded with the PDF curve stretching from $-\infty$ to $+\infty$. The greater the number of samples that are accumulated, the more likely it is that some of those samples are low probability events far from the mean. For example, assume that test equipment is used to measure the peak-to-peak limits of unbounded Gaussian Jitter for an HSS link. The longer the period over which the measurement is taken, the greater the number of signal transitions being observed, and the larger the peak-to-peak jitter value reported. For a truly Gaussian jitter source, the measured jitter approaches infinity as the measurement period is increased.

A somewhat more useful measurement for an unbounded Gaussian distribution is the root-mean-square (RMS) value, which represents the samples of the distribution that are within one σ of the mean. Unlike the peak-to-peak value, the RMS value converges to a fixed value determined by σ as the number of samples increases. As noted previously, 68.26% of the samples are in this range. Conversely, 31.74% of the samples are not in the range specified by the RMS value.

The peak-to-peak value of the measured jitter in the previous example tended toward infinity as the measurement time increased, and therefore this was not a useful measurement to use as a design constraint. The RMS value of the measured jitter converged on a useful value, but since more than 31% of the data transitions have more deviation than the RMS value, this value is also not an appropriate design constraint.

What is needed is an appropriate "worst case" limit for the jitter such that any jitter beyond this limit is sufficiently improbable. The specified *bit error rate* (BER) for the link imposes such a limit. The BER is defined as the ratio of the number of bit errors (which in this context are assumed to be caused by jitter events in the tails of the Gaussian distribution) to the total number of bits transmitted. By excluding events in the tails of the Gaussian distribution from consideration, a bounded (rather than unbounded) Gaussian distribution results.

A bounded Gaussian distribution is represented mathematically by the bounded equation:

$$p(\tau, \sigma) = K \cdot \frac{1}{\sqrt{2\pi}} \cdot \frac{1}{\sigma} \cdot e^{-\frac{\tau^2}{2\sigma^2}} \qquad \text{if} \quad \tau \le \tau_{\max} \qquad (8.5)$$

$$= 0 \qquad \text{if} \quad \tau > \tau_{\max}$$

where K is the normalization constant for the PDF, selected so that $p(\tau, \sigma)$ integrates to 1 over the range of $-\infty$ to $+\infty$.

Given this PDF, the corresponding CDF specified by $P(z)$, and a target BER limit, the probability of an event beyond the range specified by the BER must be constrained such that:

$$\text{BER} \ge [1-(P(z) - P(-z))] \qquad (8.6)$$

Q is related to the target BER by the following equation [6]:

$$\text{BER} = \frac{1}{2} \cdot \text{erfc}\left(\left(\frac{Q}{\sqrt{2}}\right)^2\right) = \frac{1}{Q\sqrt{2\pi}} \cdot e^{-\left(\frac{Q}{\sqrt{2}}\right)} \qquad (8.7)$$

where erfc(x) is the inverse of the error function erf(x): erfc(x) = 1 - erf(x), and the above equation results from the expansion of this function.

Given Q, the peak-to-peak (z_{p-p}), peak (z_{peak}), and RMS (z_{rms}) values of z are related as follows:

$$z_{peak} = Q \times z_{rms}$$
$$z_{p-p} = 2 \times z_{peak}$$

(8.8)

Some values of Q (to two decimal places) corresponding to various values of BER are shown in Table 8.1. A BER of 10^{-12} corresponds to $Q = 7.03$; a BER of 10^{-15} corresponds to $Q = 7.94$.

Table 8.1   Q factors for different BER targets

BER	Q	BER	Q
10^{-3}	3.09	10^{-10}	6.36
10^{-4}	3.72	10^{-11}	6.71
10^{-5}	4.27	10^{-12}	7.03
10^{-6}	4.75	10^{-13}	7.35
10^{-7}	5.20	10^{-14}	7.65
10^{-8}	5.61	10^{-15}	7.94
10^{-9}	6.00	10^{-16}	8.22

Gaussian distributions, both bounded and unbounded, are useful for modeling various types of jitter. Some of these jitter sources result from random events, while others are not random but are not correlated with the data (and therefore can be modeled with a bounded Gaussian distribution).

8.1.2 Dual-Dirac Distribution

Samples modeled by a dual-Dirac probability density function have equal probability of occurring at each of two values of x, as shown in Fig. 8.2. The mean (μ) of the function lies at the midpoint between the two values of x, and the PDF is nonzero for $x = (\mu \pm A)$. The dual-Dirac function is represented mathematically by the equation:

$$p(x, A) = 0.5 \, [\delta((x - \mu) - A) + \delta \ ((x - \mu) + A)]$$

(8.9)

where $\delta(x)$ is the impulse response of x:

$$\delta(x) = \begin{cases} \infty, & \text{when } x = 0 \\ 0, & \text{when } x \neq 0 \end{cases} \quad \text{and} \quad \int_{-\infty}^{\infty} \delta(t) \, dt = 1 \quad (8.10)$$

In Fig. 8.2, $\mu = 0$, $A = 1$.

The Dual-Dirac PDF is useful for modeling certain forms of deterministic jitter. For lack of a better model, it is also often used as a general model for correlated deterministic jitter.

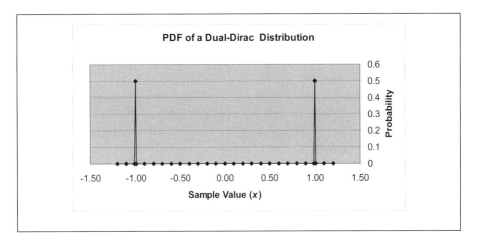

Fig. 8.2 Probability density function of a dual-dirac distribution

8.2 Jitter

One of the most important considerations in the performance of high-speed serial links is jitter. *Jitter* is defined as the deviation in arrival time of a signal from its ideal or expected arrival time. The amount of jitter present on a serial link is related to the achievable BER. If the jitter on a serial link is excessive, data errors result and overall link performance is degraded, possibly to unacceptable levels.

This chapter is only concerned with jitter where the deviation occurs sufficiently fast that the CDR circuit in the serial link cannot track the deviation. As was discussed in Sect. 4.1.2.5, *wander* and *skew* can also be considered to be forms of jitter, but with the variation in arrival time changing at much lower frequencies. Wander and skew components of jitter are therefore tracked by the CDR and do not affect the signal integrity of the serial link, although they may affect FIFO design in the protocol logic.

8.2.1 Jitter Components

This section describes the various components of jitter and the associated terminology. This topic is complicated by the fact that different standards use different terminologies. The following discussion uses one of the more common sets of terms for jitter types, and also provides alternative terminology where appropriate.

As was defined in Sect. 1.4.3, the *total jitter* (TJ) of the signal is the overall jitter as seen at the point of measurement. Total jitter can be measured directly on hardware and is calculated by determining the ideal bit time minus the actual eye width. Total Jitter is generally specified as either a peak or peak-to-peak value. As will be described in Sect. 8.2.3, jitter is statistical in nature and the value is related to the target BER of the serial link. Hardware measurements of TJ must be taken over a sufficient length of time to ensure a valid measurement for the target BER. The measurement must also be based on sufficiently random data to ensure all possible pattern-dependent effects have been observed. A typical oscilloscope providing a ± 3 sigma histogram of eye width values is not an accurate portrayal of the eye width for a 10^{-12} BER, and using such a measurement understates the jitter of the system.

Traditionally, jitter is subdivided into the following two components: *deterministic jitter* (DJ), and *random jitter* (RJ). Each of these categories accumulates differently in the link and results in different requirements for compliance and budgeting schemes.

Deterministic jitter. This is the amount of the total jitter for which the jitter distribution is non-Gaussian. Deterministic jitter is always bounded in amplitude and is created by specific, identifiable causes. The terminology used in [1] for DJ is *high probability jitter* (HPJ). Four types of jitter are typically included as part of the DJ component:

- *Duty cycle distortion* (DCD) results from the difference in width between a logic "0" and a logic "1." This element of DJ is the result of a driver circuit that has rise and fall times that are not equal. Another cause of DCD results when a DC voltage offset is present between the true and complement legs of the differential signal. Pulse width shrinkage due to passive or active components of the channel may also be a factor. DCD is sometimes called *pulse width distortion*.

- *Data dependent jitter* (DDJ) includes timing variations that result from nonclocklike serial data waveforms as they propagate through a channel with bandwidth limitations. Given knowledge of the preceding and subsequent bits of the transmission, the DDJ component of the jitter is predictable, and therefore may be corrected through equalization. DDJ is also called *pattern dependent jitter* or *intersymbol interference* (ISI).

- *Periodic jitter* (PJ) is jitter which has a single fundamental harmonic plus possible multiple even and odd harmonics. PJ results from various electromagnetic noise sources in the system such as power supply noise and crosstalk from periodic signals. Clock signals are periodic signals which cause crosstalk that results in PJ on the victim signal.

- *Sinusoidal jitter* (SJ) is jitter which has a single fundamental harmonic and no additional harmonics. Sinusoidal jitter is generally defined in the context of applied SJ for jitter tolerance testing of a receiver device. For this reason it is generally considered separate from the periodic jitter which arises from sources in the system.

- *Bounded uncorrelated jitter* (BUJ) includes all components of non-Gaussian jitter which are not included in the various components listed above. Nonclock crosstalk aggressor signals operating at a baud rate that is synchronous to the baud rate of the victim signal produce jitter on the victim which is non-Gaussian and is also not correlated with data on the victim signal.

Random jitter (RJ). This is the amount of the total jitter which conforms to a Gaussian jitter distribution. The terminology used in [1] for RJ is *Gaussian jitter* (GJ). Random jitter is caused by semiconductor imperfections and quantum effects, as well as certain types of crosstalk. Two types of jitter are typically included as part of the RJ component:

- *Uncorrelated unbounded Gaussian jitter* (UUGJ) is the component of RJ for which the jitter distribution is a true Gaussian distribution. This component results from the imperfections in the semiconductor crystal lattice, the thermal vibrations of the conductor atoms, and many other small contributors. Measured over time, the peak-to-peak value grows as the measurement time increases.

- *Correlated bounded Gaussian jitter* (CBGJ) is the component of RJ for which the jitter distribution is Gaussian, but the amplitude is bounded and correlates with the signal amplitude being transmitted. Crosstalk aggressor signals operating at a baud rate that is asynchronous to the baud rate of the victim produce jitter on the victim which may be approximated as a bounded Gaussian distribution, and is included in the CBGJ component.

Table 8.2 illustrates the taxonomy of the jitter components which make up TJ, classifying these components as either having bounded or unbounded magnitudes, and whether they are correlated or uncorrelated to the data being sent.

If all of the jitter sources were deterministic in nature, the extreme values of the timing variations could be calculated, and the absolute worst case and best case timing variation could be calculated with confidence. However, the random elements of jitter make it impossible to determine hard limits for timing variations; rather the timing limits need to be expressed in terms of probabilities of timing variation outside the defined limits.

Table 8.2 Jitter taxonomy

Total jitter (TJ) (at BER of interest)	Deterministic jitter (DJ)	Data dependent jitter (DDJ)	Bounded	Correlated
		Duty cycle distortion (DCD)		
		Sinusoidal jitter (SJ) (applied)		Uncorrelated
		Bounded uncorrelated jitter (BUJ) (including PJ)		
	Random Jitter (RJ)	Correlated bounded Gaussian jitter (CBGJ)		
		Uncorrelated unbounded Gaussian jitter (UUGJ)	Unbounded	

8.2.2 Deterministic Jitter

This section describes the various components of DJ in detail. It is important to understand the components of DJ in the system since equalization features may be incorporated into the HSS core design to compensate for correlated DJ components.

8.2.2.1 Duty Cycle Distortion

Duty cycle distortion, also known as pulse width distortion, results from various types of asymmetry in the electrical signals that are being transmitted on the serial link.

Fig. 8.3 demonstrates one possible source of DCD, where one leg of the differential pair has a DC offset compared to the other leg. The offset shifts the point at which the differential signals cross, and results in asymmetry between the width of the "0" and "1" bits.

Another common cause of DCD within a high speed serial link is asymmetry between the rise and fall times of the transmitter circuit. Fig. 8.4 illustrates a waveform where the fall time of the differential waveform is faster than the rise time. The result is that the bit width of the "1" is reduced, and the bit width of the "0" is expanded.

A probability density function can be developed for DCD by recognizing that this jitter is characterized by two mean values (μ_1, μ_2), where one of these means is associated with the deviation of the rising edge of the signal, and the other mean is associated with the deviation of the falling edge. Any given signal edge has a 0.50 probability of being a rising edge (and therefore having a jitter of approximately μ_1), and a 0.50 probability of being a falling edge (and therefore having a jitter of approximately μ_2). The dual-Dirac PDF is appropriate for modeling this distribution.

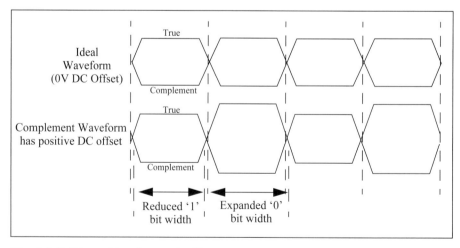

Fig. 8.3 DCD resulting from DC offset

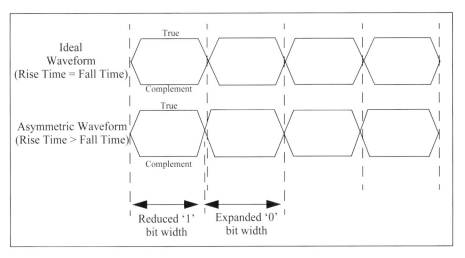

Fig. 8.4 DCD resulting from rise/fall time asymmetry

For any jitter distribution, the overall mean (μ) is zero, and therefore:

$$\mu = (\mu_1 + \mu_2)/2 = 0$$

$$A = (\mu_1 - \mu_2)/2 = DJ_{DCD}$$

where DJ_{DCD} is the peak-to-peak value of the DCD jitter component, and μ and A are characteristics of the dual-Dirac PDF defined in (8.9). The resulting PDF for DCD is therefore defined by the following equation [5]:

$$DCD(x) = 0.5\left[\delta\left(x - \frac{DJ_{DCD}}{2}\right) + \delta\left(x + \frac{DJ_{DCD}}{2}\right)\right] \qquad (8.11)$$

Since DCD jitter is correlated with the data being transmitted, equalization circuits can compensate for this jitter component.

8.2.2.2 Data Dependent Jitter

Data dependent jitter refers to the timing variations caused by the bandwidth limitations of the channel being traversed by the signal. This type of jitter, also known as pattern dependent jitter or intersymbol interference (ISI), is observed on nonclocklike waveforms since the frequency spectrum of such signals is continually changing. DDJ leads to varying amounts of signal attenuation and phase delay when exposed to the frequency response characteristics of a typical transmission channel. Such effects are not observed on clock waveforms which have constant frequency components.

Virtually all data channels exhibit increasing insertion loss as the frequency of the signal increases. More expensive channel components (exotic materials and more costly design and fabrication techniques) can mitigate these effects to a degree, but high baud rates still expose these signals to a significant amount of signal attenuation resulting in DDJ contribution.

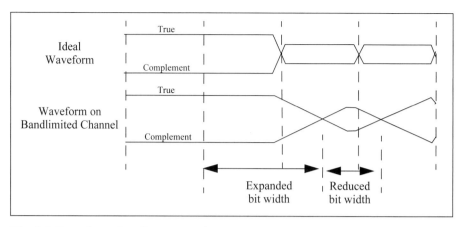

Fig. 8.5 Data dependent jitter example

An example of these effects is shown in Fig. 8.5. Instances where multiple "1" or "0" bits are sent in a row allow the signal to settle near rail voltages, while higher transition densities cause signals to reverse before reaching these voltage limits. When the signal starts near the voltage rails, it takes longer for the signal to transition. In the figure, a long string "1" bits is followed by a "010" sequence. The bit transition from "1" to the first "0" bit slides the crossover point well into the bit window, while the transition from "0" back to "1" occurs somewhat faster. Although it may appear that the waveform is exaggerated in this figure, in reality this is not atypical at higher baud rates for an uncompensated channel.

DDJ is dependent on the data pattern and the frequency response of the channel; no general equation for the PDF exists. Since DDJ jitter is correlated with the data being transmitted, equalization circuits can compensate for this jitter component.

8.2.2.3 Periodic Jitter

Periodic jitter refers to the timing variations caused by various electrical noise sources within the system which are characterized by a fixed frequency spectrum. Such noise sources have a single fundamental frequency component and may also include harmonic frequencies. The jitter induced on the serial data link is correlated to the spectral components of the noise source, but is not necessarily correlated to the data pattern being transmitted on the victim signal.

Power supply noise is an obvious potential source of periodic jitter. However, the spectral components of power supply noise are usually well below the cutoff frequency for the CDR circuit design, and therefore this noise is generally tracked by the CDR circuit. Nevertheless, it is prudent for system designers to take steps to minimize the amount of periodic jitter in the system, since most systems must meet EMI requirements dictated by government regulations. (Such regulations are defined by the Federal Communications Commission for equipment sold in the United States.)

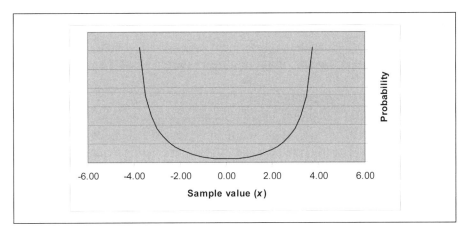

Fig. 8.6 Probability density function for a single sinusoidal aggressor PJ

Clock signals distributed through the system are a more significant source of periodic jitter. Clock signals have a single fundamental frequency component and this frequency is generally above the cutoff frequency for the CDR circuit design. Square waves additionally generate harmonics of the fundamental, with the amplitude of the harmonics increasing as the rise/fall times become faster.

When periodic signals become a crosstalk aggressor for a noise victim serial link, periodic jitter results. If the clock signal frequency is related to the serial data rate, then the periodic jitter is a non-Gaussian jitter source. This may occur, for instance, if the reference clock distribution network is acting as a crosstalk aggressor on a serial data link which is transmitted using this reference clock. Any clock that is frequency locked to this reference clock is also a potential non-Gaussian periodic jitter source.

The PDF corresponding to PJ depends on the periodic waveform of the crosstalk aggressor. However, a useful mathematical model can be developed by assuming the periodic waveform corresponds to the Sinusoidal Jitter case of a single sinusoidal signal. The waveform for this is:

$$x = A\cos(\omega t + \phi_0) \tag{8.12}$$

The corresponding PDF for periodic jitter due to this jitter source is [5]:

$$PJ(x) = \frac{1}{\pi\sqrt{1 - (x/A)^2}} \quad -A \leq x \leq A \tag{8.13}$$

This function is graphed in Fig. 8.6 for $A = 4$. The dual-Dirac function may be used as an approximation for this function, resulting in the equation:

$$PJ(x) = 0.5[\delta(x-A) + \delta\ x+A\ ]\ (\qquad\qquad) \tag{8.14}$$

The above equation is valid for PJ caused by a single sinusoidal signal. However, PJ is often caused by multiple signals which may or may not be sinusoidal, and are often asynchronous with respect to each other and to the victim link. As the number of signals contributing to PJ increases, the PDF for PJ resembles a bounded Gaussian distribution [5].

Most sources of PJ result in jitter which is not correlated to the data pattern being transmitted on the link of interest. Equalization circuits are not capable of compensating for uncorrelated jitter.

8.2.2.4 Bounded Uncorrelated Jitter

Bounded uncorrelated jitter, also called uncorrelated bounded High Probability Jitter (UBHPJ) in [1], refers to timing variations that are bounded in nature but are not directly related to the data pattern that is being sent on the high speed link. These variations can be caused by crosstalk noise coupling from adjacent channels which are transmitting asynchronous data. Since these noise sources do not have any correspondence to the data that is being sent on the serial data link of interest, the resulting jitter appears random with respect to the serial data pattern.

Crosstalk aggressors for adjacent channels can appear Gaussian or non-Gaussian. If the data on the crosstalk aggressor is truly unrelated and is asynchronous to the baud rate of the crosstalk victim, then the jitter may be approximated with a bounded Gaussian distribution and included in the CBGJ component. Otherwise, the jitter is likely to be non-Gaussian and is included as BUJ. Note this distinction between CBGJ and BUJ in [1] is somewhat arbitrary; some texts consider all crosstalk sources as BUJ [5].

The PDF for BUJ is modeled using a bounded Gaussian distribution:

$$\mathrm{BUJ}(x) = \frac{1}{\sqrt{2\pi}} \cdot \frac{1}{\sigma} \cdot e^{-\frac{\tau^2}{2\sigma^2}} \quad \text{if } \tau \leq \tau_{max} \qquad (8.15)$$

$$= 0 \quad \text{if } \tau > \tau_{max}$$

where $\tau = (x - \mu)$, representing the difference between the amount of jitter of a given edge transition (x) and the ideal timing of the signal (μ). BUJ(x) is the probability density function for the BUJ component.

Equalization circuits are generally not capable of compensating for Bounded Uncorrelated Jitter. If the crosstalk aggressor is known (as in the case of a transmitter device acting as a crosstalk aggressor for an adjacent receiver device), then the data pattern on the aggressor may be used by a crosstalk cancellation circuit to compensate for crosstalk at the receiver. Otherwise, the crosstalk contribution generally cannot be removed.

8.2.3 Random Jitter

The jitter taxonomy specified in Table 8.2 uses the definitions in [1], which describe random jitter as composed of UUGJ and CBGJ. Other texts have defined CBGJ as part of BUJ, and limit the definition of random jitter to unbounded components [5].

Equalization is useful for mitigation of jitter which correlates to the data pattern being transmitted. Since random jitter is due to factors for which no such correlation exists, equalization is of no use in mitigating random jitter. It is therefore critical that the HSS circuits be designed to minimize random jitter to the greatest extent possible.

8.2.3.1 Uncorrelated Unbounded Gaussian Jitter

Uncorrelated unbounded gaussian jitter refers to timing variations that are approximated by a Gaussian distribution function and are unbounded in range. This type of jitter is caused by thermal vibrations of the semiconductor crystal structures, material boundaries that have less than perfect valence electron mapping due to semiregular doping density and process anomalies, thermal vibrations of conductor atoms, and other smaller contributing factors.

The PDF for UUGJ is similar to (8.1):

$$\mathrm{UUGJ}(x) \,=\, \frac{1}{\sqrt{2\pi}} \cdot \frac{1}{\sigma} \cdot e^{-\frac{\tau^2}{2\sigma^2}} \tag{8.16}$$

where $\tau = (x - \mu)$, representing the difference between the amount of jitter of a given edge transition (x) and the ideal timing of the signal (μ). UUGJ(x) is the probability density function for the UUGJ component.

Because UUGJ is unbounded, it is generally specified as an RMS value (see Sect. 8.1.1). Alternatively, specification of a BER allows UUGJ to be considered as part of CBGJ. As was discussed in Sect. 8.1.1, the BER specification permits discarding the tails of the Gaussian distribution, and thus bounding the jitter value. For protocol standards which specify RJ and an associated BER, the RJ is generally assumed to be CBGJ and is modeled as described in the next section.

8.2.3.2 Correlated Bounded Gaussian Jitter

Correlated bounded gaussian jitter refers to timing variations that are bounded in nature, and appear as a Gaussian distribution with respect to the data pattern that is being sent on the high-speed link. These variations may be caused by crosstalk noise coupling from adjacent channels which is transmitting unrelated data patterns. If the data on the crosstalk aggressor are truly unrelated and is asynchronous to the baud rate of the crosstalk victim, then the jitter may be approximated with a bounded Gaussian distribution and included in the CBGJ component. Otherwise, the jitter is likely to be non-Gaussian and is included as BUJ.

The PDF for CBGJ is modeled using a bounded Gaussian distribution:

$$\text{CBGJ}(x) = \frac{1}{\sqrt{2\pi}} \cdot \frac{1}{\sigma} \cdot e^{-\frac{\tau^2}{2\sigma^2}} \quad \text{if } \tau \leq \tau_{max} \qquad (8.17)$$

$$= 0 \quad \text{if } \tau > \tau_{max}$$

where $\tau = (x - \mu)$, representing the difference between the amount of jitter of a given edge transition (x) and the ideal timing of the signal (μ). CBGJ(x) is the probability density function for the CBGJ component.

8.2.4 Total Jitter and Mathematical Models

The total jitter that is expected to be observed on a link is a combination of the deterministic jitter and the random jitter components. The exact position of a given instance of a data edge may be predicted to some degree by the data pattern being transmitted as based on the deterministic jitter components. However, the random jitter component of the signal adds some degree of uncertainty as to the exact position. At higher baud rates, the DJ components usually dominate the total jitter, and may result in the data eye being completely closed at the input to the receiver device. Fortunately, equalization can compensate for many of these DJ components. The RJ contribution to the total jitter is generally of much lower magnitude, and equalization cannot compensate for this component.

To see how DJ and RJ is combined to determinate the total jitter, the PDF associated with each type of jitter is needed. Total jitter (TJ) is the mathematical convolution of these jitter distribution functions.

In the prior sections of this chapter, bounded Gaussian distributions were used to model several uncorrelated jitter components, including:
- Periodic jitter, assuming multiple sources contribute to PJ such that the overall jitter conforms to a bounded Gaussian distribution
- Bounded uncorrelated jitter
- Correlated bounded gaussian jitter and
- Uncorrelated unbounded gaussian jitter, assuming the distribution is truncated by assuming a BER

These jitter components can be modeled as a combined Gaussian Jitter (GJ) with a bounded Gaussian distribution as defined by the following equation:

$$\text{GJ}(x) = \frac{1}{\sqrt{2\pi}} \cdot \frac{1}{\sigma} \cdot e^{-\frac{\tau^2}{2\sigma^2}} \quad \text{if } \tau \leq \tau_{max} \qquad (8.18)$$

$$= 0 \quad \text{if } \tau > \tau_{max}$$

In the prior sections of this chapter, the dual-Dirac distribution function was shown to be a reasonable model for the DCD component of deterministic jitter, as well as for a single sinusoidal PJ component. The distribution function for DDJ is less clear. However, the dual-Dirac function is often used to model the overall deterministic jitter (DJ) as is described in [1,3,5]. As is noted in [5], this is an assumption with yields a reasonable approximation and keeps the math simple.

Deterministic jitter is defined by dual-dirac distribution function:

$$DJ(\tau, W) = \frac{\delta(\tau - \frac{W}{2})}{2} + \frac{\delta(\tau + \frac{W}{2})}{2}$$ (8.19)

where W is the peak-to-peak amplitude (or width) of the DJ component.

The probability distribution function for total jitter (TJ) is formed by convolution of (8.18) and (8.19):

$$TJ(\tau, W, \sigma) = \frac{1}{2\sqrt{2\pi}} \cdot \frac{1}{\sigma} \cdot \left[e^{-\frac{\left(\tau - \frac{W}{2}\right)^2}{2\sigma^2}} + e^{-\frac{\left(\tau + \frac{W}{2}\right)^2}{2\sigma^2}} \right]$$ (8.20)

Equation 8.20 is plotted in Fig. 8.7 for the cases of $W = 3$ and $W = 4$. The characteristics of the dual-Dirac model for the DJ component produce peaks in the TJ PDF at $x = \pm W / 2$. The roll-off of the curve from these points is determined by the Gaussian Jitter component and the standard deviation.

A data eye is constructed by examining two consecutive edges of the data, separated by 1 UI, where UI = the unit interval (bit width). Adding the TJ PDF for a data edge at $x = 0$ to the TJ PDF for the next consecutive data edge at $x = UI$ (and scaling so that the result integrates to 1 over the range $\pm\infty$), the resulting equation is:

$$f(\tau, W, \sigma, UI) = \frac{1}{4\sqrt{2\pi}} \cdot \frac{1}{\sigma} \cdot \left[e^{-\frac{\left(\tau - \frac{W}{2}\right)^2}{2\sigma^2}} + e^{-\frac{\left(\tau + \frac{W}{2}\right)^2}{2\sigma^2}} + e^{-\frac{\left(\tau - UI - \frac{W}{2}\right)^2}{2\sigma^2}} + e^{-\frac{\left(\tau - UI - \frac{W}{2}\right)^2}{2\sigma^2}} \right]$$ (8.21)

The PDF represented by (8.21) is plotted in Fig. 8.8 for the range $x = 0 - 10 (= 1 \text{ UI})$. The timing deviation associated with the data edge which would ideally fall at $x = 0$ forms the left-hand portion of the plot, while the timing deviation of the edge at $x = 10$ forms the right-hand portion of the plot. From this PDF it should be obvious that the eye width depends on both the bit width and the shape of the TJ PDF.

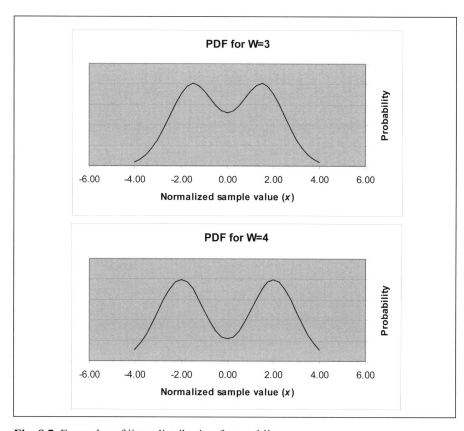

Fig. 8.7 Examples of jitter distribution for total jitter

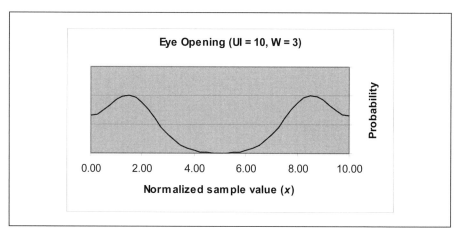

Fig. 8.8 TJ PDF of two consecutive bits (eye width)

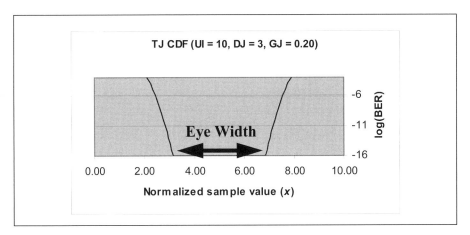

Fig. 8.9 TJ CDF of two consecutive bits (bathtub curve)

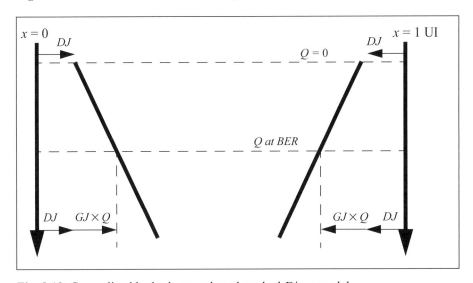

Fig. 8.10 Generalized bathtub curve based on dual-Dirac model

PDF in Fig. 8.8 is integrated to form the CDF plotted in Fig. 8.9. The TJ PDF for the bit centered at $x = 0$ is integrated over the range x to $+\infty$ to form the left-hand curve in Fig. 8.9, while the TJ PDF for the bit centered at $x = 10$ is integrated over the range $-\infty$ to x to form the right-hand curve. The resulting plot shows the envelope for the total jitter as a function of the BER.

The left and right curves in Fig. 8.9 form a *bathtub curve*, where the difference between the two sides of the bathtub is the eye width for a given target BER. As the BER becomes sufficiently small, the sides of the bathtub curve in Fig. 8.9 can be approximated by straight lines. The slope of the sides of the bathtub curve is related to the Q of the circuit as was defined in Sect. 8.1.1. Equation 8.7 specifies the relationship between Q and BER.

Fig. 8.10 illustrates a general bathtub curve which normalizes x on a scale of 0–1 UI, and plots the y-axis as a function of Q rather than as a function of *BER*. At $Q = 0$, the eye width is entirely defined by the DJ component of the jitter. As the Q is increased (downward movement on the y-axis), the GJ component causes the eye to narrow. This peak value of *GJ* is a function of Q as defined in (8.8). Lower BER requires a higher Q, and results in less eye width. The slope of the eye wall lines are therefore:

$$Q_{\text{left}} = -Q_{\text{right}} = (\tau \times \text{DJ}) \times \frac{1}{\text{GJ}} \qquad (8.22)$$

where $\tau = (x - \mu)$, DJ is the peak value of the deterministic jitter, and GJ is the RMS value of the Gaussian jitter. (Equation 8.22 assumes the eye walls are symmetrical. In some systems this may not be valid, and in a more general case Q_{left} and Q_{right} may have different slopes.) A BER of 10^{-12} corresponds to $Q = 7.03$; a BER of 10^{-15} corresponds to $Q = 7.94$. Additional Q values for various BERs were given in Table 8.1.

Note that this mathematical model for total jitter is based upon the assumption that the dual-Dirac Model is an appropriate model for deterministic jitter. Remember that this model assumes the timing of each sampled edge deviates from the ideal timing by either $+(\text{DJ}/2)$ or $-(\text{DJ}/2)$, with no values in between. As might be expected, a realistic PDF for DJ is likely to have many sampled edges which are between these values. Analysis in [5] indicates that using the dual-Dirac model to model the deterministic jitter of the link tends to overestimate the amount of *DJ* that will be present in a real system. This model is therefore appropriate to estimate worst case jitter, but characterization testing should not expect this model to correlate with hardware measurements.

8.2.5 Jitter Budgets

An example of a jitter budget is provided in Table 8.3. The chosen example is the jitter budget for the CEI-11G-LR interface as specified in [1]. This jitter budget is laid out in tabular form, with columns containing the contributions from various types of jitter, and with rows containing the contributions at various stages of the serial link.

Jitter contributors in this table are categorized as *uncorrelated* and *correlated* based on whether the jitter can be correlated to the data pattern. Gaussian (or random) jitter consists of an *uncorrelated unbounded Gaussian jitter* component and a *correlated bounded Gaussian jitter* component, and are listed in the corresponding columns. Correlated components of deterministic jitter, including DCD and DDJ components, are contained in the *correlated bounded high probability jitter* (CBHPJ) column. Remaining uncorrelated components of deterministic jitter, including PJ and BUJ components, are contained in the *uncorrelated bounded high probability jitter* (UBHPJ) column.

Table 8.3 CEI-11G-LR informative jitter budget [1]

Source	Uncorrelated jitter		Correlated jitter		Total jitter			
	Un-bounded Gaussian	Bounded high prob.	Bounded Gaussian	Bounded high prob.	Gaussian	Sinu-soidal	High prob.	Total
Abbreviation	**UUGJ**	**UBHPJ**	**CBGJ**	**CBHPJ**				
Unit	UIpp	UIpp	UIpp	UIpp	UIpp	UIpp	UIpp	UIpp
Transmitter	0.150[a]	0.150[a]			0.150		0.150	0.300
Channel			0.230	0.400				
Receiver input	0.150	0.150	0.230	0.400[a]	0.275		0.550	0.825
Equalizer				-0.300				
Post Equalizer	0.150[a]	0.150[a]	0.230	0.100[a]	0.275		0.250	0.525
DFE penalties				0.100				
Clock and Sampler	0.150	0.100		0.100				
Budget	0.212	0.250	0.230	0.300	0.313	0.050	0.550	0.913

Note:
[a]These values are normative values in [1]

[b]Due to receiver equalization, it reduces the ISI as seen inside the receiver. Thus this number is negative

[c]It is assumed that the eye is closed at the receiver, hence receiver equalization is required

Contributions of the UUGJ and CBGJ columns are combined in the *Gaussian* column of the *total jitter*, while contributions of the UBHPJ and CBHPJ columns are combined in the *high probability* column. Note that while the High Probability jitter is combined by adding the component jitter values, the Gaussian jitter must be combined using an RMS summation. This is because the probability of independent Gaussian events all having worst case values is extremely unlikely. On the other hand, the RMS summation of UUGJ and CBGJ produces a Gaussian jitter value corresponding to a BER consistent with the component numbers.

The SJ column reserves a portion of the jitter budget for applied SJ as part of jitter tolerance testing. This jitter component does not exist in an operational system.

Finally, the *Gaussian*, *sinusoidal*, and *high probability* jitter totals are summed to produce an overall *total jitter*.

The table assumes the jitter contribution of the transmitter device is entirely uncorrelated while distortion in the channel is entirely correlated. The sum of the jitter contributed by the transmitter device and the channel is seen at the receiver input. Note that the total jitter indicates very little eye opening at the receiver input. Given jitter penalties introduced by receiver logic, it would be impossible to receive this signal without equalization. As shown in the table, equalization provides a negative contributor to the CBHPJ component of the jitter budget. This equalization benefit offsets most of the jitter penalties of the receiver logic.

Jitter penalties for the receiver include *clock and sampling* penalties, *DFE penalties*, and *post equalizer* penalties. *Clock and sampling* penalties account for reference clock jitter and other uncertainties in the signal sampling point introduced by the CDR circuit. The equalization benefit assumed an ideal DFE circuit with infinite precision of tap weights which can be programmed to precisely cancel post-cursors; *DFE penalties* account for non ideal features of a realizable DFE implementation, including quantized tap weights and circuit imperfections. Circuit imperfections introduced in the receiver after the DFE are accounted for as part of *post equalizer* penalties.

Jitter contributors are totalled on the *Budget* line of the table. The *total jitter* on this line is 0.913 UI. Any value less than 1.0 indicates that the eye is open and data can be received. The extent to which this number is less than 1.0 indicates margin built into the specification.

8.2.6 Jitter Tolerance

Jitter tolerance is the ability of the receiver to successfully recover data in the presence of jitter. Jitter tolerance measurements represent the amount of jitter that is allowable at any given frequency while maintaining a specified BER.

Fig. 8.11 illustrates an typical receiver jitter tolerance mask. This mask specifies the amount of applied SJ as a function of frequency. Jitter tolerance testing is performed by sweeping the frequency of the applied SJ, and adjusting the amplitude as needed to conform to the mask. If the receiver continues to receive data and meet the specified BER, then the receiver device conforms to the jitter tolerance specification. To measure jitter tolerance of the receiver device, the amplitude of the applied SJ is increased until the specified BER is no longer achieved.

The applied SJ amplitude for the high frequency portion of the curve is specified by the jitter budget for the interface. The jitter budget in Table 8.3 for the CEI-11G-LR interface specifies this amplitude as 0.050 UI. Jitter amplitudes are generally described in terms of UI of the serial data stream.

For lower jitter frequencies, the CDR sampling point tracks the jitter rather than having to find a sampling point with sufficient margin to tolerate the jitter. This is reflected in the curve in Fig. 8.11 by the applied SJ being increased for frequencies below the baud rate divided by 1667. "Jitter" in lower frequency ranges was discussed in Sect. 4.1.2.5, where the terminology of *skew* and *wander* was introduced.

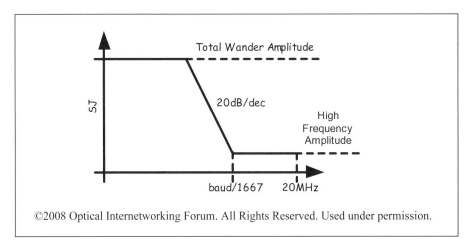

Fig. 8.11 Jitter tolerance mask example

Measuring the jitter tolerance of receiver devices, and confirming that this jitter tolerance exceeds the specifications required for standards compliance, is a necessary part of receiver characterization testing. Receiver characterization testing was described in Sect. 7.5.2.1.

8.3 Spice Models

The traditional approach to signal integrity analysis uses circuit simulation to determine whether the serial data signal meets necessary electrical characteristics at the receiver device. Since the general subject of this text includes HSS applications, the discussion of signal integrity would not be complete without including traditional approaches. However, as will be noted, execution times limit the extent to which signal integrity analysis can be exhaustively performed using circuit simulation. While some approaches to mitigate this are discussed, it is generally not practical to analyze signal integrity for baud rates of 5 Gbps and above using Spice simulations. This is especially true when the HSS receiver employs complex equalization. Statistical approaches are discussed in Sect. 8.4 which are more appropriate at higher baud rates.

If signal integrity analysis is going to employ circuit simulations, Spice models of the HSS core are needed to support such simulations. These models may be based on extracted device level models, or alternatively may be a behavioral model. The advantages and disadvantages of each model type is discussed in this section.

8.3.1 Traditional Spice Models

Traditional Spice models are based upon extracted device level models. These models are created by well-proven device and parasitic extraction programs. These extraction programs are verified for each new circuit technology during the technology qualification process to ensure that accurate models are produced.

The traditional Spice models for HSS cores are very much like the Spice models of standard I/O cells. Two major differences between HSS Spice models and standard I/O cell models are

- The HSS models are intended to operate at much higher frequencies and because of this they are much more complex than most standard I/O models and include many more control nodes.

- The Spice models for HSS cores only model the external interface circuitry so that some of the control nodes on the Spice model correspond to internal signals in the hard core. The user must consult relevant documentation on the Spice model, and determine the appropriate values for these control nodes based on the application.

The added complexity of the HSS Spice models means that they are more complex to integrate into a testbench due to the added connections that must be correctly set to make the model operate properly. Sometimes the internal signals corresponding to these control nodes are based upon a combination of HSS core parameters (either register values on input pin values). In such cases it may be necessary to calculate values for the control nodes based upon how the application is expected to configure the HSS core. In general, the transmitter models are much more complex than the receiver models due to the fact that a greater portion of the transmitter circuit is included in the Spice models.

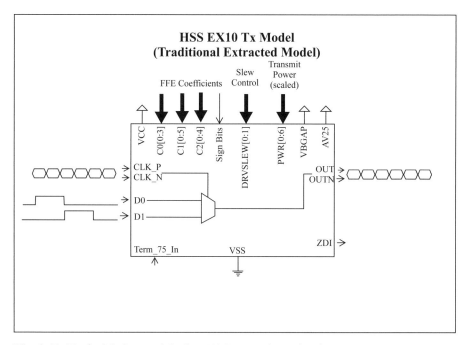

Fig. 8.12 Typical Spice model of an HSS transmitter circuit

Fig. 8.12 depicts a typical extracted Spice model for an HSS transmitter circuit with typical model nodes, including differential clock input pins, multiplexed data input pins, serial data output pins, pins to set preemphasis coefficients, sign pins for the preemphasis coefficients, slew control pins, transmitter output power pins, termination selection pins, internal power supply pins, and so forth. The settings of these nodes affect the characteristics of the transmitted signal.

Although the function of the data and clock pins is obvious, the proper values for the preemphasis, slew, output power, and termination selection nodes depend on the configuration set by the application and correspondence between this configuration and the internal control nodes of the core. If this is the Spice model for an HSS EX10 core, these control nodes do not necessarily correlate directly to register values. For example, the *PWR*[0:6] pins in Fig. 8.12 are partially based on the *transmit power register* in Table 2.6, but also must be scaled based on the *transmit tapx coefficient register* settings. Intervening logic in the HSS transmitter which generates the *PWR*[0:6] signals from various register values is not modeled in the Spice model in the interest of minimizing the complexity of the model, and the user of the model must make up for this by manually determining the proper values.

The Spice model is generally limited to modeling the analog external interface portion of the transmitter. Referring to the HSS EX10 transmitter block diagram in Fig. 2.4, this included the *driver/equalizer* and *JTAG* blocks. All of the parallel to serial conversion logic, BIST logic, clock generators, etc., in Fig. 2.4 are excluded for the Spice model. This is sufficient to support simulation of the interaction between the core circuitry and the external serial data channel.

Assuming the required signal characteristics for the input to the HSS receiver slice are specified, then the Spice model for the receiver only needs to provide an accurate model for the receiver termination and load characteristics. Referring to the HSS EX10 receiver block diagram in Fig. 2.7, the *VGA Amp*, *signal detect* block, and the JTAG receivers are included in the Spice model and are sufficient to support signal integrity simulations assuming an open eye exists at the receiver input. Other circuits are excluded from the Spice model.

It should be noted that at higher baud rates the signal input to the receiver may not have an open eye. The DFE circuit would need to be included in the Spice model to verify that equalization allows the signal to be properly received. However, circuit simulations which include the DFE circuits would have prohibitively long execution times.

8.3.2 Hybrid Spice/Behavioral Models

Extracted Spice models have a long history of use in signal integrity analyses, and as such most chip designers are very familiar with the use of these models and the accuracy of the analysis results. However, the increased complexity of these models leads to some negative consequences for a typical signal integrity analysis. First, while this type of model does a very good job of representing the deterministic performance of the actual core, the random elements of performance are much more difficult to predict and model. If the random elements are accurately included in the model, typical random variations mean that only a very few bits per time interval are affected in such a way as to cause problems in the data stream. The net result is that a *very* large number of bits would need to be simulated to successfully develop the required signal integrity statistics and thereby ensure the target BER is achieved. Simulating the transmission of a large number of bits through a complex Spice model is not a good approach for simulation efficiency, and results in very long simulation execution times. Furthermore, a large number of scenarios may need to be run to determine optimal power and equalizer settings. As a result, weeks of simulation may be required to perform signal integrity analysis for a single serial data channel. When one considers tolerances on the channel model as well as the number of different channels that are in a typical application, the turn around time for the signal integrity analysis quickly becomes unrealistic.

Hybrid behavioral Spice models are a second class of models for HSS cores. In this case, the model is still a Spice model and is built with Spice constructs, but the model is not generated through the use of an automated device and parasitic extraction program. Such models are usually coded manually. This type of Spice model is a simplified model which contains sufficient functionality to demonstrate the behavioral characteristics of the actual hardware.

The advantage of a simplified model is that the simulation run time is significantly reduced when compared to that of a full extracted version of the model. The model remains an Spice model containing nodes for each of the necessary signal pins, but the detailed device models and extracted parasitics are no longer included. In their place, appropriate circuitry to model important functional characteristics is instantiated within the model. Since this is not a device-for-device match to the hardware, the model may be simplified by removing all references to internal nodes and limiting the Spice model nodes to only those pins that would be recognized by the chip designer at the core level. The external view of the hybrid model appears much the same as that of the fully extracted model as shown in Fig. 8.13, but the contents of the model are modified to make the model run much faster.

While the simplifications described above offer significant simulation time advantages over that of fully extracted Spice models, there are some drawbacks to this type of model as well. First, there is no comprehensive design automation software to generate this type of model. Some portions of the process may be automated, but the model designer must still identify each of the circuit characteristics that is required to be incorporated into the model, and then develop an overall model that accurately describes those characteristics.

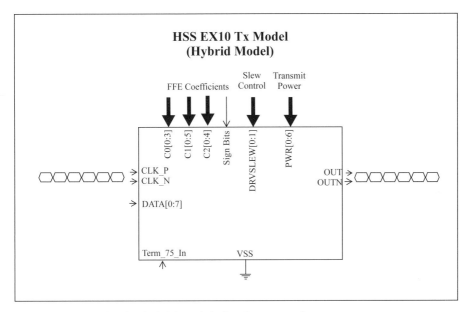

Fig. 8.13 Typical Spice hybrid model of an hss transmitter

Subsequent to creation of the hybrid model, the functional behavior of the model must be verified through extensive testing and correlation to hardware measurements. Any differences found between hardware and model behavior require changes to the model to correct the simulations. This dependence on hardware results means that the final iterations of the model may not be available until late in the development cycle. Once a final model is generated and has been through the entire verification process, the accuracy of the hybrid behavioral model should be equivalent to that of a model generated through netlist extraction.

A successful hybrid model for an HSS core can significantly improve the simulation time for a single simulation scenario, but the wide array of control nodes on the cores can still result in a large number of simulation scenarios being required to determine the optimal settings for a given channel.

8.3.3 Spice Simulation Matrices

HSS Spice models are typically used to simulate the interaction of the core circuitry with the serial data channel. There is a large matrix of variables that are typically investigated during the signal integrity analysis of an HSS link. The following parameters were defined for the HSS EX10 core in Chap. 2, and are of significance to the signal integrity of the link.

Transmitter Power Level. The HSS EX10 transmitter defined a *Transmit power register* in Table 2.6 which allowed provisioning of the transmitter launch amplitude. For the HSS EX10, this parameter is a 7-bit value supporting 128 different amplitude settings.

Transmitter slew rate settings. The HSS EX10 transmitter defined a *slow slew control* in the *transmit driver mode control register* in Table 2.6 which allowed provisioning of the transmitter slew rate. For the HSS EX10, this parameter is a 3-bit value with five valid settings. Various industry standards define minimum slew rates which may require particular settings of this parameter for compliance.

Transmitter preemphasis settings. The HSS EX10 transmitter incorporated a feed forward equalizer (FFE). Coefficients for this equalizer were provisioned using the *transmit tapx coefficient registers* as defined in Table 2.6. The number of bits in these equalizer coefficient values vary, but in almost all cases a large matrix of possible preemphasis settings exists. (The HSS EX10 has three FFE taps with a total of 15 bits leading to 32,768 possible settings.) Prior knowledge of the preemphasis effects (or a good simulation plan) could reduce the size of the matrix significantly, but there would still be a significant number of simulations that could be needed to arrive at an optimal setting.

Transmitter termination values. Although this feature was not provided on the HSS EX10 core, some HSS cores provide more than one option for termination impedance. In such cases, the appropriate value to use would generally be dictated by the interface standard. However, there may be cases where the signal integrity engineer wants to explore which of various options provides the best signal integrity.

Transmitter AC/DC coupling. The HSS EX10 transmitter supported either AC or DC coupling, as provisioned by the *HSSTXACMODE* pin defined in Table 2.1. This pin is generally tied based on the coupling method used by the channel. However, there may be cases where the signal integrity engineer wants to explore which of these coupling schemes provides the best signal integrity.

Transmitter data rate. The data rate is a significant factor in link operation. Some serial link applications only need to operate at a single data rate, while others require support for multiple data rates depending upon the specific platform in which they are deployed.

Transmitter data pattern. There are a number of encoding methods for data transmitted across serial links. Scrambling, 8B/10B, and 64B/66B were discussed for the various protocols in covered in Chap. 5. There are also a number of test patterns defined by various protocols for compliance testing. The specific characteristics of each pattern drive different performance levels on the serial links. Chip designers sometimes use Spice models to investigate the performance trade-offs associated with various data encoding methods and compliance test patterns.

Transmitter process/voltage/temperature (PVT) settings. HSS cores, like every other circuit on a chip, are affected by environmental operating conditions (transistor junction temperature and power supply voltage), and by manufacturing process parameter variations. Spice models allow specification of these parameters. The signal integrity engineer must explore several combinations of values for these parameters to ensure that any chip which was manufactured within normal process variation constraints, and operating within specified environmental conditions will in fact operate within the specified BER of the system. Generally the signal integrity engineer, working with the chip manufacturer, define a number of *PVT corners* which must be checked.

Receiver AC/DC coupling. The HSS EX10 transmitter supported either AC or DC coupling, as provisioned by the *HSSRXACMODE* pin defined in Table 2.1. Similar to the pin associated with the transmitter, this pin is tied based on the coupling method used by the channel. However, there may be cases where the signal integrity engineer wants to explore which of these coupling schemes provides the best signal integrity.

Receiver termination values. As was discussed for the transmitter, some HSS cores provide more than one option for termination impedance. In such cases, the appropriate value is generally dictated by the interface standard. However, there may be cases where the signal integrity engineer wants to explore which of various options provides the best signal integrity.

Receiver process/voltage/temperature (PVT) settings. As was discussed for the transmitter, receiver operation is affected by manufacturing process parameters, junction temperature, and power supply voltage. Receiver operation must therefore be simulated at various *PVT corner* conditions. Note that, given the transmitter and receiver devices of the link are on different chips, the PVT conditions of the transmitter and receiver are likely to be different. Therefore simulations must include all of the various combinations of transmitter PVT and receiver PVT conditions.

The results from all of these simulations are compared to the required signal characteristics specified for the HSS core at the BER of interest. The required eye height and width must be verified at the input pins of the receiver. Fig. 8.14 shows a typical channel configuration that might be simulated, along with the raw serial data pattern at the receiver and the corresponding eye diagram.

HSS cores that support data rates in excess of 5 Gbps may expect a closed eye at the receiver input. Such cores depend on receiver equalization functions to open the eye, which (as discussed previously) are generally not modeled in the Spice model. At these data rates, statistical simulation approaches are often used in place of Spice simulations. Such simulations run faster than Spice, can automate the determination of optimal settings for equalization variables, and can accurately account for random variations in the core and channel, including crosstalk effects. The next section discusses this in more detail.

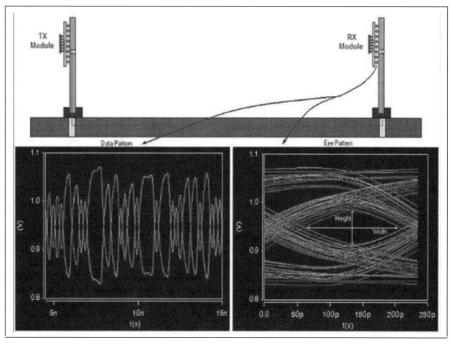

Fig. 8.14 Typical Spice simulation results

8.4 Statistical Approach to Signal Integrity

Circuit simulation using Spice models provides an accurate measure of the resulting signal waveform given the input conditions which are simulated. However, unless the input conditions are varied over a statistically representative range, Spice simulations do not guarantee that the BER of the system meets the specification. A BER of 10^{-b} implies that the signal may violate the specified eye mask no more often than once every 10^{b} bits. Many, many data patterns must be simulated under a variety of crosstalk and noise conditions to ensure the resulting signal meets this requirement. Lengthy simulations are required to guarantee a BER of 10^{-12} and simulations are prohibitively long for higher BER specifications.

At higher baud rates, intersymbol interference (ISI) becomes a key source of signal distortion. ISI is a component of data dependent jitter (DDJ), and is primarily caused by frequency response limitations of the channel. At higher baud rates, ISI effects stretch over multiple bit intervals, with the waveform of the current bit being affected by previous bits. In many cases the eye at the receiver is closed due to ISI, and receiver equalization is required to compensate. The receiver equalization must be included in the analysis in such cases to ensure the signal can be received properly and that the BER specification is met. In a Spice circuit simulation environment, this would be computationally prohibitive.

The OIF common electrical I/O (CEI) Implementation Agreement [1] was discussed in Sect. 5.2.5. This standard specifies normative channel requirements, and a statistical analysis approach was developed in conjunction with the development of this standard to verify compliance of the channel with these requirements. This approach is described in detail in [3], and was subsequently published in [1] as the normative method of determining channel compliance. An open source software tool called *StatEye* is available from [4], and implements this analysis. Other software tools also exist which implement similar statistical approaches, including the IBM HSSCDR tool described in Sect. 8.4.2.

8.4.1 Analysis Approach

Statistical approaches to signal integrity analysis can produce reliable results with significantly less computation than circuit simulation approaches. Transmitter jitter generation is statistically modeled using the dual-Dirac model for deterministic components, and using the Gaussian model for random components. Measured frequency response models are used for components of jitter due to the channel. The resulting analysis can project jitter behavior and the corresponding eye opening of the signal for a given BER at either the input to the receiver or at the output of the receiver equalization circuit. Statistical signal analysis can model complex receiver equalization circuits in the simulation without significant computational penalties. Additionally, software used to perform statistical analysis often includes algorithms which can determine the optimal settings for both transmitter and receiver equalization circuits.

The statistical analysis approach presented in this section is also described in [1] and [3], and is representative of the analysis performed by this class of software tools.

8.4.1.1 Pulse Response

Fig. 8.15 illustrates an example of input and output signals of a channel. The channel in this example is modeled by a simple RC network. The input data pattern is "011010011100," and an ideal input signal is assumed in the figure. The voltage level achieved by each bit of the output waveform depends on whether the values of prior bits were the same or different. This type of data dependent jitter is called *intersymbol interference*. While this example modeled the channel with an RC network, realistic channels generally also have inductance and impedance mismatches. The addition of inductance potentially causes ringing of the response signal, and impedance mismatches cause reflections which may lag the bit transition by up to several bit times.

The pulse response of a channel is defined as the received pulse for an ideal square wave launched into the channel, where the pulse width of the square wave is one unit interval. This response is calculated either by convolving the pulse with the impulse response of the channel, or by multiplying the Fourier spectrum of the ideal transmitted square wave with the channel response and taking the inverse Fourier transform, as described in [1]. The resulting receive pulse is illustrated graphically as shown in Fig. 8.16.

The amplitude of the receive pulse at discrete baud-spaced intervals in Fig. 8.16 are called *cursors*. The cursor corresponding to the maximum signal amplitude is labelled c_0. Cursors prior to this reference point are called *precursors*, and are labelled c_n, where $n < 0$. Cursors after this reference point are called *postcursors*, and are labelled c_n, where $n > 0$. Generally, only precursors which are within a few bit times of the main signal are significant, and only postcursors which are within twice the propagation time of the channel are significant. The $R(\tau)$ matrix represents this channel response:

$$R(\tau) = \begin{bmatrix} r_{-\frac{m}{2}}(\tau) & \cdots & r_{-1}(\tau) & r_1(\tau) & \cdots & r_{\frac{m}{2}}(\tau) \end{bmatrix} \quad (8.23)$$

where $r_n(\tau)$ are the cursors of the pulse response at sample point τ, and m is the number of cursors considered over a range that is symmetrical with respect to the sample point.

The ideal square wave launched into the channel is distorted by the channel due to ISI. If the ISI sufficiently distorts the signal, as is common at higher baud rates, the signal eye at the receiver may be closed. When this occurs, the channel cannot be analyzed as a stand-alone component to determine interoperability within the system, and equalization circuits must be included in the analysis.

8.4.1.2 Component Models

Fig. 8.17 illustrates the three fundamental components of any link: the transmitter device (including any FFE and/or other equalization), the channel interconnect (modeled as a channel frequency response), and the receiver device (including any DFE and/or other equalization).

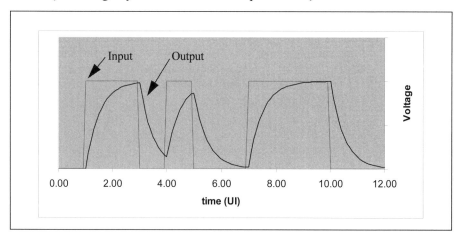

Fig. 8.15 Channel pulse response

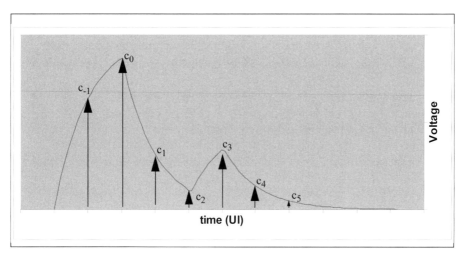

Fig. 8.16 Receive pulse representation

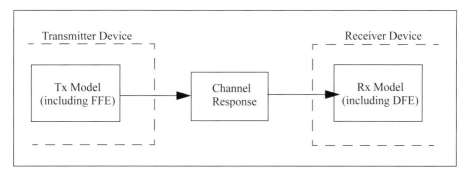

Fig. 8.17 Serial link components

These components must each be modeled, and then must be analyzed together to determine whether the overall link is operable. While an ideal square wave launched into the channel may produce a closed eye, the output of a transmitter device which incorporates equalization is not an ideal square wave. The signal emphasis injected by the transmit equalizer can partially cancel the ISI characteristics of the channel response. Similarly, even if the signal eye is closed at the input of the receiver device, equalization in the receiver may still be able to correctly receive the serial data.

Transmitter Model. A block diagram of a generalized transmitter device model is shown in Fig. 8.18. This consists of a data generation stage, transmitter equalization (usually an FFE), and stages which add losses associated with the transmitter driver stage and device package. In general, jitter (both deterministic and random jitter) introduced at the transmitter also must be considered. For the analysis described in this section, transmit jitter is considered as part of the sampling jitter in the receiver model.

Data Generator. This stage generates the bit sequences that are used by the analysis. The length of the bit sequences is determined by the number of cursors m in (8.23). Given m cursors of the channel pulse response, data sequence n is represented by the following matrix:

$$D_n = \begin{bmatrix} d_{n,1} & d_{n,2} & \cdots & d_{n,m} \end{bmatrix} \tag{8.24}$$

where each $d_{n,b}$ ($b = 1-m$) is either "-1" or "$+1$." Given that $N = 2^m$ bit sequences are possible, the resulting matrix is

$$D = \begin{bmatrix} D_1 \\ D_2 \\ \cdots \\ D_N \end{bmatrix} = \begin{bmatrix} d_{1,1} & d_{1,2} & \cdots & d_{1,m} \\ d_{2,1} & d_{2,2} & \cdots & d_{2,m} \\ \cdots & \cdots & & \\ d_{N,1} & d_{N,2} & \cdots & d_{N,m} \end{bmatrix} \tag{8.25}$$

where D is the matrix defining all possible bit sequences of length m.

In a scrambled system, each of the possible bit sequences has an equal probability of occurring. In a system using a block code, the probability associated with each bit sequence is weighted based upon its frequency of occurrence in the block code, and some bit sequences may have zero probability. The probability associated with each bit sequence is

$$P_d = \begin{bmatrix} p_d(1) & p_d(2) & \cdots & p_d(N) \end{bmatrix} \tag{8.26}$$

where $p_d(n)$ is probability associated with bit sequence n occurring, and where $n = 1-N$. The summation of all of the $p_d(n)$ must equal 1.

Transmitter Equalization. HSS devices almost universally employ transmitter equalization to compensate for distortions introduced by the channel. This transmitter equalization is usually an FFE as was described in Sect. 1.3.2. Fig. 8.19 illustrates a 3-tap FFE similar to the FFE associated with the HSS EX10 core that was described in Sect. 2.2.3.

The transmitted signal level at the output of this FFE at any given time is determined by the current bit being transmitted, as well as the bit before and after the current bit. Each of these bits is multiplied by the an equalizer coefficient, and all of these results are summed together. For a 3-tap FFE, eight different output levels are possible for a fixed set of coefficient values (as shown in Fig. 8.19). Statistical analysis must determine the link response given a transition between any pair of these output levels.

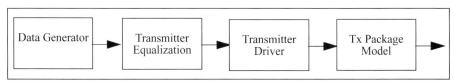

Fig. 8.18 Transmitter device model

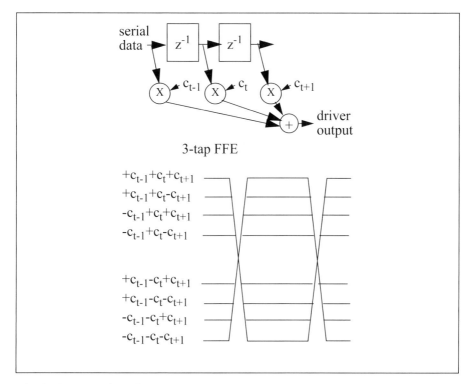

Fig. 8.19 Transmitter filter function

Transmitter equalization modifies the D_n matrix specified in (8.24). The resulting D'_n matrix is formed as follows:

$$D'_n = \begin{bmatrix} d'_{n,1} & d'_{n,2} & \cdots & d'_{n,m} \end{bmatrix} \tag{8.27}$$

where each element of this matrix is

$$d'_{n,b} = \sum_t (c_t \delta(d_{n\,b+t} - 1) - c_t \delta(d_{n\,b+t} + 1)), \tag{8.28}$$

In (8.28), the summation is performed over the range of t corresponding to the taps of the FFE, where c_t is the coefficient associated with FFE tap t. If $d_{n,b+t}$ is "+1," then the FFE tap adds to the amplitude; and if $d_{n,b+t}$ is "-1" the FFE tap subtracts from the amplitude. Similarly, the D matrix in (8.25) becomes:

$$D' = \begin{bmatrix} D'_1 \\ D'_2 \\ \cdots \\ D'_N \end{bmatrix} = \begin{bmatrix} d'_{1,1} & d'_{1,2} & \cdots & d'_{1,m} \\ d'_{2,1} & d'_{2,2} & \cdots & d'_{2,m} \\ \cdots & & \cdots & \\ d'_{N,1} & d'_{N,2} & \cdots & d'_{N,m} \end{bmatrix} \tag{8.29}$$

Transmitter Losses. There are two sources of signal loss associated with the transmitter which must be considered.

The first of these is the insertion loss resulting from bandwidth limitations of the transmitter driver stage. A real transmitter is not capable of generating an ideal NRZ pulse. The Tx_{21} term used in (8.30) defines a low-pass filter which band-limits the transmitter output to a realistic level. The OIF common electrical I/O (CEI) specifies a single pole filter with a corner frequency at 3/4 of the baud rate [1], which is a sufficient model for CML-style circuits.

The second of these is the transmitter return loss due to impedance mismatches associated with the device package. The Tx_{22} term used in (8.30) specifies the package return loss as a function of frequency.

The following matrix defines $Tx\omega$ which represents the combined losses of the transmitter:

$$Tx\langle\omega\rangle = \begin{bmatrix} 1 & Tx_{21}\langle\omega\rangle \\ 1 & Tx_{22}\langle\omega\rangle \end{bmatrix} \tag{8.30}$$

The form of this equation may be recognized as an S-parameter matrix. S-parameter matrices will be described in more detail in the next section.

Channel Response. The frequency response characteristics of the channel are typically measured, and the measured data is used for frequency domain analysis of the link.

Fig. 8.20 illustrates use of a 4-port vector network analyzer (VNA) to measure the frequency response characteristics of the channel for a differential signal. One of the differential ports of the VNA is connected to the differential pair at one end of the channel, and the other differential port is connected to the other end of the channel.

The VNA applies signals of various frequencies to one end of the channel and measures the response signal on each of the ports (including the driving port). Frequencies are tested at regular step intervals starting at a very low frequency and continuing to a frequency higher than the intended baud rate for the channel. Both differential and common mode signals are generated, and both differential and common mode response is measured. All combinations of stimulus ports and response ports are tested. The resulting frequency response data is organized into scattering parameter matrices (commonly called S-parameters), as shown in Fig. 8.20. The nomenclature used to reference the various frequency response matrices is

S_{RSji}

where: R = response type (C = common mode, or D = differential)
        S = stimulus type (C = common mode, or D = differential)
        j = output port (1 or 2)
        i = input port (1 or 2)

The S_{DDxx} are the more relevant parameters in this discussion and are used by subsequent analysis. These S-parameters characterize the transfer function of the channel for differential signals.

- S_{DD11} is the input differential return loss
- S_{DD21} is the input differential insertion loss
- S_{DD22} is the output differential return loss
- S_{DD12} is the output differential insertion loss

The return loss characteristics contribute to signal reflections and the insertion loss characteristics contribute to signal attenuation.

The other S-parameter quadrants are also potentially relevant to the system designer, but are not used by the statistical signal analysis described in this chapter. The S_{DCxx} quadrant characterizes common mode to differential conversion, and is an indication of EMI susceptibility. The S_{CDxx} quadrant characterizes differential to common mode conversion, and is an indication of EMI radiation. The propagation of common mode signals is described by the S_{CCxx} quadrant, and is not of concern for links using a properly designed differential receiver device.

Fig. 8.21 illustrates measurement of the frequency response between the primary differential channel and another nearby channel. This nearby channel is a potential crosstalk aggressor. Two cases are shown in the figure: far-end crosstalk (FEXT) is measured with the VNA connected to the end of the crosstalk channel that is furthest from the connection to the primary channel. Near-end crosstalk (NEXT) is measured with the VNA connected to the end of the crosstalk channel that is nearest to the connection to the primary channel. Which measurement is of significance depends on where the drivers are on each of these channels.

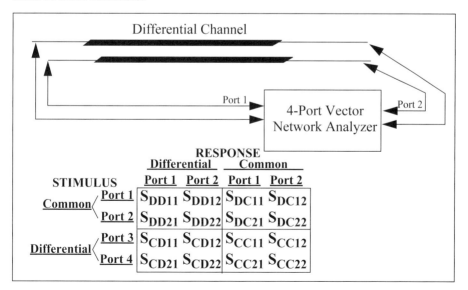

Fig. 8.20 S-parameter measurement using a VNA

Coupling of stimulus from the crosstalk aggressor channel onto the primary differential channel causes noise on the signal which is a key source of jitter. For any given differential channel, any number of NEXT and FEXT aggressor channels may exist which have significant potential to degrade the signal integrity of the channel. Measurements should be taken for all significant crosstalk aggressors, and should be included in the channel analysis.

Receiver Model. A block diagram of a generalized receiver device model is shown in Fig. 8.22. This model introduces loss associated with the device package, compensation due to receiver equalization, and the effects of jitter (both deterministic and random jitter). In the analysis described by the next section, the jitter introduced in this model covers impairments due to both jitter generation by the transmitter and sampling jitter introduced by the CDR circuit in the receiver.

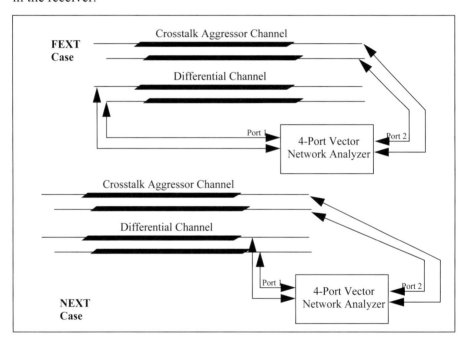

Fig. 8.21 S-parameter crosstalk measurement

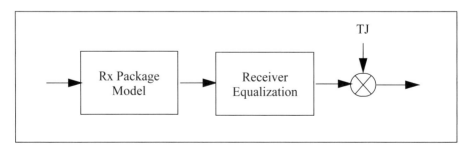

Fig. 8.22 Receiver device model

Receiver Package Model. As was the case for the transmitter, the receiver device package introduces impedance discontinuities which result in return loss. The Rx_{11} term used in (8.31) specifies this package return loss as a function of frequency. This term is captured in matrix form, where $Rx(\omega)$ represents the combined receiver losses:

$$Rx\langle\omega\rangle = \begin{bmatrix} Rx_{11}\langle\omega\rangle & 1 \\ 1 & 1 \end{bmatrix} \tag{8.31}$$

The form of this equation is once again an S-parameter matrix.

The transmitter loss matrix, and the receiver loss matrix are convolved with the channel response, giving a combined transfer function for the channel:

$$
\begin{aligned}
Tr\langle\omega\rangle &= Tx\langle\omega\rangle \otimes \begin{bmatrix} S_{11}\langle\omega\rangle & S_{21}\langle\omega\rangle \\ S_{12}\langle\omega\rangle & S_{22}\langle\omega\rangle \end{bmatrix} Rx\langle\omega\rangle \\[2mm]
&= \begin{bmatrix} 1 & Tx_{21}\langle\omega\rangle \\ 1 & Tx_{22}\langle\omega\rangle \end{bmatrix} \otimes \begin{bmatrix} S_{11}\langle\omega\rangle & S_{21}\langle\omega\rangle \\ S_{12}\langle\omega\rangle & S_{22}\langle\omega\rangle \end{bmatrix} \begin{bmatrix} Rx_{11}\langle\omega\rangle & 1 \\ 1 & 1 \end{bmatrix}
\end{aligned}
\tag{8.32}
$$

where $S_{m,n}$ is the measured 4-port differential S-parameters for the channel. $Tx<\omega>$ and $Rx<\omega>$ were defined by (8.30) and (8.31), respectively.

Receiver Equalization. At higher baud rates, HSS devices typically employ receiver equalization to cancel post cursors of the channel pulse response. This receiver equalization is usually a decision feedback equalizer (DFE) as was described in Sect. 1.3.2. Fig. 8.23 illustrates a 5-tap DFE similar to the DFE associated with the HSS EX10 core that was described in Sect. 2.3.2.

A block diagram of a receiver equalizer with five baud-spaced DFE taps is shown in Fig. 8.23. The receiver model must model the receiver sample point function (which controls the threshold at which the input signal is sampled by the DFE), and the equalizer function of the receiver. Bit values of the previous n samples are multiplied by equalizer coefficients and summed to the input signal to affect the decision as to whether the input bit is a 0 or 1. In this manner, the DFE is capable of equalizing up to n postcursors of the input signal, where n is the number of DFE taps.

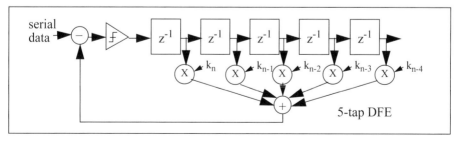

Fig. 8.23 Receiver filter function

The k coefficients of the DFE in Fig. 8.23 form the matrix:

$$K = \begin{bmatrix} 0 \ldots 0 - k_{n-4} \ldots - k_n \ 0 \ \ldots \ 0 \end{bmatrix} \qquad (8.33)$$

where K represents the response of the DFE, and coefficients k_1-k_n are the coefficients of an n-tap DFE. This matrix is the same size as that of the $R(\tau)$ channel response matrix defined in (8.23), with entries corresponding to matrix positions τ labelled $-m/2$ to -1, and $+1$ to $+m/2$. Matrix K in (8.33) contains DFE coefficients k_1-k_n in positions of the matrix corresponding to τ in the range of $1-n$, and 0 in all other positions.

Sample Jitter. The analysis presented in the next section incorporates Data Dependent Jitter (DDJ) by analyzing all possible waveforms propagating through the channel with the transfer function defined in Eqn (8.32). Crosstalk is also incorporated into this analysis. Additional sources of jitter in the system include:

- Transmitter jitter generation, and
- Sampling jitter in the receiver

Deterministic jitter generation in the transmitter is primarily the result of jitter on the clock reference of the transmitter circuit. Dominant forms of deterministic jitter are therefore the result of duty cycle distortion, and periodic jitter. The dual-Dirac model is appropriate to model these types of jitter. In addition, Gaussian jitter results from semiconductor imperfections and quantum effects. Therefore, the equation for total jitter (TJ) in (8.20) provides the PDF for the transmitter jitter generation.

Sampling jitter in the receiver CDR circuit results from similar root causes. The clock reference for this circuit is subject to DCD and PJ, and Gaussian jitter again results from semiconductor-related effects. The equation for total jitter in (8.20) also provides the PDF for the sampling jitter.

The analysis described in the next section injects jitter into the analysis of the received signal using the PDF of the total jitter as defined (8.20). This approach is intended to model both the transmitter jitter generation and the sampling jitter, as described above.

8.4.1.3 Statistical Eye Analysis

Each pulse response waveform at the receiver input is analyzed as illustrated in Fig. 8.24 and as described by the following steps:

Determining Channel Response. The first step in the analysis procedure is to form the channel response matrix for $R(\tau)$ as defined by (8.23).

The channel response is determined by the transfer function for the channel, $Tr(\omega)$, as defined by (8.23). The pulse response of the channel is plotted as shown by the waveform on the left side of Fig. 8.24. The position of the c_0 cursor is chosen arbitrarily on the pulse response waveform, and baud-spaced precursors and postcursors are determined from this arbitrary reference point in the manner described in Sect. 8.4.1.1. These cursors define the $r_n(\tau)$ elements of the $R(\tau)$ matrix.

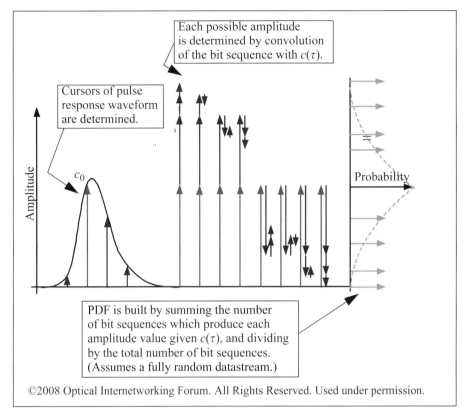

Each possible amplitude is determined by convolution of the bit sequence with $c(\tau)$.

Cursors of pulse response waveform are determined.

c_0

Probability

PDF is built by summing the number of bit sequences which produce each amplitude value given $c(\tau)$, and dividing by the total number of bit sequences. (Assumes a fully random datastream.)

Fig. 8.24 Probability density function for pulse response

Determining the DFE Response. Once the channel response matrix $R(\tau)$ has been determined, static DFE tap coefficient values are selected to cancel postcursors of the channel response. An n-tap DFE can negate up to n postcursors. The DFE tap coefficient values determined in this step are used to form the DFE response matrix K as defined in (8.33).

The channel response, in the presence of receive equalization, becomes the equalized cursors defined by:

$$C(\tau) = R(\tau) - K$$

$$= \begin{bmatrix} r_{-\frac{m}{2}}(\tau) & \cdots & r_{-1}(\tau) & r_1(\tau) - k_1 & \cdots & r_b(\tau) - k_b & r_{b+1}(\tau) & \cdots & r_{\frac{m}{2}}(\tau) \end{bmatrix}$$

$$= \begin{bmatrix} c_{-\frac{m}{2}}(\tau) & \cdots & c_{-1}(\tau) & c_1(\tau) & \cdots & c_b(\tau) & c_{b+1}(\tau) & \cdots & c_{\frac{m}{2}}(\tau) \end{bmatrix}$$

(8.34)

where $r_n(\tau)$ are the cursors of the pulse response at sample point τ, as defined by (7.14), and coefficients k_i (i in the range of $1-b$) are the coefficients of a

b-tap DFE. The nomenclature $c_n(\tau)$, where $c_n(\tau) = r_n(\tau) - k_n$ is used to designate elements of the $C(\tau)$ matrix.

Given an ideal DFE, the k_i (i in the range of $1-b$) coefficients are selected such that the resulting $c_n(\tau)$ elements are zero. Real DFE circuits do not have infinite precision and range, and therefore these $c_n(\tau)$ elements are minimized but may be nonzero.

Building the Probability Density Function. Given a fully random datastream with no transmitter equalization and a limited number of cursors, the cursors superimpose on each other with equal probability.

This is illustrated in Fig. 8.24. The possible signal amplitudes are formed by superimposing the c_0 cursor with each possible combination of $+c_n(\tau)$ or $-c_n(\tau)$ values for the remaining cursors. Each combination of superimposed cursors represents the signal amplitude for the corresponding bit sequence of 1's and 0's. Each bit sequence is equally likely to occur for a fully random datastream, and a PDF for the corresponding signal amplitude can be built by creating a histogram of amplitude values and normalizing the result. The resulting PDF is illustrated on the right side of Fig. 8.24.

Mathematically, the amplitude of the signal for a given sequence of bits and a given sample point (τ) is:

$A_n = d_n \times C(\tau)$

Considering all possible combinations of $d = \{ -1, +1 \}$, the number of bit sequences for which A_n has a given value is

$$A = \sum_n d_n \times C(\tau)$$

and normalizing this (by dividing by the number of patterns) gives the probability density function for a given sample point (τ). For a channel with an ideal transfer function, the signal at the c_0 sample point would have an amplitude of 1, and all precursors and postcursors would have an amplitude of 0. In the presence of ISI, other amplitudes have nonzero probability.

Assuming a fully random datastream and no transmitter equalization, the PDF of the ISI for a given sample point τ, is therefore:

$$p(\text{ISI}, \tau) = \frac{1}{2^m} \sum_{n=1}^{2^m} \delta[(d_n \cdot C(\tau)) - \text{ISI}] \qquad (8.35)$$

This equation sums the number of bit sequences d_n for which the convolution of the bit sequence with the equalized channel response $c(\tau)$ results in a given value of *ISI*. The number of such patterns is divided by the total number of patterns to determine the probability of this value of ISI.

If transmitter equalization is used, this alters the d_n matrix as specified by (8.26). Also, if the datastream is not fully random, then each bit sequence has a probability of occurrence $p_d(n)$ as defined in (8.27).

A more generic form of (8.35) is therefore:

$$p(\text{ISI}, \tau) = \sum_{n=1}^{2^m} p_d(n) \; \delta[(d'_n \cdot C(\tau)) - \text{ISI}] \qquad (8.36)$$

where (8.36) calculates the probability of a given value of ISI for a given sampling point τ. The matrix of these probability values for all possible values of ISI forms the probability density function associated with sampling point τ. Similarly, probabilities and corresponding crosstalk PDFs may be generated for the crosstalk pulse response (using S-parameters measured as described in Fig. 8.21).

Note that simplifications of the above algorithm are described in [3].

Varying the Sampling Point. The arbitrary choice of c_0 in effect chooses a sampling point for the CDR circuit in the receiver. Equation 8.36 calculates the PDF of the ISI given this sampling point. Additional PDFs may similarly be calculated for other values of c_0. By repeating the process in Fig. 8.24 and building PDFs for different c_0, and then weighting these PDFs based on the TJ PDF in (8.20), jitter is incorporated into the analysis.

The CDR sampling point is assumed to be nominally centered, but with some jitter around the ideal sampling point. Some of this jitter is due to jitter sources in the CDR circuit, and some of this jitter is the result of jitter generation in the transmitter. As previously noted, the PDF for the total jitter from the combination of these sources is defined in (8.20).

The pulse response PDFs for the forward channel, the crosstalk PDFs, and the PDF for the sampling jitter may therefore be combined to form a joint probability density function (p_{joint}). This calculation involves convolving the crosstalk PDFs with the forward channel PDFs, and multiplying this result by the PDF for the sampling jitter. This PDF is calculated as follows:

$$p_{\text{joint}}(\text{ISI}, \tau) = \int_{-\infty}^{\infty} \{[p_{\text{xtalk}}(\text{ISI } \tau + \upsilon + w) \otimes p_{\text{fwd}}(\text{ISI } \tau \mp \upsilon)] \cdot p_{\text{jitter}}(\upsilon, w, \sigma)\} d\upsilon \qquad (8.37)$$

where:

$P_{\text{fwd}}(\text{ISI}, \tau)$ is the probability density function of the ISI of the forward channel (from (8.36))

$P_{\text{xtalk}}(\text{ISI}, \tau)$ is the probability density function of the crosstalk (determined in a similar manner to $P_{\text{fwd}}(\text{ISI}, \tau)$, but using S-parameters for the crosstalk channel response as measured in Fig. 8.21)

$P_{\text{jitter}}(\tau, w, \sigma)$ is the dual-Dirac PDF of the transmit and sampling jitter (from (8.20))

The resulting p_{joint} probability density function incorporates the effects of forward channel response, crosstalk channel response, and jitter.

Plotting the Results. Using the combined joint probability density function (p_{joint}) defined by (8.37), the PDF for the signal amplitude at various values of τ is plotted across a 1 UI range as shown in Fig. 8.25(a). Points of these PDFs corresponding to similar BERs are connected as shown in Fig. 8.25(b) to form eye contours. As BER is reduced, the points being connected on the PDF curves move downward on the lower tail of each curve, reducing the amplitude of the signal. In addition, the point at which this eye contour crosses the zero line determines the eye width, and reducing the BER also results in less eye width.

The joint PDF in (8.37) can be integrated to produce the corresponding Cumulative Distribution Function, and can be plotted as a bathtub curve as shown in Fig. 8.25(c). As was described in Sect. 8.2.4, the bathtub curve plots eye width as a function of BER. The slope and y-intercept of the bathtub curve can be used to approximate the decomposition of the jitter into deterministic jitter and Gaussian jitter components as described in Fig. 8.10.

The *statistical eye* shown in Fig. 8.25(d) is determined using the eye contours in Fig. 8.25(b). These contours are cut off at the zero line and plotted on both sides of the decision threshold axis to produce an equivalent receiver eye. The statistical eye shown in the figure is specified for different levels of probability, or circuit Q. The relationship between Q and BER was discussed in Sect. 8.1.1.

The statistical eye for a given Q and corresponding BER indicates the bounds of the eye width and amplitude corresponding to this probability level. Given the statistical eye for a BER of 10^{-12} ($Q = 7.04$), for example, the signal remains outside the contour of this statistical eye opening most of the time, but strays into this contour with a frequency of once every 10^{12} bits. If the eye contour for this BER is sufficiently open for the receiver to correctly receive the signal, then the corresponding BER is 10^{-12} or better.

Generally, the minimum acceptable eye opening is defined by the interface standard or by the HSS receiver vendor. Using measured channel S-parameters and models for the transmitter and receiver, the resulting statistical eye opening is determined. If this eye is at least as open as the minimum acceptable eye opening for the BER of interest, then the channel design meets require-ments. To the extent that the eye is more open than required, margin exists in the system.

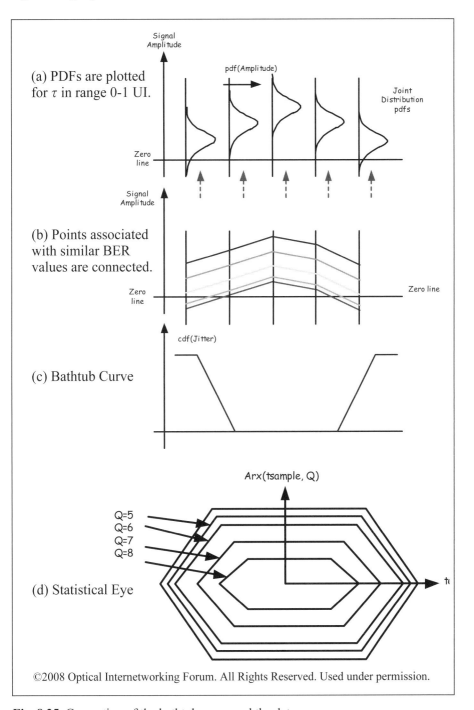

©2008 Optical Internetworking Forum. All Rights Reserved. Used under permission.

Fig. 8.25 Generation of the bathtub curve and the data eye

8.4.2 HSSCDR Software

There are a number of software tools which perform the statistical signal analysis described in the previous section. One or the other of these tools will likely be used by the signal integrity engineer analyzing the system design of the serial data channels associated with HSS cores. It is therefore instructive to provide some description of the data entry and output reports associated with this class of tools.

An open source software tool called *StatEye* is available from [4], and implements the analysis described in Sect. 8.4.1. Another example of such a tool is the IBM HSSCDR software tool, used to analyze serial links for IBM ASIC chips. HSSCDR is used as an example for the descriptions in this section, primarily because the description of the graphical user interface data entry is more straightforward. Although the details vary, the types of data which must be entered to run HSSCDR are representative of the data entry for other statistical-based signal integrity analysis tools including StatEye. Also, the report outputs described are representative of any statistical-based signal integrity analysis tool. All such software tools use statistical methods to calculate the signal eye shape.

The entry screen for the IBM HSSCDR software tool is shown in Fig. 8.26. The entry screen has four columns, corresponding to the transmitter model, channel model, and receiver model for the serial link being analyzed, and a column to define report outputs that are to be generated. To illustrate basic concepts, entry fields are described generally in this section.

8.4.2.1 Transmitter Entry

The entry fields associated with the transmitter model are in the left column of the entry screen in Fig. 8.26.

Core. This field selects the specific IBM HSS core at the transmit end of the serial link.

Technology. This field selects the IBM ASIC process technology for the chip containing the HSS transmitter. In conjunction with the *Core* field, this uniquely selects one of the built-in transmitter models.

Options. This field allows selection of the operating mode for the transmitter. Selections typically provide for enabling/disabling use of the transmitter FFE, predefined FFE coefficient settings, selection of the transmitter amplitude level, etc.

Corner. This field selects worst case, nominal, or best case process, voltage, and temperature conditions for analysis. Analysis should be performed for all process corners since results may vary.

Package. This field selects one of several package models to be used for the transmitter chip package.

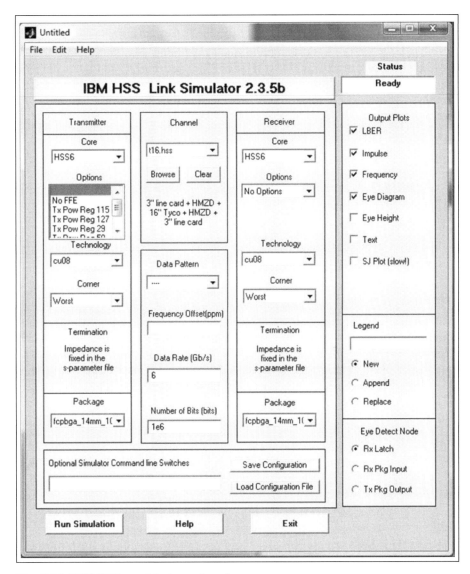

Fig. 8.26 HSSCDR graphical user interface

8.4.2.2 Channel Entry

The entry fields associated with the channel model are in the second column from the left of the entry screen in Fig. 8.26.

Channel. This field selects a command file which defines S-Parameter format options and loads S-parameters for the channel to be analyzed.

Data Pattern. This field selects the data pattern to be analyzed. Scrambled data is fully random and all bit sequences are equally likely. Systems using block coding or scrambled block coding constrain the allowed bit sequences and alter the probabilities associated with each bit sequence as was described by (8.26).

Frequency Offset. For a plesiosynchronous clocked system, this field selects the frequency offset between the reference clocks of the Tx and the Rx device, specified in parts per million (ppm). This is used to calculate the PDF for CDR sampling. More tightly constrained tolerances reduce sampling jitter and improve link performance.

Data Rate. This field selects the baud rate of the signal to be analyzed.

Number of Bits. This field selects the number of data bits to be simulated. Higher numbers result in more statistical accuracy, but increase execution time. HSSCDR uses a default value of 2,40,000 bits which is found to be sufficient for most cases.

8.4.2.3 Receiver Entry

The entry fields associated with the receiver model are in the second column from the right of the entry screen in Fig. 8.26.

Core. This field selects the specific IBM HSS core at the receive end of the serial link.

Technology. This field selects the IBM ASIC process technology for the chip containing the HSS receiver. In conjunction with the *Core* field, this uniquely selects one of the built-in receiver models. Note that the transmitter chip and the receiver chip in a real system may use different HSS cores or even different ASIC technologies.

Options. This field allows selection of the operating mode for the receiver. Selections may provide for enabling/disabling use of the receiver DFE, etc.

Corner. This field selects worst case, nominal, or best case process, voltage, and temperature conditions for analysis. Results should be performed for all process corners since results may vary. Note that the transmitter chip and the receiver chip in a real system may not be operating at the same process corner. All combinations of process corners for the transmitter and receiver chips should be analyzed.

Package. This field selects one of several package models to be used for the receiver chip package.

8.4.2.4 Output Selection

The entry fields associated with output plots and reports are in the right column of the entry screen in Fig. 8.26.

Output Plots. This field selects which of various output plots and reports are to be generated. These are discussed below.

Legend. This field specifies the label used in output plots and reports.

New/Append/Replace. This field determines whether this analysis is appended to or replaces prior analysis in the log file.

Eye Detect Node. This field determines whether output plots and reports are generated based on the signal at the receiver output, the receiver input, or the transmitter output.

The following plots and reports can be generated by the HSSCDR software:

Log Bit Error Rate (LBER) Plot. An example of this plot is shown in Fig. 8.27. The eye width (*x*-axis) is plotted as a function of BER (*y*-axis). This is the bathtub curve output of the statistical signal analysis.

Eye Diagram Plot. An example of this plot is shown in Fig. 8.27. This plots the eye diagram at the node selected by *eye detect node* selection.

Impulse Response Plot. An example of this plot is shown in Fig. 8.28. This plots the impulse response of the channel as defined by the S-Parameters for the channel.

Frequency Response Plot. An example of this plot is shown in Fig. 8.28. This plots the S_{DD21} insertion loss for the channel as defined by the S-parameters for the channel.

Eye Height Plot. This plots the cumulative distribution function for the eye height at receiver output.

Text Output File. This file contains the results of analysis in a text form. Contents of this file are specific to the software tool, and are beyond the scope of this text.

Sinusoidal Jitter (SJ) Plot. This file plots jitter tolerance as a function of sinusoidal jitter.

8.4.2.5 Filter Coefficient Optimization

As has been discussed previously, the HSS transmitter may include an FFE which must be modeled by the transmitter model described in Sect. 8.4.1.2. Likewise, the HSS receiver may include a DFE which must be modeled by the receiver model also described in Sect. 8.4.1.2. Signal integrity analysis results are dependent on these filter functions being tuned to provide optimal results for a given channel.

As was discussed in Sect. 8.4.1.3, the optimal values for DFE coefficients are determined by selecting values which negate postcursors in the pulse response waveform at the input of the receiver. This algorithm is straightforward, and is an integral part of most software tools used to perform statistical eye analysis, including the HSSCDR software.

Optimizing FFE coefficients is less straightforward. Generally, these coefficients are optimized by performing analysis for various coefficient values until the "best" eye opening is achieved at the receiver output. The criteria used to determine the "best" eye opening may vary: algorithms exist which attempt to maximize the eye amplitude, and other algorithms exist which attempt to maximize the eye width. In point of fact, both eye amplitude and eye width contribute to the resulting BER, and therefore some algorithms calculate the overall BER of the eye and use this as a basis for determining the "best" eye opening.

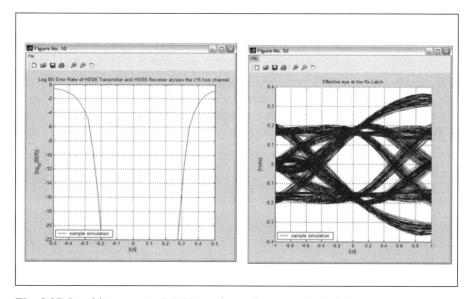

Fig. 8.27 Log bit error rate (LBER) and eye diagram output plots

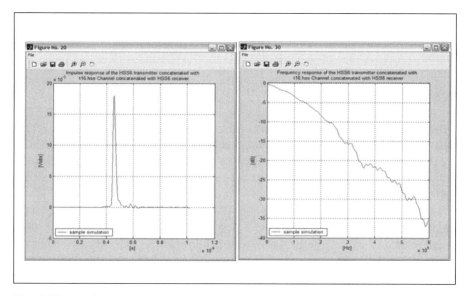

Fig. 8.28 Impulse response and frequency response output plots

Many software tools which perform statistical signal analysis, including HSSCDR, incorporate an algorithm for determining optimal FFE coefficient values. FFE coefficients to be used for analysis may either be forced by the user to specific values, or may be automatically determined. One important use of such software tools is to determine the optimal FFE coefficient values to be used when testing the hardware, and when using the serial link in the system environment.

8.4.2.6 Pass/Fail Criteria

The end goal of using HSSCDR (or any other signal integrity analysis tool) is, of course, to predict whether or not the serial link will operate correctly when all of the channel impairments in an actual application are considered.

The software does not produce a PASS/FAIL result for a simulation. The user must use the program to investigate all of the process corners and environment conditions, including channel variations and crosstalk contributions, and then interpret the results of simulations to determine whether adequate margin exists for the application.

Many of the interface standards discussed in Chap. 5 specify requirements for eye dimensions that must be met at the receiver to claim compliance with the standard. Statistical eye analysis allows users to determine whether or not these criteria are met.

8.5 References and Additional Reading

The following standards documents are applicable to topics in this chapter:

1. "Common Electrical I/O (CEI) - Electrical and Jitter Interoperability agreements for 6G+ bps and 11G+ bps I/O", OIF-CEI-02.0, Optical Internetworking Forum (http:\\www.oiforum.com), Feb. 28 2005.

2. "ANSI INCITS TR-34-2004: INCITS Technical Report for Information Technology - Fibre Channel - Methodologies for Jitter and Signal Quality Specification (MJSQ)", American National Standards Institute, Inc., International Committee for Information Technology Standards, Jan. 1 2004.

The following reading is recommended for more information regarding statistical signal integrity analysis methods and StatEye software:

3. "Channel Compliance Testing Utilizing Novel Statistical Eye Methodology", Anthony Sanders, Mike Resso, John D'Ambrosia, IEC, Designcon, 2004.

4. Open source StatEye Software and additional documentation is available at http\\www.stateye.org.

The following reading is recommended for more information regarding signal integrity analysis in general:

5. "Jitter, Noise, and Signal Integrity at High Speed", Mike Peng Li, Prentice Hall, 2007.

6. "Dwdm Network Designs and Engineering Solutions", Ashwin Gumaste, Tony Anthony, Cisco Press, 2003.

8.6 Exercises

(1) Use (8.1) to calculate $p(\tau, \sigma)$ for x given a Gaussian distribution with μ and σ as specified below:

(a) $x = 2.0, \mu = 3.0, \sigma = 1.5$ (b) $x = 5.0, \mu = 2.0, \sigma = 0.5$

(c) $x = 4.0, \mu = -1.0, \sigma = 1.5$ (d) $x = 1.5, \mu = 1.5, \sigma = 0.75$

(2) Given a Gaussian distribution of x, calculate the probability that x is within $1.5\ \sigma$ of μ.

(3) Use (8.5) to calculate $p(\tau, \sigma)$ for x given a bounded Gaussian distribution with μ, σ, and τ_{max} as specified below:

(a) $x = 2.0, \mu = 3.0, \sigma = 1.5, \tau_{max} = 4.5$

(b) $x = 5.0, \mu = 2.0, \sigma = 0.5, \tau_{max} = 1.5$

(c) $x = 4.0, \mu = -1.0, \sigma = 1.5, \tau_{max} = 5.5$

(d) $x = 1.5, \mu = 1.5, \sigma = 0.75, \tau_{max} = 1.0$

(4) Assume an unbounded Gaussian distribution where z_{rms} has the values specified below. Given the specified BER, what is the z_{p-p} value?

(a) $z_{rms} = 0.02$ UI, BER $= 10^{-12}$ (b) $z_{rms} = 0.020$ UI, BER $= 10^{-15}$

(c) $z_{rms} = 30$ ps, BER $= 10^{-9}$ (d) $z_{rms} = 0.10$ UI, BER $= 10^{-12}$

(5) Use (8.9) to calculate $p(x, A)$ for x given a dual-Dirac distribution with μ and A as specified below:

(a) $x = 2.0, \mu = 3.0, A = 1.5$ (b) $x = 1.5, \mu = 3.0, A = 1.5$

(c) $x = 0.5, \mu = -1.0, A = 1.5$ (d) $x = -0.75, \mu = 0, A = 0.75$

(6) What is the probability of x being in the specified range given a dual-Dirac distribution with μ and A as specified below:

(a) $1.0 < x < 2.0, \mu = 3.0, A = 1.5$ (b) $1.0 < x < 5.0, \mu = 3.0, A = 1.5$

(c) $-2.0 < x < 0.0, \mu = -1.0, A = 1.5$ (d) $x = 0.75, \mu = 0, A = 0.75$

(7) For each of the jitter types below, specify which PDF is used to model this component, state whether this model is optimistic or pessimistic, and state whether equalization can compensate for this type of jitter.

(a) DCD   (b) PJ   (c) BUJ

(d) UUGJ   (e) CBGJ

(8) What type of jitter results from each of the following system contributors:

(a) Transistor device effects

(b) Channel transfer function

(c) Crosstalk from a clock signal

(d) Crosstalk from a data signal (not in the same clock domain)

(9) Use (8.20) to calculate TJ(τ, W, σ) of x given total jitter distribution with $\mu=0$, and with σ and W as specified below:

(a) $x = 0.2$, $\sigma = 0.02$, $W = 0.15$ (b) $x = 0.2$, $\sigma = 0.03$, $W = 0.08$

(a) $x = 0.4$, $\sigma = 0.03$, $W = 0.08$ (b) $x = 0.15$, $\sigma = 0.02$, $W = 0.25$

(10) For each case of σ and W in exercise 9, calculate the eye width given:

(a) BER $= 10^{-9}$ (b) BER $= 10^{-12}$ (c) BER $= 10^{-15}$

(11) Does the jitter budget in Table 8.3 still result in an operational link assuming the following hypothetical cases:

(a) If the UUGJ of the transmitter doubles?

(b) If the UBHPJ of the transmitter increases by 0.100 UI, and the receive equalization is also improved to provide an additional –0.080 UI of jitter compensation.

(12) Figure 8.11 describes the SJ generated as part of a jitter tolerance test.

(a) Draw a figure similar to Fig. 8.11 with jitter amplitudes and frequencies labelled. Assume a CEI-11G-LR link operating at 11.1 Gbps with the jitter budget in Table 8.3 and the skew/wander budget in Table 4.3.

(b) Is SJ typically encountered in a real system? Explain.

(c) At lower frequencies the SJ in part (a) of this question is several UI. Why does the link work even with this much jitter?

(13) Using the HSS EX10 description in Chap. 2, speculate as to which I/O pins and register bits may affect each of the control inputs to:

(a) The Spice model for the HSS EX10 transmitter in Fig. 8.12 which was produced using traditional circuit extraction methods.

(b) The hybrid Spice model for the HSS EX10 transmitter in Fig. 8.13.

(14) Spice simulation is to be performed to verify signal integrity of a link which uses HSS EX10 cores on both the transmitter and receiver end. The transmitter power level, slew rate, and preemphasis settings have already been determined. The data coding, data rate, coupling, and termination have also been specified for the link. The Spice models support selection of best case, nominal, and worst case PVT conditions. Create a simulation matrix which indicates the Spice runs that must be performed to validate signal integrity under all PVT conditions.

(15) A block code is devised which has a maximum run length of 2 bits. Given $m = 5$ cursors of the channel response are significant, devise a D matrix as described by (8.25) which represents all possible bit sequences allowed by this block code.

(16)	Assume a 3-tap FFE with the coefficients $c_{t-1} = 0.15$, $c_t = 0.80$, and $c_{t+1} = -0.05$. Given the D matrix from exercise 15, construct a D' matrix as described by (8.29) which represents the equalized bit sequences.

(17)	Given S-parameters of a channel, in general $S_{DD21} \neq S_{DD12}$ and $S_{DD11} \neq S_{DD22}$. Explain why this is the case.

(18)	Draw a diagram of a channel consisting of a transmitter chip on one circuit board driving a link to a receiver chip on a different circuit board, and connected through a backplane. Label the various points on this channel where impedance discontinuities may occur.

(19)	The HSS EX10 core described in Chap. 2 contained both transmitters and receivers. The transmitter slices and receiver slices are physically alternated on this core. For this core configuration, which type of crosstalk (FEXT or NEXT) is more likely to be a significant factor for signal integrity?

(20)	A system designer chooses to use simplex cores to implement an interface to reduce link signal integrity issues due to NEXT. The resulting links still have FEXT which must be considered in the analysis. Why does this system designer prefer to deal with the FEXT instead of the NEXT?

(21)	In (8.32) the channel S-parameters are convoluted with the S-parameters for the transmitter and receiver package models. This equation reflects the fact that the chip packages are part of the channel interconnect between the driver circuit and the receiver circuit. Explain why it is not easier to simply include the transmitter and receive packages in the S-parameter measurements for the channel.

(22)	Given the output plot of the bathtub curve shown in Fig. 8.27, what is the approximate eye width for each of the following BER values?

	(a) BER $= 10^{-9}$	(b) BER $= 10^{-12}$		(c) BER $= 10^{-15}$

(23)	Given the output plot of the S_{DD21} channel response shown in Fig. 8.28, what is the loss at each of the following frequencies?

	(a) 4 GHz		(b) 2 GHz			(c) 1 GHz

(24)	Many systems use simulations (using StatEye, HSSCDR, or a similar tool) to determine a set of FFE coefficients that works for all links regardless of how cards are populated on the backplane. What is the disadvantage of this approach as opposed to dynamic training as was described for the Backplane Ethernet standard described in Chap. 5?

Chapter 9
Power Analysis

In Chap. 1, HSS cores were introduced as being the result of increases in silicon density outpacing increases in the pin densities of chip packaging technologies. Increases in silicon circuit density have also outpaced advances in the ability of chip packages to dissipate heat. Predicting and controlling the power dissipation of the chip design has become an increasingly important part of chip design.

The multiplexing and demultiplexing functions of HSS cores operate at very high frequencies, and the associated signals have high activity factors. High frequencies typically require higher voltages in order to provide the necessary circuit performance, and higher voltages increase power dissipation. The nature of the function performed by HSS cores dictates that these cores are power hungry devices.

The power consuming circuits in the HSS core can be categorized into two basic types: digital logic circuits and nondigital logic circuits. Power in nondigital logic circuits can be subcategorized as follows: AC (active) power, DC (leakage) power, and DC quiescent power. Digital Logic AC power is a function of the applied voltage (V_s), the frequency (f) of operation, and the activity factor (AF) for the given circuit. DC leakage power is a function of the chip process technology, transistor threshold voltage (V_t) temperature, and the supply voltage (V_s). DC quiescent power is a function of the circuit design and results from the amount of continuous power that is needed to sustain the circuit (for example, to operate an amplifier). The factors that affect power dissipation are discussed in this chapter, along with methods the designer can use to control the power dissipation of the HSS core.

9.1 Digital Logic Circuits

The discussion of power dissipation for digital logic circuits can be broken into the categories of AC (active) power and DC (leakage) power.

9.1.1 Digital Logic Active or AC Power

AC power can be derived several different ways and for this discussion can start with the well-known physics textbook power relationship

$$P_s = I_s \cdot V_s, \tag{9.1}$$

where P_s is the power delivered from the power supply, I_s is the supply current, and V_s is the supply voltage. Another well-known relationship from physics textbooks is the definition of capacitance:

$$C = \frac{q}{V_c}, \tag{9.2}$$

D. R. Stauffer et al., *High Speed Serdes Devices and Applications,*
© Springer 2008

where C is the capacitance value, q is the charge on the capacitor, and V_c is the voltage across the capacitor.

Equation (9.2) can be rewritten as:

$$V_c = \frac{q}{C}. \tag{9.3}$$

Assuming C is a constant and differentiating (9.2) with respect to time results in

$$\frac{d}{dt}V_c = \frac{d}{dt}q \cdot \frac{1}{C} \tag{9.4}$$

and by definition:

$$I = \frac{dq}{dt}. \tag{9.5}$$

Combining (9.4) and (9.5) results in

$$\frac{dv}{dt} = \frac{I}{C}. \tag{9.6}$$

The energy stored on a capacitor can be assumed to have the following relationship:

$$E_c = \frac{1}{2}C \cdot V_c^2, \tag{9.7}$$

where C is the capacitor value, and V_c is the voltage on the capacitor. This relationship can be derived from the previous physics relationships as is shown below. The energy stored on a capacitor is simply a function of the static voltage on the capacitor V_c and capacitance value C, and is not a function of the time it takes to charge the capacitor.

Assuming a capacitor of value C is being charged with a current I, the voltage on the capacitor rises at a constant rate as defined by (9.6). This equation can be rewritten as:

$$dv = \left(\frac{I}{C}\right)dt. \tag{9.8}$$

Integrating Eqn 9.8 with respect to t and assuming $V(0) = 0$ results in

$$V_c(t) = \left(\frac{I}{C}\right) \cdot \int_0^t dt = \left(\frac{I}{C}\right)t. \tag{9.9}$$

Assuming a constant current I, the energy to charge this capacitor after a time T (in seconds) is:

$$E(T) = \int_0^T P(t)dt = I \cdot \int_0^T V_c(t)dt \tag{9.10}$$

Substituting $V_c(t)$ from (9.9) into (9.10) results in:

$$E(T) = \int_0^T \left(\frac{I}{C}\right) \cdot t \, dt = I \cdot \frac{I}{C} \cdot \frac{t^2}{2}\Big|_0^T = \frac{T^2}{2} \cdot \frac{I^2}{C} \tag{9.11}$$

From (9.9), the voltage on the capacitor after T seconds is

$$V_c(T) = \left(\frac{I}{C}\right) T.$$ (9.12)

Combining Eqn 9.11 and Eqn 8.38, and rearranging the result:

$$E(T) = \frac{C}{C} \cdot \frac{T^2}{2} \cdot \frac{I^2}{C} = \frac{C}{2} \cdot \left(T \cdot \frac{I}{C}\right) \cdot \left(T \cdot \frac{I}{C}\right) = \frac{1}{2} \cdot C \cdot V_c^2.$$ (9.13)

which is equivalent to (9.7). This equation also implies that it does not matter how much time (T) is required to charge the voltage (V_c), the energy stored on it is the same. Although this analysis assumes I_c and dv/dt are constants (and this text will continue to do so for simplicity), it can also be shown that (9.7) remains valid even if these parameters are variable over time. Proof of this is beyond the scope of this text.

The above analysis leads to a simple model using a lump sum capacitance which can be used for analysis of digital logic power relationships in the CMOS chip as shown in Fig. 9.1a. The model shown in the figure consists of a power supply delivering power through a current limiting device to charge the lump sum capacitance of the network (represented by C in prior analysis). The "current limiting device" represents the transistor of the CMOS logic gate, and the "network" is the output net of the CMOS logic gate which is connected to transistor inputs of other CMOS logic gates. Assuming I_s and V_s are constant, then the power is calculated as described in (9.1), and the supply energy (E_s) provided by the power supply to charge the capacitance of the network is:

$$E_s = P_s \cdot T = I_s \cdot V_s \cdot T$$ (9.14)

Rearranging Eqn 9.6, the following equation is obtained

$$I_s = C \cdot \frac{dv}{dt}$$ (9.15)

and substituting Eqn 9.15 into Eqn 9.14 results in the equation

$$E_s = C \cdot \frac{dv}{dt} \cdot V_s \cdot T$$ (9.16)

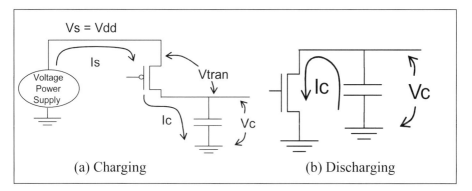

(a) Charging   (b) Discharging

Fig. 9.1 Network model for AC power

Allowing for the capacitor to charge for a time (T) until the capacitor is fully charged to the power supply voltage (V_s), and making the simplifying assumption that the rate of change of the voltage (dv/dt) is constant during this period, the following relationship results:

$$V_s = \frac{dv}{dt} \cdot T \tag{9.17}$$

Rearranging Eqn 8.43, and substituting this into (8.42)

$$E_s = C \cdot (V_s)^2 \tag{9.18}$$

The conclusion implied by (9.18) is that the energy delivered by the power supply to the network is twice the charge stored on the lump sum capacitance of the network (as was indicated by (9.13)). The difference between the supply energy (E_s) and the energy stored on the capacitor (E_c) is dissipated by the current limiting device and is released as heat.

Now consider how this model appears when the capacitance is being discharged. Fig. 9.1b shows the charge stored on the capacitance discharging through another transistor to ground. During this discharge, the transistor dissipates all of the energy stored on the capacitor as heat, and the power supply does not supply any additional energy to the network.

As the network capacitance is charged and then discharged through one complete cycle, the total energy supplied by the power supply is specified by (9.18). All of this energy is dissipated as heat during the cycle: one-half of it during the charging phase and one-half of it during the discharging phase. Assuming the network switches states at a constant rate of F cycles per second, the resulting power dissipation is

$$P_s = E_s \cdot F \tag{9.19}$$

where E_s (joules) is the power supply energy for one cycle and P_s (joules/sec or watts) is the power supply power at the specified frequency.

Substituting (9.18) for E_s results in

$$P_s = C \cdot (V_s)^2 \cdot F \tag{9.20}$$

If all of the logic gates on the chip switched at the same rate as defined by F, then (9.20) could be extrapolated to calculate supply power (P_s) for the entire chip by using a value for C which corresponds to the sum of all net capacitances on the chip. Of course, some signals in the chip switch more often than other signals, and using the same value of F for all circuits would not be appropriate. It therefore becomes useful to introduce the concept of an average Activity Factor (AF) defined as follows:

$$AF = \frac{F_s}{F_c} \tag{9.21}$$

where F_s is the frequency of the signal and F_c is the frequency of the clock associated with this logic. The Activity Factor (AF) is the ratio of these frequencies.

The Activity Factor can be calculated in the following manner: F_s is determined by the number of times the signal toggles divided by the time period (Tp) over which the measurement is taken, and divided by 2 since both a rising and falling transition are required to produce one cycle of the signal. F_c is calculated in a similar manner for the clock signal. This results in

$$AF = \frac{\left(\frac{(\Sigma TG_s)}{2 \cdot Tp}\right)}{\frac{(\Sigma TG_c)}{2 \cdot Tp}} = \frac{(\Sigma TG_s)}{(\Sigma TG_c)} \qquad (9.22)$$

where TG_s is the number of signal toggles and TGc is the number of clock toggles summed over the time period Tp. The Tp terms cancel, and the equation reduces to the ratio of toggles of the signal over toggles of the clock. Of course, a real signal does not toggle at a constant rate, but rather may have bursts of activity followed by quiescent periods. However, if toggle activity is measured over a sufficiently long period, the resulting value of F_s is representative of average activity, and is an accurate predictor of the average power consumed.

Given the relation:

$$F_s = F_c \cdot AF \qquad (9.23)$$

and substituting for F in (9.20), the result is

$$P_{s_{ac}} = C \cdot F_c \cdot AF \cdot (V_s)^2 \qquad (9.24)$$

This power analysis approach can be extended beyond that of a single signal and can be applied to all signals in a region of a chip or across an entire chip. In the context of (9.24), scope is limited to the digital logic domain of the HSS core. All net capacitances on CMOS digital signals in the HSS core can be lumped together into a single lumped capacitance term and used in (9.24). Each interconnect wire has a set of capacitance components associated with it, including wire-to-ground, wire-to-substrate, wire-to-wire, and gate-to-substrate capacitances, all of which should be included. The activity factor AF is calculated as a weighted average of the activity factors of the individual signals, where the net capacitance is used to weight this calculation.

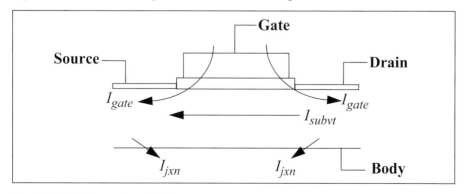

Fig. 9.2 CMOS transistor leakage currents

The C term in (9.24) is more or less proportional to the area of the digital logic in the core and affects power dissipation linearly. The F_c and AF terms also have a linear relationship to power dissipation. However, the Vs^2 term indicates that the value of V_s has a dramatic impact on power dissipation. Small reductions in V_s can substantially reduce power dissipation.

9.1.2 Digital Logic Leakage or DC Power

Digital Logic Leakage power (DC power) is proportional to the total leakage current in all of the transistors in the digital logic section of the HSS core. This leakage current in each transistor is made up of several different leakage mechanisms including:

1. Subthreshold Leakage Current (I_{subvt})
2. Gate dielectric (tunneling) Current (I_{gate})
3. Junction Leakage current (I_{jxn}), and
4. Gate Induced Drain Leakage Current (I_{gidl})

The above leakage components are illustrated in Fig. 9.2, and account for the majority of the total leakage current. Lesser contributors are not considered in this text. The total leakage current, for purposes of this discussion, is therefore the sum of these components

$$I_{leakage} = I_{subvt} + I_{gate} + I_{jxn} \quad I_{gidl} \qquad (9.25)+$$

Note that each of the leakage components in (9.25) has a different dependency on voltage and temperature.

9.1.2.1 Subthreshold Leakage Current

I_{subvt} is the leakage current between the CMOS transistor drain and source terminals when the device is turned off. I_{subvt} is determined by the magnitude of the threshold voltage and the slope of the subthreshold voltage region of the current $-$ voltage (IV) curve for the transistor.

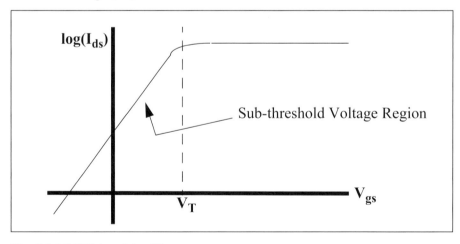

Fig. 9.3 NMOS transistor IV curve

Fig. 9.3 illustrates the ideal *IV* curve for an NMOS transistor, which plots the log of the drain-to-source current (I_{ds}) as a function of the gate-to-source voltage (V_{gs}). The subthreshold region of this curve corresponds to V_{gs} values below the transistor threshold voltage (V_t); in this region I_{ds} increases exponentially as a function of V_{gs}. Typically $I_{ds} > 0$ mA when $V_{gs} = 0$ V; the value of I_{off} is defined as I_{ds} at $V_{gs} = 0$ V.

This drain current in the sub threshold region of the curve can be calculated by the following equation [3]

$$I_{subvt} = \mu(C_d + C_{it})\frac{Z}{L}\left(\frac{kT}{q}\right)^2\left(1 - e^{\frac{-qV_{ds}}{kT}}\right)\left(e^{\frac{q(V_{gs} - V_t)}{C_\tau kT}}\right), \qquad (9.26)$$

where

V_{ds} = drain-to-source voltage on the transistor
V_{gs} = gate-to source voltage on the transistor
V_t = threshold voltage of the transistor
T = junction temperature of the transistor (in Kelvin)
k = Boltzmann's constant
q = magnitude of the electronic charge
L = transistor channel length
Z = transistor channel width
μ = low field mobility of the semiconductor material
C_d = depletion layer capacitance of the semiconductor material
C_i = insulator capacitance of the semiconductor material
C_{it} = fast interface state capacitance of the semiconductor material
$C_\tau = [\,1 + (C_d + C_{it})\,/\,C_i\,]$

The C_d, C_i, and C_{it} capacitance parameters, as well as μ, are determined by the semiconductor material. Given a particular chip fabrication process, the values of these parameters are subject to process variation, but are otherwise constants for a given chip. The Z, L, and V_t parameters are determined by the transistor design. Some variation of V_t may occur due to process variation, but these parameters are also otherwise constants for a given transistor on a given chip. Operating conditions are specified by the V_{ds}, V_{gs}, and T terms.

Equation (9.26) can therefore be rewritten as a function of the operating conditions and various constants:

$$I_{subvt} = K1(T)^2\left(1 - e^{\frac{K2(V_{ds})}{T}}\right)\left(e^{\frac{K3(V_{gs})}{T}}\right)\left(e^{\frac{K4}{T}}\right),$$

$$(9.27)$$

$$K_1 = \mu\,C_d + C_{it}\frac{Z}{L}\left(\frac{k}{q}\right)^2 \quad K_2 = \frac{-q}{k} \quad K_3 = \frac{q}{C_\tau k} \quad K_4 = \frac{-qV_t}{C_\tau k}.$$

Equation (9.27) is further reduced by substituting the conditions for I_{off}: $V_{ds} = V_s$, and $V_{gs} = 0V$. Given that $Vs >> q/kT$, the dependencies on V_{ds} and V_{gs} are eliminated and (9.27) becomes

$$I_{subvt} = K_1 T^2 \left(e^{\frac{K_4}{T}} \right) \tag{9.28}$$

Equation (9.28) implies that I_{subvt} has a strong dependence on temperature, but no dependence on Vs. However, this equation does not provide a complete picture. The dependence of (9.26) (and by extension (9.28)) on V_{ds} assumes threshold voltage (V_t) and (L) are fixed quantities for the transistor device. In reality, V_t is dependent on both V_{ds} and L. As the device electric field increases (either by increasing V_{ds} or shortening L) the drain induced barrier lowering (DIBL) effect results in a lowering of the threshold voltage, which correspondingly increases I_{subvt}. The threshold voltage also decreases as the temperature increases, decreasing the slope of the IV curve in the subthreshold voltage region, and further enhancing the I_{off} leakage current.

This effect on I_{off} is illustrated in Fig. (9.8) as an upward shift in the point at which the IV curve crosses the $V_{gs} = 0$ axis as V_{ds} is increased. Increased I_{off} can also be caused by decreasing channel length (L). Figures 9.4 and 9.5 show the relationship of I_{off} to decreasing channel length and increasing $V_{ds} = V_s$, respectively, and are based on characterization data collected for an example of an ASIC chip fabrication technology. Fig. 9.4 also shows data collected for various junction temperatures. The characterization data suggests an exponential relationship between these parameters and the corresponding I_{subvt} leakage current. For operating temperatures greater than about 50°C this leakage mechanism dominates over all others.

The empirical data suggests that I_{subvt}, when the effects of DIBL are included, is exponentially related to the power supply voltage. Also, the empirical data suggests that the I_{subvt} dependence on T is also exponential once DIBL is considered. This implies that the terms in (9.28) containing T can be replaced by an exponential dependency. The following equation results

$$I_{subvt} = K_{subvt} \left(e^{\frac{T}{T_{REF}}} \right) \left(e^{\frac{Vs}{V_{REF}}} \right) \tag{9.29}$$

where:

V_s = power supply voltage (in volts)

T = junction temperature of the transistor (in Kelvin)

V_{REF} = empirical constant (in volts) derived from transistor characterization data

T_{REF} = empirical constant (in Kelvin) derived from transistor characterization data

K_{subvt} = empirical constant (in amps) derived from transistor characterization data

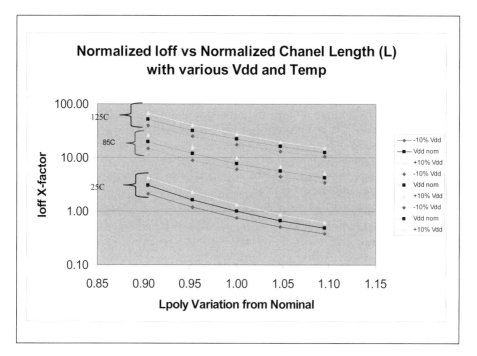

Fig. 9.4 I_{off} relationship to channel length

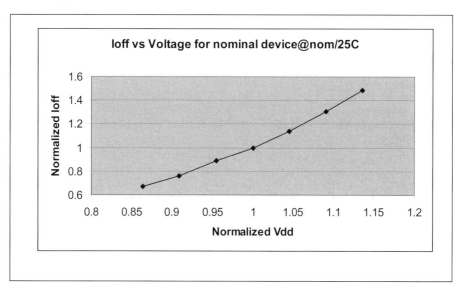

Fig. 9.5 I_{off} relationship to power supply voltage

Equation (9.29), with appropriate constants derived empirically from characterization data, is a reasonable approximation of I_{subvt} assuming that values
of V_s and T are in typical operating ranges: $V_s > 2V_t$ and $T > 0C$ (273K).
Behavior as these parameters approach zero is more complex, and other terms

in (9.26) become significant. However, given that circuit speed is important to HSS devices, lower supply voltages are generally not used, and the above restriction can be assumed. Furthermore, nominal/maximum temperatures used in calculating power dissipation for typical circuit applications generally exceed the above limit.

The above equations applied to a single transistor device, and the constants in these equations applied to specific transistor characteristics. One approach to calculating the I_{subvt} leakage current for the overall circuit would be to sum the contributions of the individual transistors in the circuit. However, note that the number of transistors turned off ($V_{gs} = 0\,V$ and $V_{ds} = V_s$) at any given time is state dependent. The form of (9.29) can be extended to model the total I_{subvt} leakage current for the HSS core by assuming that the constants in (9.29) are derived using characterization data which measures power dissipation for the HSS core rather than individual transistors. This is represented by the following equation

$$Ic_{subvt} = Kc_{subvt}\left(e^{\frac{T}{Tc_{REF}}}\right)\left(e^{\frac{Vs}{Vc_{REF}}}\right),\qquad(9.30)$$

where:
V_s = power supply voltage for the core (in volts)
T = junction temperature of the core (in Kelvin)
Vc_{REF} = empirical constant (in volts) derived from HSS core characterization
     data
Tc_{REF} = empirical constant (in Kelvin) derived from HSS core
     characterization data
Kc_{subvt} = empirical constant (in amps) derived from HSS core
     characterization data

The constants in (9.30) differ from those in the prior equations in that these constants apply to the overall HSS core logic rather than individual transistors. Kc_{subvt}, Vc_{REF}, and Tc_{REF} are not calculated, but rather are empirically derived through measuring power dissipation as part of characterization testing of the HSS core in a laboratory environment. As such, these coefficients are inherently based on the average transistor characteristics and the average number of transistors for which $V_{gs} = 0\,V$ applies at any given point in time (thereby take into account the dependencies on circuit state). Equation (9.30) is therefore a reasonable approximation of the Ic_{subvt} leakage current for the digital logic in the HSS core (and for nondigital circuits as well if such logic is included in the characterization test per Sect. 9.2.2).

9.1.2.2 Gate Dielectric or Tunneling Current

I_{gate} is the leakage current between the gate electrode and the substrate. This current results from a tunneling leakage mechanism whereby the electrons directly tunnel from the silicon surface to the gate through the forbidden energy gap of the SiO_2 dielectric layer. When the gate voltage is high with respect to the channel this leakage current adds to the I_{on} current, however, the additional current must be sourced from the node driving the gate. When the gate voltage is low with respect to the channel, this leakage is sourced from the channel and flows to ground.

The I_{gate} leakage current has a weak dependence on temperature and an exponential dependence on power supply voltage (V_s). This leakage current is also exponentially dependent on oxide thickness, with thinner oxide thickness resulting in larger I_{gate} leakage. Gate tunneling current is more pronounced in thinner oxide device gates (around 12 Å), and less of an issue with transistors designed for low power applications which usually employ thicker gate oxides (around 18 Å).

If the HSS core (or the chip) implements a stand-by mode in which portions of the core are powered down while selected FETs remain active to hold state values, then the gate tunneling current imposes a lower limit on stand-by power dissipation. Gate tunneling currents sourced by active FETs continue to tunnel through the gates of inactive FETs while the core is in stand-by mode. Conversely, the target stand-by power dissipation specification for the core imposes a limit on the number of FETs which may be active while the core is in stand-by mode.

9.1.2.3 Junction Leakage Current

I_{jxn} is the leakage current between the cathode and the anode of the reverse biased junction of the drain (and source) to the well (body) terminals of the transistor. This leakage component is the result of band-to-band tunneling as illustrated in Fig. 9.6. When the channel (either the drain or the source) of the transistor is biased at a higher voltage with respect to the substrate, electrons (for an n-channel device) tunnel from the valence band of the p-region to the conduction band of the n-region. For a p-channel device, holes tunnel in the opposite direction.

The reverse-biased junction of a CMOS transistor is shown in Fig. 9.7. The source and drain of the transistor are assumed to be at a higher voltage than the substrate in this figure. Under these conditions, leakage currents result from electrons (or holes) tunneling across the reverse-biased junctions into the substrate.

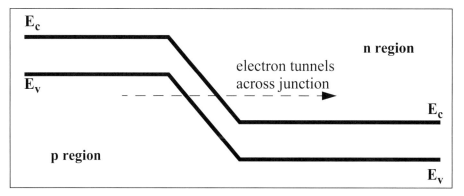

Fig. 9.6 Tunneling through a reverse biased pn junction

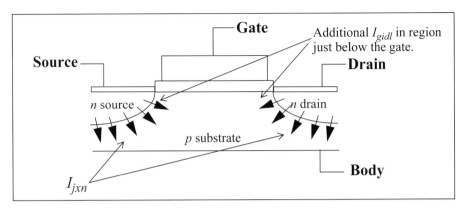

Fig. 9.7 CMOS transistor junction leakage currents

This leakage current is an exponential function of the voltage from the drain (or source) to the body terminal (V_{db} or V_{sb}) of the transistor [2]. Assuming $V_d = V_s$ and $V_b = 0$, then

$$V_{db} = V_{sb} = V_s$$

and an exponential dependence once again exists on V_s.

9.1.2.4 Gate Induced Drain Leakage Current

I_{gidl} is the additional leakage current, also caused by band-to-band tunneling, between the reverse biased junction and the channel of the transistor in the presence of the vertical gate field. This gate field can result in an amplification of the standard reverse bias junction leakage. As is shown in Fig. 9.8, at some $V_{gs} < 0\,\text{V}$ point the I_{ds} current increases with increasingly negative V_{gs} voltages. As V_{ds} increases, the trough of this curve moves toward $V_{gs} = 0\,\text{V}$, resulting in a delta between the ideal I_{jxn} leakage and the measured leakage.

In Fig. 9.7, I_{gidl} results at the $p- \; n$ junctions immediately below the gate. Without an electric field, I_{jxn} current would be uniform across the entire $p- \; n$ junction. However, the electric field induced by the gate causes additional current to flow in the immediate region below the gate.

I_{gidl} is weakly temperature dependent, but exponentially dependent on voltage in a similar manner to I_{jxn}. Worst case leakage current occurs when $V_d = V_s$.

I_{gidl} is primarily a concern with low power technologies where the threshold voltage has been increased to limit the I_{subvt} component. Such technologies reduce leakage currents, but the higher threshold voltage reduces switching speeds of the transistor devices. HSS cores typically do not use low power technologies since this conflicts with requirements to achieve high baud rates.

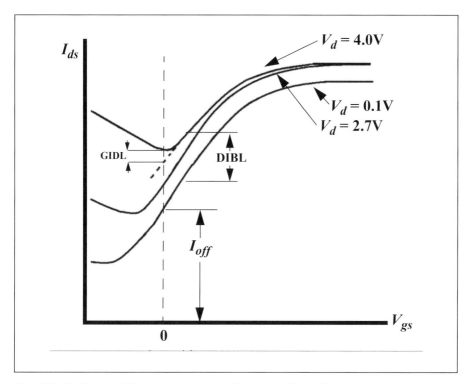

Fig. 9.8 N-Channel IV curve defining various tunneling effects

9.1.2.5 Total Leakage Power

Equation (9.30), repeated here, was noted as a reasonable approximation for total Ic_{subvt} of the HSS core

$$I_{c_{subvt}} = K_{c_{subvt}} \left(e^{\frac{T}{Tc_{REF}}} \right) \left(e^{\frac{Vs}{Vc_{REF}}} \right)$$

In general, the Ic_{subvt} contribution to leakage current dominates over the other leakage contributors. (I_{gidl} can be significant for high V_t devices used in low power technologies, but as noted previously this is not usually a concern for HSS cores.) Therefore, (9.30) is a reasonable approximation for the overall leakage current for the HSS core

$$I_{c_{leak}} \cong K_{c_{leak}} \left(e^{\frac{T}{Tc_{REF}}} \right) \left(e^{\frac{Vs}{Vc_{REF}}} \right), \tag{9.31}$$

where the Kc_{leak}, Vc_{REF}, and Tc_{REF} constants in this equation are the result of characterization testing of the HSS core considering all leakage currents, and not just the Ic_{subvt} leakage component.

The total power dissipation associated with (9.31) is therefore

$$Pc_{\text{leak}} \cong Ic_{\text{leak}} \times Vs = Kc_{\text{leak}} \left(e^{\frac{T}{Tc_{REF}}} \right) \left(e^{\frac{Vs}{Vc_{REF}}} \right) \times V_s \quad (9.32)$$

Once again the value of V_s has a dramatic impact on power consumption. Small reductions in V_s can substantially reduce power dissipation. Maximum junction temperature also has a significant impact on the power dissipation, and circuit operation at lower temperatures can substantially impact power dissipation.

9.2 Non Digital Logic Circuits

As was the case for digital logic circuits, the discussion of power dissipation for nondigital logic circuits can also be broken into the categories of AC (active) power and DC (leakage) power. In addition, DC quiescent power also contributes to the power dissipation for nondigital logic circuits.

9.2.1 AC (Active) Power

Active power for nondigital logic circuits results from similar switching activity of the signals to that of digital logic circuits, and is calculated in a similar manner. Eqation (9.24) described this calculation, and can be applied to digital or nondigital logic circuits (or the combination thereof) based on the scope of the C and AF parameters used in the equation.

9.2.2 DC (Leakage) Power

Leakage power for nondigital logic circuits results from similar mechanisms to that of digital logic circuits, and is calculated in a similar manner. Equation (9.32) described this calculation, and can be applied to digital or nondigital logic circuits (or the combination thereof) based on the scope of the characterization testing used to determine the Kc_{leak}, Vc_{REF}, and Tc_{REF} values.

9.2.3 Quiescent Power

The quiescent current (I_q) in a nondigital logic circuit is defined as the DC current drawn by the circuit while the circuit is in a quiescent state with no activity. By definition, power is related to I_q as follows

$$Ps_q = I_q \cdot V_S, \quad (9.33)$$

where Ps_q is the power delivered by the power supply, I_q is the steady-state circuit current, and V_s is the supply voltage. I_q results from DC current paths in a nondigital (analog) circuit, and does not include any leakage current. Some DC steady-state current paths approximate the behavior of current sources for which power dissipation is proportional to V_s. Other DC steady-state current paths approximate the behavior of resistances for which power dissipation is proportional to Vs^2. Power dissipation due to quiescent current paths is therefore modeled by the following equation

$$P_{s_q} = (K_{1_q} \cdot Vs) + K_{2_q} \cdot Vs^2 \tag{9.34}$$

where constants $K1_q$ and $K2_q$ are empirically determined for a given HSS core.

9.3 HSS Power

This section combines the power equations developed previously, and discusses topics related to power dissipation for the overall HSS core.

9.3.1 HSS Power Equation

The power in the HSS core is the sum of all types of power consuming circuits discussed previously, including digital and nondigital circuits; AC power, DC leakage power, and DC quiescent power. Equation (9.35) summarizes this relationship

$$P_{hss} = Ps_{ac} + Pc_{leak} + Ps_q \tag{9.35}$$

where P_{hss} is the total power dissipated and consumed from the supply by the HSS core. The Ps_{ac} and Ps_{leak} terms are calculated as described in (9.24) and (9.32), and calculated such that AC power and leakage power for all digital and nondigital circuits of the HSS core are included in these terms. The Ps_q term for nondigital circuits is calculated using (9.34).

When the underlying equations for these terms are substituted into (9.35), the following equation results

$$\begin{aligned} P_{hss} = (C \cdot F_c \cdot AF \cdot V_s^2) + \\ (K1_q \cdot V_s) + (K2_q \cdot Vs^2) + \\ (Kc_{leak} \cdot \exp(T/Tc_{REF}) \cdot \exp(V_s/Vc_{REF}) \cdot V_s) \end{aligned} \tag{9.36}$$

Equation (9.36) can further be rearranged as follows:

$$\begin{aligned} P_{hss} = [(C \cdot F_c \cdot AF) + K2_q] \cdot V_s^2 + \\ [(Kc_{leak} \cdot \exp(T/Tc_{REF}) \cdot \exp(V_s/Vc_{REF})) + K1_q] \cdot V_s \end{aligned} \tag{9.37}$$

where P_{hss} is a function of V_s, T, F_c, plus various empirically derived coefficients. Equation (9.37) is strongly dependent on V_s, with linear, quadratic, and exponential terms associated with the power supply voltage.

While (9.37) does combine all of the power dissipation factors into a single equation, this is not the end of the story. Often, multiple sets of coefficients may be supplied for (9.37). These various sets of coefficients may reflect the following depending on the applicability to a given HSS core:

- HSS cores consist of some number of transmitter, receiver, and PLL slices as shown in Fig. 9.9. Coefficients are generally specified separately for each slice to facilitate scaling the calculation of P_{hss} to arbitrary core configurations.

- In cases where the HSS core requires more than one power supply, coefficients are specified separately for each power supply.

- Coefficients are generally specified for various chip fabrication process points (nominal, worst case, best case).

- If the HSS core includes modes of operation where power dissipation is reduced, coefficients are generally specified for each of these modes.

Generally, the coefficients associated with P_{hss} calculations are provided to the chip designer for each slice of the HSS core. This allows the chip designer to perform analysis to determine the amount of power required for each transmitter, receiver, and PLL slice. The HSS power dissipated on the chip is calculated by multiplying each of these power calculations by the corresponding number of core slices.

Other factors in the above list are described in more detail below.

9.3.2 Multiple Power Supplies

Many HSS cores require more than one power supply input. It is necessary to calculate the power dissipation separately for each supply using (9.37). The coefficients used in this equation are specific to the supply for which power is being calculated.

9.3.2.1 Logic Power Supply (Vdd)

All HSS cores have a primary power supply input which is used to supply power to digital logic, as well as portions of the nondigital logic. Most of the power consumed by the HSS core is generally drawn from this supply. This supply is usually shared with other digital logic on the chip, and therefore may include significant amounts of noise.

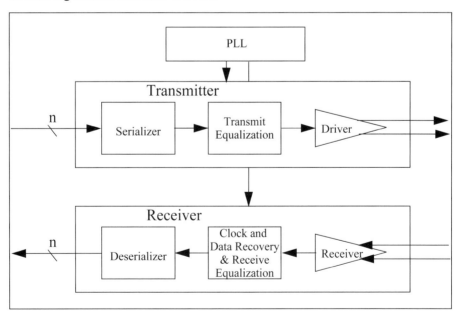

Fig. 9.9 Basic block diagram of typical high-speed serdes

Because of the strong dependence of P_{hss} on supply voltage V_s in (9.37), power dissipation is minimized by using the lowest possible *Vdd* voltage. AC and leakage power terms in (9.37) dominate power dissipation associated with this supply; quiescent power is not a significant factor for the Vdd supply unless this supply is used to power termination networks.

9.3.2.2 Analog Power Supply (AVdd)

Some HSS circuits may have a separate analog power supply input. Analog circuits may require a supply voltage higher than the voltage of the Vdd supply, or may require a power supply with less noise than would be present on the Vdd supply. Using a separate AVdd supply provides noise isolation from the Vdd supply, and provides a higher supply voltage where it is needed. Using the higher voltage only where needed optimizes power dissipation.

Dominant terms in (9.37) for the AVdd supply depend on which circuits in the HSS core use this supply. As with Vdd, selecting the lowest possible voltage for AVdd is desirable to minimize power dissipation.

9.3.2.3 Termination Supply (AVtt)

It is sometimes desirable to provide a separate power supply input for biasing termination impedance networks. The higher the amplitude of the differential signal, the higher the AVtt voltage must be to avoid clipping the signal. Such clipping can damage transmitter and receiver devices in addition to impacting signal integrity, and therefore should be avoided.

HSS cores sometimes bias termination networks by tieing AVtt to Vdd internal to the core, while other designs may provide separate AVtt supply inputs. In the latter case, AVtt is tied to a voltage as required by the application, which may be the same or greater than the Vdd voltage.

Since the AVtt supply is generally used to bias output drivers and resistive termination networks, quiescent power terms in (9.37) dominate for this supply. If the core internally ties AVtt to the Vdd supply, then this power is included in the equation for Vdd as increased quiescent power. As with the other supplies, selecting the lowest possible voltage for the AVtt supply is desirable to minimize power dissipation.

Note that a termination supply on a receiver device is sometimes referred to as the AVtr supply. In a DC coupled application, AVtt and AVtr would generally be tied to a common Vtt power supply.

9.3.3 Chip Fabrication Process

The values of the various coefficients in (9.37) are strongly influenced by chip fabrication process variation. These coefficients are usually defined for a nominal chip fabrication process, resulting in calculations which yield nominal power values. A given chip may consume more or less power depending on the actual manufacturing process conditions.

It is useful to also provide coefficients associated with worst case (maximum power) chip fabrication process. Calculations using these coefficients predict the maximum possible power dissipation. Power dissipation on a given chip may be significantly less than this value.

9.3.4 Mode-Dependent Power

9.3.4.1 Functional Mode Dependencies

Some HSS cores support various modes of operation which affect power dissipation. The power dissipation for the HSS EX10 transmitter slice described in Chap. 2 is affected by the *Transmit Power Register* setting defined in Table 2.6. Driving higher amplitudes on differential signals results in higher internal capacitive load, and as such Ps_{ac} increases. Various sets of coefficients for (9.37) are supplied for common settings of the *Transmit Power Register*.

The power dissipation for the HSS EX10 receiver core defined in Chap. 2 is affected by the *DFE/non-DFE Mode Select* setting in the *Receive Configuration Mode Register* defined in Table 2.7. When *non-DFE* mode is selected, the DFE circuit is bypassed and the activity factor of the associated logic is significantly reduced. This results in lower power dissipation for the receiver slice. When *DFE-3* mode is selected, portions of the DFE circuit are bypassed, resulting in higher power dissipation than *non-DFE* mode, but still less than *DFE-5* mode where the entire DFE is being used. Equation (9.37) coefficients are supplied for the receiver slice for each functional mode.

The features noted above are commonly recognized to influence power dissipation on most HSS core designs which support these features. Depending on the design of the HSS core, other configuration parameters may also result in significant effects on power dissipation.

9.3.4.2 Power Down Modes

Many HSS cores include the capability to selectively power down portions of the core. While power down mode controls may disable power distribution in some HSS core designs, the HSS EX10 core defined in Chap. 2 uses the more common approach of gating clocks to the affected circuits. The activity factor of the affected logic is forced to zero, thus eliminating Ps_{ac} power dissipation term. Leakage and quiescent power dissipation is generally not affected when clocks are gated.

There are a number of control signals on the HSS EX10 core, as defined in Table 2.1–2.3, which selectively power down portions of the core. Some of these control signals are used to disable unused channels, while others are used to implement a *PCI Express* dynamic power management scheme as was described in Sect. 5.5.4. The PCI Express link power states are defined as follows:

- L0: Normal Operation, Active Transmit and Receive
- L0s: Power Saving, Transmit and/or Receive Idle
- L1: Standby, Transmit and Receive in Sleep Mode
- L2: Powered Down, Tx and Rx Powered Off, Beacon Enabled

Table 9.1 summarizes the power control pins of the HSS EX10 and their typical application.

Table 9.1 HSS EX10 power mode core pin definitions

Pin name	Typical use	Description
PLL signals		
HSSPDWNPLL	Power down cores on unused interfaces	HSS PLL Power Down 0=normal operation, 1=power down the HSS PLL Slice
HSSSTATEL2	PCI Express power state L2	Power down signal which powers off part of the PLL slice in compliance with implementation of a PCI Express L2 link state. Also forces power down of transmitter and receiver slices 0=normal operation 1=core is in L2 link state
Transmitter signals		
TXxPWRDWN	Power down unused Tx channels on a core	Transmit Power State: Power down signal which powers off the Transmitter slice. 0 = normal operation, 1 = Power down
TXxSTATEL1	PCI Express power state L1	Transmit Power State: Power down signal which powers off the Transmitter slice in compliance with implementation of a PCI Express L1 link state. 0=normal operation, 1=transmitter is in L1 link state
TXxELECIDLE	PCI Express power state L0s	Transmit Electrical Idle: Forces transmit serial data to an electrical idle signal level. 0 = normal operation, 1 = electrical idle state
Receiver signals		
RXxPWRDWN	Power down unused Rx channels on a core	Receive Power State: Power down signal which powers off the Receiver slice. 0 = normal operation, 1 = power down
RXxSTATEL1	PCI Express power state L1	Receive Power State: Power down signal which powers off the Receiver slice in compliance with implementation of a PCI Express L1 link state. 0=normal operation, 1=transmitter is in L1 link state

Table 9.1 HSS EX10 power mode core pin definitions

Pin name	Typical use	Description
RXxSIGDETEN	Power down signal detect circuit on channels which do no use RXxSIGDET	Signal Detect Enable 0=Signal Detect power control using *Signal Detect Power Down* bit in *SIGDET Control Register*, 1=Signal Detect circuit powered on

9.3.5 Power Dissipation Breakdown

The discussion of HSS power dissipation is not complete without some discussion of the magnitude of the relative contributions of various HSS core components and power terms to the overall HSS core power dissipation. As is shown in Fig. 9.9, the HSS core is made up of three slices: the Phase Lock Loop (PLL) slice, the transmitter (TX) slice, and the receiver (RX) slice.

For purposes of this discussion, assume a typical full duplex HSS core consisting of one PLL slice, four transmitters, and four receivers. Also assume that this core has a separate AVdd supply used to supply power to analog circuits in the PLL slice, and separate AVtt and AVtr supplies to power termination networks. The AVtt supply powers the output driver stage of the transmitter. The following relative contributions were determined through analysis of examples of actual cores with a configuration similar to that of the HSS EX10 core defined in Chap. 2.

Considering this typical core configuration, the PLL dissipates approximately 10% of the total core power, the four Tx slices dissipate 40% of the core power, and the four Rx slices dissipate the remaining 50% of the total core power.

Within the PLL slice, around 34% of the PLL power is dissipated in digital circuits and 66% is dissipated in the analog circuits. A further breakout of the PLL digital circuit power shows that 75% of the power dissipation is AC power and the remaining 25% is leakage. The breakout of the PLL analog circuit power shows that only 20% of the power dissipation is dynamic (AC) power and the remaining 80% is quiescent power.

Within the Tx slice, around 23% of the Tx power is dissipated in digital circuits and 77% is dissipated in the analog circuits. A further breakout of the Tx digital circuit power shows that 80% of the power dissipation is AC power, and the remaining 20% of the power is leakage. The breakout of the Tx analog circuit power shows that only 20% of the power dissipation is dynamic (AC) power and the remaining 80% is quiescent power. Approximately 60% of the power for the Tx slice is supplied by the AVtt supply, most of this being the quiescent power component. The remaining power is supplied by the Vdd power source.

Within the Rx slice, around 48% of the Rx power is dissipated in digital circuits and 52% is dissipated in the analog circuits. A further breakout of the Rx digital circuit power shows that 77% of the power dissipation is AC power and the remaining 23% of the power is leakage. The breakout of the Tx analog circuit power shows that only 20% of the power dissipation is dynamic (AC) power and the remaining 80% is quiescent power. Most of the power for the Rx slice is supplied by the Vdd power source, with almost no power dissipated by the AVtr supply.

These relative power contributions should not be taken as absolute values, but rather as a representative aid to the chip designer as to where attention can be focused to save power. For example, the large quiescent power contributor in the Tx can be reduced by reducing the AVtt supply voltage.

9.4 Reducing Power Dissipation

This section discusses general techniques for reducing power dissipation of the HSS design, and of the system using the HSS core.

9.4.1 Power Concerns for the HSS Core Design

The architecture of the HSS design is probably the largest contributor to the power dissipation of the core. More complex design architectures require more logic gates and more circuit transistors, and correspondingly result in greater power dissipation. Of course, complex design architectures may be the necessary result to meet other requirements such as baud rate or to provide the feature set necessary for a particular protocol standard. Performing trade-offs between features, baud rate, and power dissipation is a necessary part of the design process for any HSS design.

However, HSS core architecture and design topics are not the subject of this text. In general, minimizing the number of logic gates and analog circuits (for a given clock frequency) results in the lowest power dissipation. This text assumes appropriate trade-offs have been made by the design process, and focuses on approaches for minimizing the power dissipation given a particular design architecture.

9.4.1.1 Clock Frequencies

All stages of an HSS core datapath must process bits at the target baud rate. However, these bits can be processed serially, or can be processed in parallel at a lower clock frequency. Except for the differential serial data output stage of the HSS transmitter and the differential receiver stage of the HSS receiver, all circuits within the HSS core can be designed to any arbitrary datapath width in order to reduce the clock frequency of the circuit.

Equation (9.24), repeated below, states that the AC power of logic is linearly related to both C and F_c of the circuit:

$$P_{s_{ac}} = C \cdot F_c \cdot AF \ (V_s)^2 \qquad \qquad (9.38)$$

Assume that two options exist for logic block. One option processes an n-bit data path using a clock of frequency f. The other option processes a $2n$-bit data path using a clock frequency of $f / 2$. Given the data being processed is the same, the value of AF can be assumed to be equivalent for both circuits. It is also assumed that V_s is the same for both circuits. The first logic block uses a higher clock frequency while the second design likely requires more logic gates (increasing the value of C).

From a power dissipation standpoint, the optimal choice between these two circuits depends on the implementation details and the resulting logic size. If the second implementation (which must process a $2n$-bit data path) contains twice the number of logic gates of the smaller logic block, then the Ps_{ac} of the two implementations is equivalent. In addition, since Pc_{leak} is also proportional to the number of logic gates, the implementation using $F_c = f$ is the proper choice for optimal power dissipation in this example.

In general, a wider datapath operating at a lower frequency reduces power dissipation if the implementation of this circuit is such that the number of logic gates does not grow at the same rate as the increase in datapath width. The power dissipation of the two circuits must meet the following equation to justify the larger (slower) circuit:

$$C_1 \cdot F_c + Pc_{leak} < C_2 \cdot nF_c + \frac{C_2}{C_1} Pc_{leak} \tag{9.39}$$

where C_1 is the capacitance associated with the circuit with the wider datapath, C_2 is the capacitance associated with the circuit with the higher clock frequency, F_C is the frequency of the clock for the circuit with the wider datapath, and n is the ratio of the clock frequencies for the two circuits. This equation assumes AF and V_s are the same for both circuits, and assumes Pc_{leak} scales linearly with the logic size (and capacitance). This equation can be rearranged as follows:

$$C_1 < C_2 \cdot \left[\frac{nC_1 F_C + Pc_{leak}}{C_1 F_C + Pc_{leak}} \right] \tag{9.40}$$

If $Pc_{leak} \ll Ps_{ac}$, then this equation reduces to:

$$C_1 < n \cdot C_2. \tag{9.41}$$

There are two cases which must be considered where the trade-off suggested by the above equation is not valid.

First, the derivation of this equation assumed the power supply voltage (V_s) is equivalent for both circuits. If reducing the clock frequency allows the circuit to use a lower supply voltage, then the reduced V_s^2 term more than compensates for the increased number of circuits.

Second, the above discussion regarding datapath logic blocks (and circuits) does not necessarily apply to initialization, control, and status logic. Logic with low activity factors already has minimal contribution of Ps_{ac} to power dissipation. However, there is additional benefit to reducing the frequency of clock signals themselves. The lowest possible clock frequency should be used in order to minimize power dissipation of this logic.

9.4.1.2 Clock Gating

Clock signals by definition have an inherent activity factor of 1, and are therefore a key contributor to Ps_{ac} in any logic design. If a logic block is not needed for some period of time, the clock signal can still cause significant power dissipation even if AF = 0 for all nonclock signals. Gating the clock to the circuit provides substantial power reduction.

Datapath logic in the HSS core processes data on every clock cycle, and clocks to such logic cannot be gated during normal operation. However, clocks can be gated off for power-down modes. Also, operational modes may exist where clocks may be turned off for some circuits. Examples of such modes for the HSS EX10 core include:

* *Non-DFE* or *DFE3* selections for the *DFE/non-DFE Mode Select* setting in the *Receive Configuration Mode Register* defined in Table 2.7. These modes gate the clock to portions of the DFE logic of the receiver slice.

* Setting *Signal Detect Power Down* in the *Signal Detect Control Register* defined in Table 2.7. This powers down the signal detection circuit for applications which do not use this function.

In general, if a function contributes significantly to power dissipation and may not be needed in some applications, then it is worthwhile to provide the user with the capability to turn off the function. Such capabilities are generally implemented by gating off the associated clocks to the logic block, which results in AF = 0, and eliminates Ps_{ac} power dissipation.

Initialization, control, and status logic blocks are also candidates for gating the clock off when the circuit is not performing useful work.

9.4.1.3 Multiple V_t Logic Circuits

The threshold voltage (V_t) of transistors used to implement a logic gate has an influence on both AC power dissipation (Ps_{ac}) and the leakage power dissipation (Pc_{leak}). Higher transistor V_t levels result in lower leakage currents, but also impact circuit timing. The slower slew times of such circuits increase the period over which the gate draws current while switching states, and thereby increases Ps_{ac} .

Often, logic gates are available in an ASIC technology which use a variety of transistor devices with different V_t levels. Logic gates using a lower V_t are used in circuits where timing performance is of greatest concern, and higher V_t logic gates are used in slower circuits to reduce leakage currents.

Note that it is *not* the case that higher V_t devices should be used universally in slower circuits. Two cases should be considered:

First, in circuits with higher activity factors, the increase in Ps_{ac} associated with such devices may more than offset the Pc_{leak} savings. Synthesis tools with power optimization capabilities are generally used to select logic gates for initialization, control, and status circuits in the HSS core. Higher V_t devices are generally used in such logic except where faster circuits are needed on critical timing paths. Datapath circuits must function at higher speeds and have relatively high activity factors; these circuits rarely use high V_t devices.

Additionally, if use of higher V_t devices forces the use of a higher power supply voltage in order to meet circuit performance requirements, then the increased power dissipation associated with the higher supply voltage offsets any advantage of using the higher V_t devices. Synthesis tools optimize for a specified V_s and do not consider potential advantages of reducing the power supply voltage. If the potential exists to lower the power supply voltage, and restricting the circuit to lower V_t devices facilitates this, then this is almost always the optimal approach.

9.4.1.4 Multiple Power Supplies

As was discussed earlier in this chapter, some HSS cores require multiple power supply inputs. The logic power supply (Vdd) is typically used to supply power to most circuits in the HSS core. This supply is also used by most of the logic on the chip, and as such the chip designer specifies the Vdd voltage to be as low as possible to minimize power dissipation.

Additional power supplies may be required by the HSS core if:
- Supply voltages higher than Vdd are required by some circuits
- Critical analog circuits require isolated power supplies with minimal noise

These additional power supplies generally use a supply voltage which is at least equal to, and often greater than, the voltage of the Vdd supply. Higher supply voltages substantially increase power dissipation. For this reason, the HSS core designer should power as many circuits as possible using the Vdd supply. Circuits powered by additional power supplies should be minimized, especially if it is expected that these supplies will use higher supply voltages.

9.4.2 *Power Dissipation Concerns for the Chip Designer*

For the chip designer, minimizing power dissipation of the HSS core requires utilizing power savings features of the HSS core to the extent that the application permits. Relevant considerations are listed below, however, many of these items are probably obvious to most readers at this point.

Turn off Unused Channels. HSS cores generally include multiple transmitter and receiver slices. Depending on the number of links required for a given chip and the granularity of the number of links available per HSS core, some chips may have transmitter and/or receiver slices which are never used. Also, some chips may have multiple modes of operation, and some of these modes may not use some of the HSS links. In these cases, unused HSS transmitters and receivers should be powered down using the appropriate control inputs.

Minimize Power Supply Voltages. Chip and system designers should always use the minimum possible power supply voltage(s) as permitted by the application. As was obvious from earlier topics in this chapter, V_s has a dramatic impact on power dissipation.

Minimize the Baud Rate. Generally, baud rates are dictated by the application. However, lower baud rates do reduce power dissipation. Using lower baud rates during periods when less bandwidth is required is one approach for reducing power dissipation.

Minimize Transmit Amplitude. The differential amplitude (V_{diff}) of the transmitted signal affects the quiescent power of the transmitter driver stage. Lower amplitudes dissipate less power. Transmit amplitude is provisionable on most HSS cores. Minimum signal amplitude may be constrained by protocol specifications and/or by signal integrity requirements for the link, however, the minimum possible amplitude should always be used if power dissipation is a concern.

Select Operational Modes to Reduce Power. The HSS core may support provisioning of operational modes such that unnecessary functions are turned off to reduce power. For the HSS EX10 core, this included *Non-DFE* or *DFE-3* selections for the *DFE/non-DFE Mode Select* setting in the *Receive Configuration Mode Register* defined in Table 2.7. If either the *Non-DFE* or *DFE-3* selections are sufficient to meet signal integrity requirements of the link, then these modes can be used and power dissipation of the receiver is reduced.

Power Down States. Some applications may experience periods when no useful data is being sent or received on the serial data interface. During such periods, links can be partially or completely powered down to reduce power dissipation. The PCI Express application defines protocols associated with entering and exiting power down states during such periods, and utilizes power state control signals on the HSS core as was described in Sect. 9.3.4.2.

9.5 References and Additional Reading

The following reading is recommended for more information regarding topics in this chapter:

1. "Thermal and Power Management of Integrated Circuits", A. Vassighi and M. Sachdev, Springer, Berlin, 2006.
2. "ULSI Devices", C.Y. Chang and S.M. Sze, Wiley, New york, 2000.
3. "Solid State Electronic Devices", B. G. Streetman and S. Banerjee, Prentice Hall, New Jersy, 2000.

9.6 Exercises

1. For the driving CMOS logic inverter circuit in the figure below:
    (a) Highlight the portions of the circuit which compose the lump sum capacitance C which stores charge driven by the inverter.
    (b) Indicate the current path when the inverter charges the this lump sum capacitance.
    (c) Indicate the current path for discharging this lump sum capacitance.

(d) If more circuits are driven by this inverter, how does this affect power dissipation?

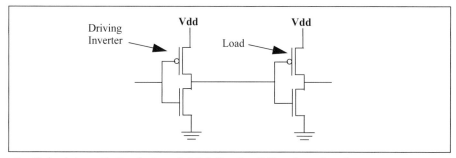

2. Calculate activity factors (AF) for the following signals:

   (a) 100-MHz clock signal.

   (b) Data signal transmitting a 00110011... repeating pattern.

   (c) Data signal transmitting a 00001111010100110011... repeating pattern.

   (d) Control signal which is set during initialization and is not changed.

3. Using your answer to Chap. 7 Exercise 16, calculate activity factor for the output of a Pseudorandom Bit Sequence (PRBS) generator which uses the following polynomial: $G(x) = x^7 + x^6 + 1$.

4. Calculate AC power dissipation (Ps_{ac}) for each of the following cases:

   (a) 100-MHz clock buffer driving a 10 pF load and a 1.2 V supply.

   (b) 100-MHz clock buffer driving a 10 pF load and a 0.8 V supply.

   (c) 200-MHz clock buffer driving a 20 pF load and a 1.2 V supply.

5. Calculate AC power dissipation (Ps_{ac}) for each of the following cases:

   (a) Logic gate driving a 50 pF load with an activity factor of 0.1 and powered by a 1.0 V supply.

   (b) Logic gate driving a 25 pF load with an activity factor of 0.2 and powered by a 1.0 V supply.

6. Calculate total AC power dissipation (Ps_{ac}) for a circuit composed of the following logic blocks assuming $Vs = 1.2$ V:

Logic Block	Clock Frequency	Activity Factor	Lump Capacitance
Block #1	100 MHz	0.3	172 nF
Block #2	333 MHz	0.5	88 nF
Block #3	333 MHz	0.01	250 nF
Block #4	25 MHz	0.5	205 nF

7. The circuit described in Exercise 6 is modified such that logic block #2 is partitioned into a voltage island on the chip for which $V_s = 0.8$ V. Recalculate Ps_{ac} for a circuit given this change.

8. Logic gates are available in an ASIC technology which use transistor devices with one of two V_t levels. In order to reduce leakage current in the circuit, should the higher V_t or lower V_t logic gates be used? Explain your answer in terms of its affect on the calculation of I_{subvt}.

9. A logic gate must operate across a temperature range of 0–100°C. Which of these temperature limits results in the lowest I_{subvt}? Compare I_{subvt} at each of these temperature limits. (Note: K = C + 273.15.)

10. Assume the driving CMOS logic inverter circuit in Exercise 1 is driving a logic "1" to the load circuit. This means the p-FET transistor of the inverter is conducting current in an *on* state, and the n-*FET* transistor is turned *off*. Show the leakage currents through these transistors while in this state.

11. Explain using Fig. 9.3 why increasing V_t reduces I_{subvt}.

12. The *IV* curve in Fig. 9.3 is somewhat simplified; more realistic *IV* curves are shown in Fig. 9.8. Given your answer to Exercise 11 and the definition of GIDL in Fig. 9.8, why do you think GIDL is more of a concern in low power ASIC technologies which use higher V_t transistors?

13. Given all other factors are constant, how does increasing the power supply voltage from 1.0 to 1.2 V impact Pc_{leak} based on (9.32)? How is P_{hss} impacted based on (9.37)?

14. Given all other factors are constant, how does increasing the maximum junction temperature from 100 to 125°C impact Pc_{leak} for a circuit based on (9.32)? How is P_{hss} impacted based on (9.37)? (Note: K = °C + 273.15.)

15. Quiescent Power calculations for nondigital circuits have both V_s and V_s^2 terms. What types of circuits are associated with each of these terms?

16. Assume the HSS EX10 core has Vdd, AVdd, Vtt, and Vtr power supply inputs. How many sets of coefficients for (9.37) would need to be supplied to the chip designer to support power calculations for nominal and worst case power dissipation? (Ignore operating mode dependencies when answering this question.)

17. There is a desire to build a spreadsheet to calculate power dissipation for the HSS EX10 core using the sets of coefficients described in Exercise 16. However, it is also desired that the spreadsheet allow the user to select between four possible values for the *Transmit Power Register*, and the various *DFE/non-DFE Mode Select* operating modes. How many sets of coefficients for (9.37) would need to create this spreadsheet?

18. The *Transmit Power Register* in the HSS EX10 transmitter sets the V_{diff} amplitude of the transmitted signal. Which variables in (9.36) would you expect are affected by this value?

19. The *DFE/non-DFE Mode Select* operating mode programmed in the *Receive Configuration Mode Register* in the HSS EX10 receiver turns off portions of the DFE digital logic circuit by gating the clocks to this circuit. Which variables in Eqn 9.36 would you expect are affected by this operating mode?

20. Assume the HSS EX10 *TXxELECIDLE* pin reduces Tx slice power by 60%, and *TXxSTATEL1* reduces Tx slice power by 90%. Similarly, *RXxSTATEL1* reduces Rx slice power by 90%. Also, the *HSSSTATEL2* pin reduces PLL slice power by 80% in addition to forcing all Tx and Rx slices into L1 power state. Given the power breakdown for the HSS EX10 core described in Sect. 9.3.5, how much is the power dissipation reduced for each of the following PCI Express power states?

    (a) All links in L0s power state.

    (b) All links in L1 power state.

    (c) HSS core in L2 power state.

21. Alternative HSS EX10 core configurations are proposed which include different numbers of transmitter and receiver slices. Given the power breakdown for the HSS EX10 core described in Sect. 9.3.5, estimate the power dissipation for each of the configurations below relative to the four-lane full duplex HSS EX10 configuration described.

    (a) Simplex transmit core with one PLL and four Tx slices.

    (b) Simplex receive core with one PLL and four Rx slices.

    (c) Full duplex core with one PLL, eight Tx, and eight Rx slices.

22. Two versions of a datapath circuit are considered for inclusion in a 10-Gbps HSS core design. One version processes data serially while the other version processes 4-bit parallel data. The implementation of the parallel version of the circuit requires four times as much logic. Which circuit is optimal from a power dissipation perspective?

23. How does your answer to Exercise 22 change if the 4-bit parallel data circuit requires twice as much logic as the serial circuit?

24. Derive (9.40) from Eqn (9.39).

25. Given methods of reducing power dissipation described in Sect. 9.4.2, discuss how improvements to the design of the channel (i.e., printed circuit board materials, connector selection, backplane design, etc.) can lead to lower power dissipation for the system.

Chapter 10
Chip Integration

A general methodology flowchart for the chip design process is shown in Fig. 10.1 [1]. This flowchart illustrates the major steps that must be performed to transform the chip design into a manufacturable chip. (In practice, many steps are overlapped to speed up the process and are iterated as the design is altered or optimized. For simplicity, linear processing of the traditional methodology is assumed for this discussion.)

The chip design is initially expressed in some higher form by the chip designers, usually in a High-level Design Language (HDL). This design is simulated using various means to verify that the design meets applicable design specifications. The design is then processed by a synthesis tool to map the design into the available library of logic gates and to optimize the logic structure. Specialized synthesis tools add Design-For-Test structures and clock trees. The resulting gate-level netlist is verified using Static Timing Analysis software to verify prelayout timing requirements are met; Test Structure Verification software to verify testability requirements are met; and Formal Verification software to verify that the gate-level representation of the design is boolean equivalent to the original HDL design.

The gate-level netlist is the input to a physical design process which includes floorplanning, layout and wiring, and optimization. Major blocks and I/O cells are placed as part of the floorplanning stage. Logic gates are placed and nets are wired during the layout and wiring stage. The optimization stage consists of analyzing the timing of the routed design and making appropriate changes to the logic mapping or placement. Extracted resistance and capacitance characteristics of the routed wires are used in this stage by Static Timing Analysis software to obtain an accurate view of the timing characteristics of the design and verify all requirements are met. Timing characteristics may also be used to perform full-timing simulation of the gate-level design.

Once the logical and physical design of the chip is verified to meet all requirements, chip manufacturing data is generated. At this stage, it is also possible to use Automatic Test Pattern Generator (ATPG) software to generate test patterns for the chip.

All of the components used in the chip design, including HSS cores, must support the methodology used to design the chip. Significant facets of using HSS cores within this methodology are discussed in this chapter, including:

- Simulation models for HSS cores
- Test synthesis support
- Timing models and timing assertions to support static timing analysis
- Floorplanning considerations

This chapter is not intended to be a comprehensive chip design methodology discussion. HSS cores are processed in the same manner as any other monolithic logic blocks through much of the chip design methodology. However, the topics in this chapter warrant special consideration.

D. R. Stauffer et al., *High Speed Serdes Devices and Applications,*
© Springer 2008

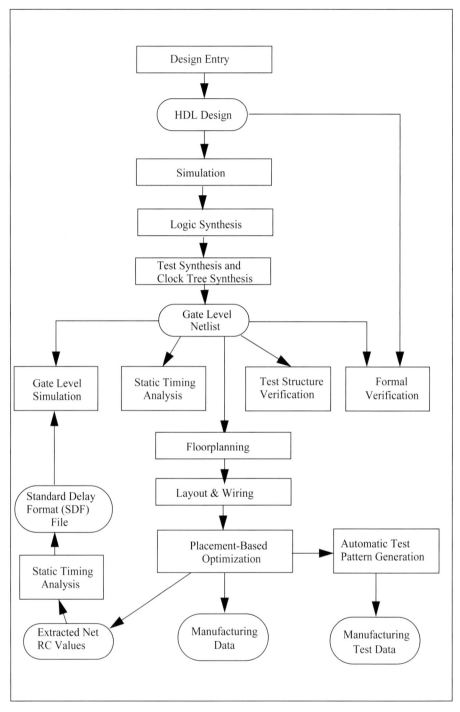

Fig. 10.1 Chip design methodology [1]

10.1 Simulation Models

Simulation is the primary method used to verify that chip designs function as intended and meet the requirements of the chip specification. The chip design and verification teams generate stimulus for chip inputs, simulate the chip design's response to the stimulus, and check that the chip outputs behave as expected. Models must exist for all components of the chip design, including HSS cores, in order to support this simulation.

The simulation model for the HSS core must accurately reflect the behavior of the hardware. To the extent that variation exists between the behavior of the HSS model and the hardware, the risk is increased that a chip design may pass its simulation testcases and then not work in hardware. However, simulating the HSS core through an accurate reset and initialization sequence is computationally expensive. Chip design and verification teams generally prefer to be able to abbreviate the reset and initialization process for most simulation runs, and limit simulation of the full sequence to only a few testcases. Also, accurately modeling the analog behavior of the core is not relevant if the resulting analog characteristics cannot be observed in the digital simulation. An Analog Mixed Signal (AMS) simulation is required in order to observe these effects. However, AMS simulations are generally not done at the chip level, and would be computationally expensive if they were performed.

10.1.1 Reset and Initialization Short Cuts

The HSS EX10 core reset sequence was described in Fig. 2.16. There are a number of wait loops in the reset sequence which generally require from hundreds of microseconds to several milliseconds to complete in hardware. A full function digital simulation would need to simulate all of these steps to initialize the HSS cores before any operational functions could be simulated. This would require significant execution time for the typical event-driven simulation software, and simulating the reset sequence more than a few times would not provide any additional coverage of chip functionality.

10.1.1.1 Power Application

The first wait loop in Fig. 2.16, "Allow time for core inputs to stabilize," exists to allow input pins to the core to stabilize after power is applied to the chip. None of the logic in the chip can be reliably reset until the power supply reaches its steady-state value and circuits are driving valid logic "0" or "1" levels. This may take many milliseconds in hardware, but this behavior is not modeled at all in a digital simulation. This step in the process may be abbreviated in a digital simulation, and HSSRESET may be asserted almost immediately after the simulation starts. Since the chip logic controls when HSSRESET is asserted, abbreviating this wait loop does not require special support from the simulation model for the HSS core. However, the chip designer should ensure the ability exists to speed up wait loops for any relevant finite state machines which may exist in the chip design or the simulation environment.

10.1.1.2 PLL Calibration and Lock

The second wait loop in Fig. 2.16, "Wait for PLL Reset Completion," exists to allow the PLL slice to achieve a lock state. After the HSS core is reset, the PLL slice executes a calibration sequence which adjusts circuit parameters until the PLL successfully locks to the reference clock input. In hardware this process may take from hundreds of microseconds to several milliseconds. While it is useful to simulate the detailed behavior of the PLL for a few testcases (to make sure chip logic responds correctly when realistic delays are involved), it is not efficient to perform this simulation for every simulation run.

To provide better efficiency for chip simulations, most simulation models of HSS cores include a mode of operation which shortens the PLL calibration and lock sequence by jumping directly to a lock state after a small arbitrary delay. This feature may be activated using one of several means, the most common of which is to provide a Verilog parameter or VHDL generic on the simulation model which selects the corresponding mode of operation.

For example, assume the simulation model for the HSS EX10 core described in Chap. 2 has a Verilog parameter *BYPASS_CAL*. When this parameter is set to "1," PLL state machine outputs are forced to a "locked" state after minimal delay, and chip simulation of transmit and receive functions commences. This results in simulation waveforms as shown in Fig. 10.2, where the calibration sequence completes only 1,644 ns after the HSSRESET input is deasserted. In this figure, the transmitters start to send data at some point before HSSPLLLOCK is asserted. Although serial data is driven onto the receiver inputs in this simulation, the HSS receivers do not attempt to deserialize data until after HSSPLLLOCK is asserted.

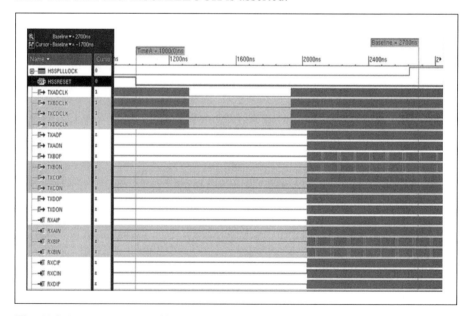

Fig. 10.2 Reset sequence with BYPASS_CAL = "1"

10.1.1.3 Rx Training

Additional wait loops in Fig. 2.16, "Wait for Rx Training Completion," exist to allow the CDR circuit to find the center of the data eye, and to allow the DFE circuit to train on the receive data and determine the optimal set of coefficients to use. This process may take several hundred microseconds in hardware, but generally is not modeled in a digital simulation. In hardware, the receive data waveform has jitter and other signal impairments; CDR and DFE circuits must compensate for these impairments. On the other hand, a digital simulation generates pristine "1" and "0" levels with no impairments. Even if the CDR and DFE circuit operation were to be modeled in the simulation model, there is no signal degradation to exercise these functions.

CDR circuits are often modeled with the equivalent digital function in HSS simulation models. However, assuming a pristine receiver input with no jitter, the CDR finds the center of the eye and starts receiving data correctly after only a small delay.

The only DFE behavior which needs to be modeled in the HSS simulation model is the DFE interaction with chip logic or system software. This includes supporting register read/write operations for DFE parameters and generating any relevant status signals. If the DFE circuit is modeled in the simulation model, then this circuit trains nearly immediately, and the resulting filter coefficients are trivial. More often, a behavioral model is used for this circuit.

As indicated above, CDR and DFE circuits should train quickly in a digital simulation environment where the receiver input does not have any signal impairments. It is therefore generally not necessary to incorporate special features into the HSS simulation model to shorten the execution of Rx Training.

10.1.2 Simulation 'X' States

When power is initially applied to a state element in a circuit (either a flip-flop, latch, or SRAM cell), the initial value of that state element is undefined. The value initializes to either a logic "0" or "1" depending on power supply sequencing, circuit conditions, and other factors. The initial value may or may not vary after every power cycle. In a well-designed circuit, the initial value of these state elements does not matter. Either the circuit is subsequently reset to force state elements to known values, or values in some state elements force values in downstream state elements to known values after a few clock cycles.

The determination as to whether a circuit is well designed is part of design verification. It is possible, for instance, to design a finite state machine for which some of the state codes are not used. Such a state machine may include a reset input to set the state to an initial value, or alternatively may be designed so that invalid states always transition to the reset state. However, a poorly designed state machine may loop between one or more invalid states if it powers up in an invalid state, and may never get properly initialized. Design verification must ensure all circuits in the device under test are initialized to known states by whatever means, and cannot power up in a state which leads to the circuit hanging in an invalid state.

A potential method of verifying that circuits cannot power up in an invalid state is to verify that the device under test can be initialized properly for all possible initial values of its state elements. This would require a simulation run for every possible combination of initial values. Given n state elements in the device under test, 2^n simulation runs are required for exhaustive verification. Unless the device under test is relatively small, exhaustive verification using this approach is prohibitive due to the number of simulation runs involved. A random sampling of initial states is possible, but would not be guaranteed to find all problems that may exist for a particular design.

Event-driven simulators have approached this problem by initially assigning a value of "X" to state elements in the simulation. Logic gates calculate output values using a signal resolution function whereby the output of the gate is a "1" or "0" only if the gate inputs cause this value to be known. For example, the output corresponding to various combinations of "0" and "1" on the inputs of an AND gate are defined by the logic function. However, if one or both of the inputs are "X", then:

"X" AND "X" = "X"
"X" AND 1 = 1 AND "X" = "X"
"X" AND 0 = 0 AND "X" = 0.

Likewise for an OR gate:

"X" OR "X" = "X"
"X" OR 1 = 1 OR "X" = 1
"X" OR 0 = 0 OR "X" = "X".

Event-driven simulation propagates signals according to these resolution functions. In a well-designed circuit the "X" states are forced to known values by reset signal(s) or by other means within a few clock cycles. If the device under test reaches a state where no "X" states remain, this proves exhaustively that the circuit cannot power up in a state where initialization is not possible. Subsequent simulations from this state may be assumed to be representative of device operation regardless of the initial power-up state of the circuit.

This approach of using "X" states to verify initialization of the device under test is a pessimistic approach. The chip designer may reset state elements or otherwise add logic to eliminate the propagation of "X" states in simulation, while the real hardware may operate normally without this additional logic. However, this additional logic is viewed as necessary design overhead to ensure the design has been exhaustively verified and avoid any problems going undetected.

This discussion is relevant to the simulation model for the HSS core in the following manner: This simulation model must be initialized within the chip to eliminate "X" states for signals within the model, and must not have "X" states driven on output signals which interfere with the initialization of other chip logic. In order to accomplish this, the chip designer may need to ensure certain inputs to the core are at known values, or to sequence input signals to the core in a specific manner.

For the HSS EX10 core, the HSSRESET input resets the HSS core to a known state and eliminates "X" states within the model. However, certain inputs to the core must be at known values (not "X") when HSSRESET is asserted. If these inputs are not at known values, clock signals within the core are "X," and the simulation model reloads "X" values.

Table 10.1 defines the requirements for the simulation model of the HSS EX10 core. Signals which gate or otherwise affect propagation of clocks internal to the simulation model are listed in this table. These signals must be driven to known values as indicated before HSSRESET is asserted or the model for this core cannot be initialized. Note that although the HSS EX10 is a tutorial example, unless the core designer goes to the effort to gate signals internally, there are similar requirements for any HSS core.

Table 10.1 Signal requirements for initializing the HSS EX10 model

Slice	Signal	Value	Description of rationale
PLL	HSSREFCLK[T,C]	Oscillating	The reference clock must be running and stable when HSSRESET is asserted
	HSSREFDIV	0 or 1	This input controls a clock divider and must be a known value to avoid "X" on internal clocks
	HSSDIVSEL[1:0]	0 or 1	This input controls a clock divider and must be a known value to avoid "X" on internal clocks
	HSSPDWNPLL	0	This input controls power to the PLL slice and must be "0," otherwise clocks are gated off
	HSSRESYNCCLKIN	0	This input gates clocks in the simulation model, and must be "0" to avoid "X" on internal clocks
	HSSSTATEL2	0	This input controls power to the PLL slice and must be "0," otherwise clocks are gated off
TX	TXxPWRDWN	0	This input controls power to the Tx slice and must be "0," otherwise clocks are gated off
	TXxSTATEL1	0	This input controls power to the Tx slice and must be "0," otherwise clocks are gated off
RX	RXxPWRDWN	0	This input controls power to the Rx slice and must be "0," otherwise clocks are gated off
	RXxSTATEL1	0	This input controls power to the Rx slice and must be "0," otherwise clocks are gated off
	RXxDATASYNC	0	This input gates clocks in the simulation model for the Rx deserialization stage, and must be "0" to avoid "X" on internal clocks

10.1.3 Modeled and Unmodeled Behavior

While it is desirable that simulation models for HSS cores accurately reflect the behavior of the hardware, it is not desirable for models to be of such detail that simulation run times are adversely impacted. Furthermore simulation models are used in a digital simulation environment where many of the analog-mixed signal functions of the core are not exercised or observable. Functions which cannot be exercised or observed need not be modeled. Behavioral models are used for these circuits to model the equivalent digital behavior and thereby improve execution times for simulations using the model. As a general rule, features and functions implemented in the digital logic of the core are fully implemented in the simulation model, whereas AMS features and functions are either implemented with a behavioral model or not modeled.

Chapter 2 provided a detailed description of the HSS EX10 core which has been used as a tutorial example throughout this text. This section continues to develop this tutorial example by describing which functions are modeled in a hypothetical simulation model for this core, and which functions are not modeled or modeled behaviorally.

10.1.3.1 Reset Sequence

The reset sequence is fully modeled in the HSS EX10 simulation model, including all control and status signals, and all state machines involved in the reset sequence and VCO calibration. The analog PLL circuits are replaced with a behavioral model which simulates the digital behavior of the circuit.

The simulation model executes the reset sequence in a similar manner to hardware, accurately modeling signal handshakes, timing, VCO calibration results, etc. As was discussed previously, it is often desirable to shorten the execution time associated with VCO calibration. The *BYPASS_CAL* parameter was defined to do this.

10.1.3.2 Data Serialization and Deserialization

The parallel data interface of both the transmitter and receiver are fully modeled. Parallel data inputs of the transmitter are serialized, and data is driven onto the serial data outputs. Since this is a digital simulation model, the serial data outputs are driven to "0" and "1" values according to the data bit being sent. The true and complement legs of the differential signal are always driven to opposite values.

Serial data inputs to the receiver are deserialized and are driven to the parallel data output of the receiver. Note that the RXxI[P,N] inputs must have opposite values to result in a valid decode of the digital bit; if the serial data inputs are the same level, the model decodes this as an "X" value. Modeling of this interface includes deriving the RXxDCLK frequency from the incoming serial data and modeling the operation of RXxDATASYNC.

10.1.3.3 Analog Signal Characteristics

The registers defined in Table 2.6 contain parameters that set analog characteristics of the transmitted serial signal, including signal amplitude and slew rate. Although these registers are included in the digital simulation model for the HSS core, they have no affect on the operation of the model.

Equalizer operation is also modeled in a limited fashion. Since the serial data output of the digital simulation model can only be "0" or "1," the waveform variations generated by the FFE are not observable in the digital simulation environment. For this reason, only the register interface is generally modeled. A behavioral model is used to model any analog circuits which may be needed to provide appropriate register readback values. (One exception to this may be inverting the polarity of the serial data output based on the sign bit for the z^0 filter coefficient. This feature is useful for simulation of some applications.)

In a similar manner to the transmitter, Table 2.7 defines registers in the receiver associated with the DFE. Only the register interface of the DFE is generally modeled. A behavioral model is used to model any analog circuits which may be needed to provide appropriate register readback values. Since the received data is an ideal signal in a digital simulation environment, any more detailed functionality would not be exercised even if it were modeled.

10.1.3.4 Power Control

Various controls on the HSS EX10 core force portions of the core into various power down states. The outward effects of these controls are modeled in the HSS EX10 simulation model. TXCTS and TXBPWRDWN have been asserted in the example shown in Fig. 10.3, forcing the corresponding serial data outputs of the respective channels to a "Z" value. TXBPWRDWN also stops clocks in the transmitter slice as evidenced by TXBDCLK remaining at a fixed level. TXCTS only shuts down the driver stage and does not stop any clocks.

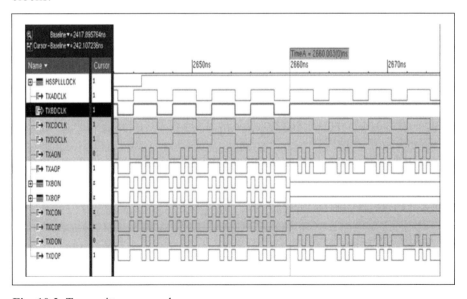

Fig. 10.3 Transmitter power down

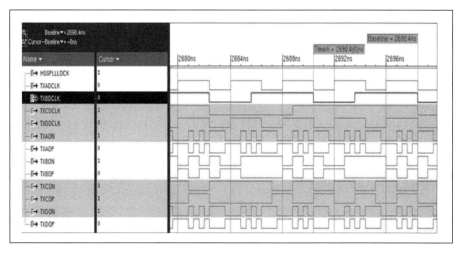

Fig. 10.4 Full, half, and quarter rates

10.1.3.5 Data Rate Selections

The full-, half-, quarter-, and eighth-rate modes of the HSS EX10 core are fully modeled in the simulation model. Figure 10.4 illustrates a case where the "A" and "D" transmitter channels are operating at full rate, the "B" channel is operating at half rate, and the "C" channel is operating at quarter rate. The figure illustrates how this affects the TXxDCLK operating frequencies. Note that while the duty cycle of the TXxDCLK is approximately 50% in full-rate mode, this is not true in all modes.

10.1.3.6 Diagnostics

All diagnostic functions which are primarily implemented in digital logic are modeled in the HSS EX10 simulation model. These include:
- PRBS generator and PRBS checker logic circuits
- Loopback paths
- JTAG 1149.1 and 1149.6 functions

Although these functions may also be used during chip manufacturing test, these functions are primarily needed for system characterization and system manufacturing test. The chip designer must verify not only the normal functional operation of the chip, but also the diagnostic test modes of the chip. It is therefore important that the simulation model for the HSS core contain support for these functions.

10.2 Test Synthesis

Manufacturing test for the HSS EX10 core was described in Chap. 7. As was discussed in that chapter, the chip is tested using some form of scan test, and the digital logic of the HSS core is tested as part of this scan test. Scan test performs DC stuck fault testing and is capable of achieving high coverage metrics using efficient test patterns generated by ATPG software. In addition, a suite of Macro Tests is applied to the HSS core to test analog circuits and electrical parametrics.

Requirements are imposed on the chip designer in order to facilitate execution of the various types of manufacturing test. Test control signals and scan chain inputs and outputs on the HSS core must be connected to the correct chip level signals. Macro test execution may require controllability and observability of various HSS core pins in order to facilitate test execution. Additionally, some HSS core pins may require preconditioned logic values during scan test, macro test, or JTAG test execution.

This mish-mash of requirements falls on the chip designer to implement. This section starts by discussing logic structures to implement various classes of requirements. Later, the concept of a *test wrapper* is discussed, which is delivered as part of the HSS core design kit, and allows connection of these logic structures to be automated.

10.2.1 Scan Test Support

The HSS EX10 core description in Chap. 2 did not include necessary pins to support manufacturing test of the core. In order to facilitate scan testing of the HSS EX10, the pins described in Table 10.2 are added to the core.

Table 10.2 HSS EX10 scan test pins

Pin name	Type	Description
SCANIN[8:0]	In	Scan chain input pins. All the flip-flops in each PLL, Tx, or Rx slice are connected as a long scan shift register when SCANGATE is asserted. The HSS EX10 core has nine slices (one PLL, four Tx, and four Rx) and therefore has a total of nine scan inputs
SCANOUT[8:0]	Out	Scan chain output pins. See description of SCANIN[8:0] for more information
SCANGATE	In	Test control pin which is asserted (= "1") when scanning the scan chains of the HSS EX10 core
TESTENABLE	In	Test control pin which is asserted (= "1") at all times when executing scan test and macro test sequences
LT	In	Test control pin which is asserted (= "1") to disable any DC leakage current paths in the core to facilitate Iddq testing
DI	In	Test control pin which is asserted (= "1") to disable all drivers on the chip
RI	In	Test control pin which is asserted (= "1") to disable all receivers on the chip
ZDI	Out	Redriven DI signal
ZRI	Out	Redriven RI signal

Scan test requires all of the flip-flops on the chip to be configured as scan shift registers when specific chip-level input signals are asserted. Table 10.2 assumes flip-flops in each slice of the core have been connected as such, and the inputs/outputs of these scan chains have been added as *SCANIN* and *SCANOUT* pins of the core, respectively. Core scan chains are embedded by the chip designer within chip scan chains. Although it would be sufficient to connect all of the flip-flops of the core into one scan chain, the lowest manufacturing cost is realized when the chip contains as many scan chains as possible (within limitations imposed by the number of package pins), and the scan chains have roughly equal numbers of flip-flops in each scan chain.

The *SCANGATE* pin controls whether or not the scan chain operates as a shift register, or whether flip-flops are connected to their functional data sources. This signal is connected to one of the chip-level manufacturing test signals and must be asserted to scan the chip.

The *TESTENABLE* pin performs design-specific functions to ensure test functions are accessible, and to disable functions that are to be excluded from testing. This pin is connected to an appropriate chip-level manufacturing test signal which is expected to be held active through all stages of manufacturing test, and which is expected to be held inactive in an operational system.

Additional test pins are defined in Table 10.2 which perform specific, but commonly required, functions:

- The *LT* pin disables DC current paths so that quiescent currents are not lumped into leakage current measurements, thus supporting a chip-level *Iddq* or *Leakage Test* as described in Sect. 7.4.1.4.
- The Driver Inhibit (*DI*) and Receiver Inhibit (*RI*) pins defined in Table 10.2 disable the drivers and receivers, respectively, in the HSS EX10 core. These signals are used during chip I/O tests, and must be connected to the equivalent chip-level manufacturing test signals. The *DI* and *RI* input pins are redriven onto the *ZDI* and *ZRI* output pins to support a daisy-chain connection of these signals to adjacent chip I/O.

It is the responsibility of the chip designer to connect scan chains of the core into chip level scan chains, and to connect the various test control signals to appropriate sources. Automation methods are considered later in this section.

10.2.2 Macro Test Support

10.2.2.1 Input Controllability

Macro tests for the HSS EX10 core were described in Chap. 7. In order to execute these tests, various pins on the HSS EX10 must be preconditioned to specific values, or must be fully controllable. The following cases may exist:
- HSS pin must be held at logic "0" for all test sequences.
- HSS pin must be held at logic "1" for all test sequences.
- HSS pin must be held at logic "0" or "1" depending on the test.
- HSS pin must be changed or toggled during the test sequence.

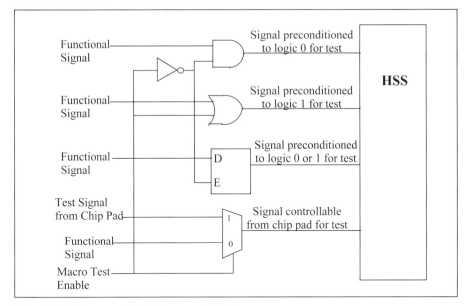

Fig. 10.5 Macro test logic for controllability

Logic to implement these cases is illustrated in Fig. 10.5. This logic assumes a *Macro Test Enable* which is asserted and held throughout all macro test sequences, and is not asserted during normal operation of the chip. As shown in the figure, this signal is used to force signals which must be preconditioned to fixed values of "0" or "1" during macro test sequences.

For cases where the signal must be preconditioned with a value that varies from test to test, a transparent latch is inserted as shown in Fig. 10.5. The *Macro Test Enable* signal is used to control the *E* input of this latch. During normal function, the latch is held *flush* such that the functional signal always controls the pin on the HSS core. During test operation, the latch is disabled so that it holds whatever value has been loaded into it. Although the scan connections are not shown in Fig. 10.5, it is assumed that this latch is stitched into the chip scan chains. The desired precondition value is scanned into the latch prior to execution of the test sequence, and then is held since the *Macro Test Enable* holds the *E* pin at a logic "0" during the test sequence.

Finally, consider the case where the HSS pin must be fully controllable during the macro test sequence. Signals which must be pulsed or toggled during the test sequence must be controlled from a chip input in order to do so. This input is usually not the normal functional path for driving the HSS pin. As shown in Fig. 10.5, a multiplexor selects between the normal functional control signal and the test signal source; the selection control for this multiplexor is the *Macro Test Enable* signal. Since the chip input pad is only used as a test signal source when testing the HSS core, this same input may be used for other purposes during other modes of operation. For instance, this pin may also be a scan input when scanning the chip, and may be a functional signal input for some unrelated function during normal chip operation.

10.2.2.2 Output Observability

The above discussion concerned controllability of HSS input pins to support macro test execution. In a similar manner, some HSS output pins require observability during the macro test sequence. Two cases exist:
- Final value of the HSS pin at the conclusion of the test must be observed.
- HSS pin must be observed dynamically during the macro test sequence.

Logic to implement these cases is illustrated in Fig. 10.6. This logic assumes the same *Macro Test Enable* signal which was defined previously. This logic also assumes a *Macro Test Complete* signal which is held at logic '0'during execution of the test sequence, and is asserted at the end of the test sequence to capture results.

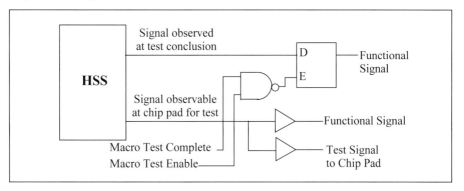

Fig. 10.6 Macro test logic for observability

For cases where the HSS pin logic value must be observed at the conclusion of the test, a transparent latch is inserted as shown in Fig. 10.6, and is used in a similar manner to the transparent latch in Fig. 10.5. During normal function, the latch is held *flush* by the *Macro Test Enable* signal such that the HSS output signal always propagates to downstream logic. During test operation, the latch E pin is controlled by the *Macro Test Complete* signal. This signal is held low during the test sequence and goes high at the end of the sequence, capturing the HSS signal output logic level in the latch. The latched value is subsequently scanned out of the chip.

If the HSS output pin must be dynamically observed during execution of the test sequence, then the signal must be observable at a chip output pad. This output is usually not the functional connection for the HSS signal. As shown in Fig. 10.6, the HSS pin is driven to both its functional sink as well as a test sink. The buffers shown in the figure are not logically required, but may be required to preserve net names for some design automation flows. The test sink, a chip output pad, may also be used for other purposes during other modes of operation. For instance, this pin may also be a scan output when scanning the chip, and may be a functional signal output for some unrelated function during normal chip operation. A multiplexor would be used to select the source of the signal driving the chip output pin based on the operational or test mode.

10.2.2.3 HSS EX10 Example

Continuing with the expansion of the HSS EX10 tutorial example, Table 10.3 defines macro test requirements for the HSS EX10 core. Preconditioning is required for many of the HSS input pins. Pins not listed in this table are gated with other signals internal to the HSS core and do not affect test execution.

During macro test, chip pads must drive the *HSSRESET*, *TXxPRBSRST*, *RXxPRBSRST*, and *RXxPRBSFRCERR* signals. Each of these signals must be driven by a chip input pad during macro test execution. This chip input may be shared by all of the HSS cores (and all channels of the same HSS core) on the chip. Four test inputs are required regardless of the number of cores on the chip.

Table 10.3 HSS EX10 macro test controllability and observability

Pin name	Type	Slice	Controllability/observability description
HSSRESET	In	PLL	Full controllability required during test sequence
HSSPLLLOCK	Out	PLL	Full observability required during test sequence
HSSPDWNPLL HSSRECCAL HSSRESYNCCLKIN HSSSTATEL2	In	PLL	Preconditioned to "0" for all tests
HSSREFDIV HSSDIVSEL[1:0] HSSPRTWRITE HSSPRTAEN HSSPRTADDR[7:0] HSSPRDATAIN[15:0]	In	PLL	Preconditioned to "0" or "1" depending on test
TXxPRBSRST	In	Tx	Full controllability required during test sequence
TXxBYPASS TXxPWRDWN TXxSTATEL1 TXxELECIDLE	In	Tx	Preconditioned to "0" for all tests
TXxTS	In	Tx	Preconditioned to "1" for all tests
TXxPRBSEN	In	Tx	Preconditioned to "0" or "1" depending on test
RXxPRBSRST RXxPRBSFRCERR	In	Rx	Full controllability required during test sequence
RXxDATASYNC RXxPHSLOCK RXxPWRDWN RXxSTATEL1 RXxSIGDETEN	In	Rx	Preconditioned to "0" for all tests
RXxACJPD[P,N] RXxPRBSEN	In	Rx	Preconditioned to "0" or "1" depending on test
RXxSIGDET	Out	Rx	Must obverse value at conclusion of the test

The *RXxSIGDET* and *HSSPLLLOCK* outputs require observability. In the case of *RXxSIGDET*, the logic value can be captured in a latch and observed at the conclusion of the test sequence. In the case of *HSSPLLLOCK*, the signal must be observed dynamically, and therefore the signal must drive a chip output during the test sequence.

10.2.3 JTAG Logic Connections

10.2.3.1 JTAG Boundary Scan Cell Connections

As was discussed in Sect. 7.1, JTAG 1149.1 and JTAG 1149.6 compliance require all input/output signals of the chip connect to Boundary Scan Cells. This places a requirement on the HSS core to provide signals to connect to the Boundary Scan Cell, which is part of the JTAG Boundary Scan Register. HSS pin connections to Boundary Scan Cells were described for the HSS EX10 core in Sects. 2.2.6 and 2.3.6.

10.2.3.2 Input Preconditioning

In addition to the signals described in Sects. 2.2.6 and 2.3.6, additional pins on the HSS core may need to be preconditioned to specific logic values for JTAG testing. The following cases may exist:
- HSS pin must be held at logic "0" for JTAG test execution.
- HSS pin must be held at logic "1" for JTAG test execution.
- HSS pin must be controlled by the JTAG TAP Controller.

Logic to implement these cases is illustrated in Fig. 10.7. This logic assumes a *JTAG Compliance Enable* is asserted and held by the TAP controller when appropriate JTAG instructions have been loaded, and is not asserted during normal operation of the chip. Logic in Fig. 10.7 preconditions and multiplexes signals in a similar manner to Fig. 10.5.

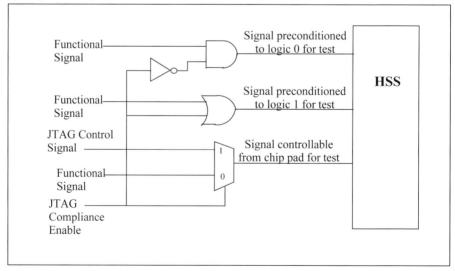

Fig. 10.7 Logic for JTAG test preconditioning

10.2.3.3 HSS EX10 Example

Continuing to expand the HSS EX10 example, Table 10.4 defines the JTAG signal connection requirements for this core. Many of the pins in this table are connected to the Boundary Scan Cells, while the rest are described below.

The *HSSJTAGCE* pin is driven by the TAP Controller based on a decode of the contents of the JTAG Instruction Register. This is the equivalent description to the *JTAG Compliance Enable* signal described in Fig. 10.7, and this signal is typically connected to the *HSSJTAGCE* pin. This signal is used internal to the HSS EX10 core to gate signals and enable JTAG test mode. Table 10.4 describes additional signals which must be preconditioned in logic outside the core by *HSSJTAGCE*.

Table 10.4 HSS EX10 JTAG test preconditioning requirements

Pin name	Type	Slice	Controllability/observability description
HSSJTAGCE	In	PLL	Driven by the JTAG TAP Controller based on contents of the JTAG Instruction Register
HSSACJPC HSSACJAC	In	PLL	Signal connections to boundary scan cells
HSSPDWNPLL	In	PLL	Preconditioned to "0" during JTAG test
HSSTXACMODE HSSRXACMODE	In	PLL	Tied to "0" or "1" based on system design
TXxBSIN TXxJTAGTS	In	Tx	Signal connections to boundary scan cells
TXxBSOUT	Out	Tx	Signal connections to boundary scan cells
TXxJTAGAMPL[1:0]	In	Tx	Tied to "0" or "1" during JTAG test. (Only used when HSSJTAGCE = "1.")
TXxPWRDWN	In	Tx	Preconditioned to "0" during JTAG test
RXxBSOUT	Out	Rx	Signal connections to boundary scan cells
RXxACJPDP RXxACJPDN	In	Rx	Signal connections to boundary scan cells
RXxACJZTP RXxACJZTN	Out	Rx	Connection to boundary scan cell
RXxPWRDWN	In	Rx	Preconditioned to "0" during JTAG test

Preconditioned signals must be driven to valid values during JTAG test. The reset sequence for the chip is *not* executed during JTAG testing of a circuit board, and therefore the state of most flip-flops on the chip is unknown. If, for example, some of these signals are driven by a programmable control register, the state of this register is unknown during JTAG test. In order to ensure values are valid during JTAG, these pins must either be tied directly to a logic "0" or

"1" level, must be connected to a chip I/O, or must be connected to logic which is controlled by a JTAG compliance signal.

10.2.4 Automation of Test Requirements

A number of requirements for test connections to the HSS core have been discussed, the implementation of which has been left as an exercise to the chip designer. This suggests opportunities for Design-For-Test (DFT) design automation software to be used in the following areas:

- DFT design automation software tools which make logical changes to the netlist to stitch scan chains, generate and connect JTAG boundary scan cells, etc.

- Additional files in the design kit for the HSS core which provide logic and connections for test signals. A *test wrapper* is discussed in this section which serves this purpose.

10.2.4.1 Design Automation Software

Software tools exist which are capable of automatically connecting some of the test signals discussed in this section.

Scan chain stitching software connects scan input and output pins of flip-flops and latches on the chip into contiguous scan chains. Such software can also stitch scan chain segments embedded in the HSS cores into the chip level scan chains. This software is generally sophisticated enough to perform stitching based on the physical locations of the elements on the chip (stitching flip-flops based on nearby neighbors), and to balance scan chains such that scan chains have roughly equal numbers of flip-flops in each chain.

Top level insertion software generates the logic for JTAG boundary scan cells and connects JTAG signals to boundary scan cells. This software is also capable of generating logic for I/O sharing. Chip input pins may have functional use during normal chip operation, may be used as a scan input, and may connect to an HSS core (or another core type) to control a signal during macro test. Likewise, chip output pins may have defined functional use, may be used as a scan output, and may be used to observe an HSS core signal during macro test.

Top level insertion software is driven by input from the chip designer supplied in the form of I/O definition statements in an I/O control file. Each I/O statement specifies the net name associated with the I/O; the I/O cell type (name of the library cell or other keyword); net names of signals associated with functional, scan, and macro test use; and net names associated JTAG boundary scan cell connections.

The base functionality of top level insertion software assumes that all of the nets referenced by the statements in the I/O control file exist at the top level of the chip. Test signals on HSS cores must be routed through the chip design hierarchy to the top level so that these signals may be connected to the appropriate sources and sinks. Alternatively, it is possible to add preprocessing to the top level insertion software such that test signals do not need to be routed through the chip hierarchy. This approach is discussed further in the next section.

10.2.4.2 Test Wrappers

Sects. 10.2.2 and 10.2.3 discussed logic which was required to be added by the chip designer around the HSS core. This logic gated signals to the HSS core to force preconditioned values, and provided controllability and observability of various HSS pins for macro test and for JTAG test. Given requirements for a specific pin, the logic associated with the pin is typically one of the cases illustrated in either Fig. 10.5, 10.6, or 10.7 (or some combination thereof). This test logic does not vary significantly from one instance of the HSS core to the next, or from one chip design to the next. This suggests that the logic can be supplied in the form of a *test wrapper* as part of the design kit for the HSS core.

A *test wrapper* is a Verilog or VHDL design which instantiates the HSS core and provides the necessary test support logic around the core. This is shown conceptually in Fig. 10.8. Inputs to the test wrapper consist of functional signals and test signals, and any preconditioning logic or multiplexor logic is included in the wrapper. The chip designer connects functional signals as needed by the application, and connects test signals to appropriate test control sources; control logic is provided by the test wrapper.

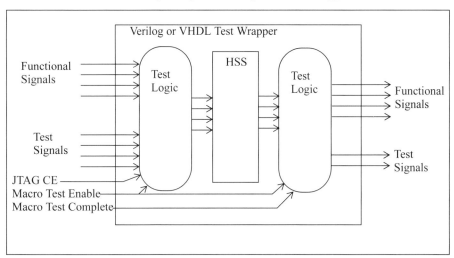

Fig. 10.8 Test wrapper concept diagram

10.2.4.3 Combining Test Wrappers with Top Level Insertion

The previous description of test wrappers assumed all test signals were connected by the chip designer. Almost universally, test signals must be connected through the design hierarchy of the chip in order to make appropriate connections at the top-level of the netlist. This produces an opportunity for further automation by combining the test wrapper concept with preprocessing functionality in the top level insertion software.

In order to implement automation, the test signal input and output ports in Fig. 10.8 are removed from the Verilog module statement (or VHDL entity statement). Instead, test signals are tied to their normal (functional) values inside the test wrapper. These nets are "tagged" in the test wrapper so that the nets are recognized by top level insertion software. "Tags" may use net attributes, special net names, special cells, or special instance names to identify the nets to the top level insertion software. Software scans the chip netlist for tagged nets, removes the tie values from these nets, and connects the nets to the appropriate sources and sinks in the top-level of the chip netlist. When this approach is used to process test wrappers, there is no longer a need for the chip designer to connect test signals through the chip netlist hierarchy.

An integrated netlist processing flow which combines I/O definition statement examples and a test wrapper with appropriate design automation software can greatly simplify implementation of the test requirements for the HSS core. What started out as a rather complex set of requirements to support various test modes becomes an almost fully automated methodology for generating and connecting the necessary test logic.

10.2.5 Running Macro Test using the JTAG Interface

The discussion of JTAG 1149.1 in Sect. 7.1.1 described an optional *RUNBIST* instruction which can be loaded into the *JTAG Instruction Register*. This instruction is sometimes utilized to execute a Macro Test sequence controlled by internal state machines within the HSS core. When the *JTAG Instruction Register* contains the *RUNBIST* instruction, signals are preconditioned to run Macro Test as needed. Signals which require controllability to either logic "0" or logic "1" can be driven from additional user-defined registers scanned through the JTAG interface; signals which must be observed at the conclusion of the test can be similarly captured in user-defined registers. Signals toggled during the test sequence are controlled by BIST state machines in the HSS core.

Using the JTAG interface to execute the Macro Test sequence is attractive because it simplifies the requirements for special test connections to the HSS core and potentially eliminates most of the logic that would otherwise need to be added by the chip designer (or included in a test wrapper). However, this approach does not reduce the complexity of the macro test implementation, but rather simply embeds the associated logic in the HSS core. If this approach is used, the Macro Test sequence is dictated by the design of the associated state machines in the HSS core. This sequence cannot be significantly altered by reprogramming chip test equipment.

Despite these limitations, Macro Test execution through the JTAG interface is widely implemented on HSS cores. This is especially true on mature core designs targeting lower baud rates.

10.3 Static Timing Analysis

Static timing analysis software is used in the methodology shown in Fig. 10.1 to exhaustively verify that timing requirements of all logic paths on the chip have been met. Static timing analysis requires a timing model for all physical blocks on the chip, including HSS cores. This *timing model* defines timing checks at the input pins of the HSS core and propagation delays associated with timing paths through the core. The format of the timing model depends upon the software being used to perform static timing analysis.

For the HSS EX10 core example, the timing model must include the following timing information:

- Propagation delays from the *HSSREFCLK[T,C]* clock inputs to all clock outputs of the core, including: *TXxDCLK, RXxDCLK, RXxRCVC16,* and *HSSRESYNCCLKOUT.*
- Pulse Width test at the *HSSREFCLK[T,C]* clock input pins.
- Setup and Hold test at the *TXxD[n:0]* pins referenced to the corresponding *TXxDCLK* output clock.
- Propagation delays from the *RXxDCLK* output clocks to the corresponding *RXxD[n:0]* output pins.
- Setup and Hold test at the *HSSRESYNCCLKIN* input pin referenced to the *HSSRESYNCCLKOUT* clock output.
- Timing tests and propagation delays associated with the register interface, including:
  - Pulse Width test at the *HSSPRTWRITE* input pin.
  - Setup and Hold tests at the *HSSPRTDATAIN, HSSPRTADDR,* and *HSSPRTAEN* input pins referenced to *HSSPRTWRITE.*
  - Propagation delays from the *HSSPRTADDR* and *HSSPRTAEN* input pins to the *HSSPRTDATAOUT* output pins.

Although the above is specific to the HSS EX10 tutorial example core, similar timing parameters for clocks and parallel data apply generically to any HSS core. Timing parameters for the register read/write interface are similar to the requirements for any core with an interface used to read/write registers.

10.3.1 Clock Timing

The timing model for the HSS EX10 core includes propagation delays associated with the timing paths from the *HSSREFCLKT/C* input pins to the *RXxRCVC16, HSSRESYNCCLKOUT, TXxDCLK,* and *RXxDCLK* outputs. Assuming the PLL is locked, these clock outputs do have a fixed frequency relationship to the reference clock, and the timing model incorporates these propagation delay paths so that timing phase properties of the reference clock may be inherited by the corresponding clock outputs. However, static timing analysis software has shortcomings in how these clocks are checked and propagated. These are described in the following sections.

10.3.1.1 Differential Clock Analysis

Static timing analysis tools do not perform sufficient analysis to ensure the signal integrity of differential signals. Therefore, the *HSSREFCLK[T,C]* signals should be primarily verified using Spice as was described in Sect. 6.4.3. *RXxRCVC16* clocks, if used, should be analyzed in a similar manner using Spice.

In some cases the chip may contain protocol logic which is clocked by a single-ended clock derived from a differential clock. In these cases, the phase and frequency associated with the reference clock does become relevant for logic connected to the single-ended portion of the clock tree. This case is illustrated in Fig. 10.9. The buffer shown in the figure is a differential clock buffer with a differential input and both single-ended and differential outputs.

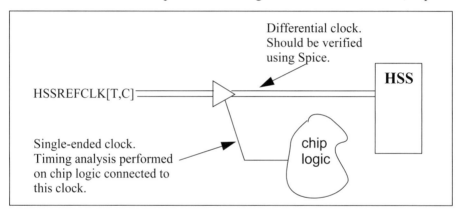

Fig. 10.9 Chip logic clocked by a differential clock

10.3.1.2 TXxDCLK / RXxDCLK Clock Outputs

For purposes of this discussion, the *TXxDCLK* and *RXxDCLK* output pins are referred to generically as the *DCLK* outputs.

Timing Adjusts

The frequency of the *DCLK* clock outputs is fractionally related to the frequency of the *HSSREFCLK[T,C]* input on the HSS EX10 core. The phase relationship between *DCLK* and *HSSREFCLK* is indeterminate.

Static timing analysis propagates the clock attributes of the *HSSREFCLK* signals through the HSS timing model to the *DCLK* outputs. Because of static timing analysis limitations, propagation through the delay path in the timing model causes the *DCLK* outputs to be treated as data signals as shown in Fig. 10.10. The arrival time of the rising and falling edges of the *DCLK* output are determined by adding the propagation delays in the timing model to the rising edge arrival time at the *HSSREFCLKT* input pin. For proper analysis to occur at downstream sinks for *DCLK*, the arrival time of the falling edge must be shifted by one-half *DCLK* cycle. Parameters must also be set on this pin to indicate the signal is, in fact, a clock.

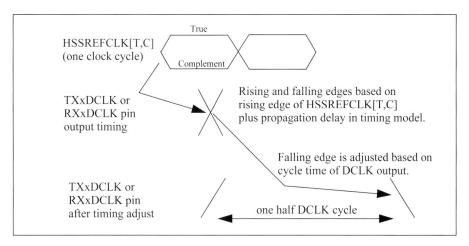

Fig. 10.10 Timing adjust for DCLK outputs

These actions are communicated to the static timing analysis software using directives, called *timing assertions*, which are written by the chip design team. The command format for timing assertions is specific to the software tool being used. The following example assumes the IBM EinsTimer software is used to perform static timing analysis of a chip containing an HSS EX10 core with the instance name *HSS_TX_CORE*. (An equivalent example using commands for Synopsys software is provided later in this section.) Consider the transmitter portion of an HSS EX10 core running at a data rate of 8.5 Gbps with an 16-bit data bus width. The cycle time of this clock is 1.882 ns, taking into account the baud rate, data width, and rate mode settings at which the HSS core is used by this application. The leading (rising) edge of the signal arrival on the *DCLK* pins should be adjusted by +0.00 ns, and the trailing (falling) edge should be adjusted by +0.941 ns (one-half of the clock cycle).

Assuming that the clock phase name arriving on the *HSSREFCLK* pins is *TXREFCLK*, and the new phase name associated with the *TXADCLK* output pin is *DCLK+*, then the following timing assertions are needed. A *DCLK+* phase definition is created using the following timing assertion:

```
et::create_clock -period 1.882 -waveform { 0.0 0.941 } -name "DCLK"
```

The timing of the *DCLK+* clock falling edge is adjusted with the *TXREFCLK@L* tag as shown in the following timing assertions:

```
et::adjust_signal -pins { "HSS_TX_CORE/TXADCLK" } -rise \
                -time 0.0 -phase "TXREFCLK@L"
et::adjust_signal -pins { "HSS_TX_CORE/TXADCLK " } -fall \
                -time 0.941 -phase "TXREFCLK@L"
et::rename_phase -pins { "HSS_TX_CORE/TXADCLK " } -phase * \
                -new_phase "DCLK+"
```

The "@L" suffix of the *TXREFCLK* phase notation in these assertions indicates that a data signal is being adjusted, and that this data signal was propagated from the rising edge arrival time on the *HSSREFCLK* input. The first timing assertion adjusts the leading edge arrival time, and the second assertion adjusts the falling edge. The last timing assertion changes the phase name on the *TXADCLK* pin from *TXREFCLK* to *DCLK*. The "+" suffix indicates that this is a clock signal (as opposed to a data signal).

As noted in Sect. 4.2.4, the *TXxDCLK* and *RXxDCLK* outputs of the HSS EX10 do not have a guaranteed phase relationship. Transmit protocol logic should use either elastic FIFOs to retime data to individual *TXxDCLK* clock domains or the *HSSRESYNCCLKIN* input to synchronize the transmitters. In the latter case, channel-to-channel skew specifications for the HSS core must be taken into account in the timing analysis.

Clock Jitter

The clock output of the HSS PLL slice is divided in the HSS core to produce the *TXxDCLK* and *RXxDCLK* outputs. Jitter on the PLL clock is a source of jitter on the *DCLK* outputs. An additional source of jitter for *RXxDCLK* results from the CDR circuit tracking the center of the received data eye. The CDR circuit that was described in Sect. 3.3.1 updated sampling phase in quantized steps; these updates cause jitter on the corresponding *RXxDCLK*.

The jitter on *DCLK* is period related and varies from one edge to the next. Because the first edge can move independently with respect to the next edge (which could also move sooner or later), the application must account for periods of shrinkage or growth in the cycle time caused by jitter. Fig. 10.11 illustrates how this jitter can affect *DCLK*. As shown, the edges of the "Jittery DCLK" may shift by ±0.05 (units are not important) relative to the ideal clock. Worst case conditions occur when consecutive edges shift in opposite directions. Period jitter in this example is 0.10.

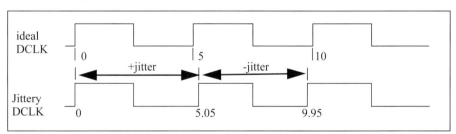

Fig. 10.11 HSS DCLK jitter

DCLK jitter is of significance since it results in cycle times which may be less than the ideal cycle time. The clock cycle time must be adjusted by the period jitter value to ensure flip-flop setup times are met.

Table 10.5 provides an example of a jitter specification for the HSS EX10 core. Using this specification, the timing assertions in the example from the previous section may be amended with the following additional assertion:

```
et::set_clock_jitter -clocks "DCLK" -jitter_late 0.071 -jitter_early 0.0 \
              -half_cycle_jitter 0.0 -pll_gate HSS_TX_CORE
```

Table 10.5 HSS EX10 *DCLK* period jitter

Channel type	Data bus width (bits)			
	8	10	16	20
TX	±71 ps	±151 ps	±71 ps	±71 ps
RX	±153 ps	±208 ps	±153 ps	±153 ps

For example, the TX core specifies period jitter as ±71 ps, which means the *DCLK* period could grow or shrink by 71 ps. This effect is modeled in the same timing run by adding a set clock jitter statement to the timing assertions

This timing assertion defines the amount of jitter on the *DCLK* clock phase. Static timing analysis software uses this value when evaluating setup tests by assuming a cycle time of: cycle time – jitter = 1.811 ns.

Duty Cycle Variation

The duty cycle of the clock must be considered to properly evaluate pulse width checks throughout the clock tree, as well as to evaluate timing for any logic clocked by the falling clock edge.

Assume a duty cycle variation from 45% of the cycle time to 55% of the cycle time is specified for the HSS EX10 core. The timing assertions in the example from the previous sections are modified as follows:

```
et::create_clock -period 1.882 -waveform { 0.0 0.941 } -name "DCLK"

et::adjust_signal -pins {"HSS_TX_CORE/TXADCLK"} -rise \
                -time 0.0 -phase "TXREFCLK@L"
et::adjust_signal -pins {"HSS_TX_CORE/TXADCLK "} -fall \
                -time 1.0351 -phase "TXREFCLK@L" -late
et::adjust_signal -pins {"HSS_TX_CORE/TXADCLK "} -fall \
                -time 0.8469 -phase "TXREFCLK@L" -early
et::rename_phase -pins {"HSS_TX_CORE/TXADCLK "} -phase * \
                -new_phase "DCLK+"

et::set_clock_jitter -clocks "DCLK" -jitter_late 0.071 -jitter_early 0.0 \
                -half_cycle_jitter 0.0 -pll_gate HSS_TX_CORE
```

The *adjust_signal* timing assertion for the falling clock edge has been split into two assertions in this example, with separate adjust times specified for *late* and *early* signal arrival. The late value is 1.0351 ns (55% of the cycle time), and the early value is 0.8469 ns (45% of the cycle time). Static timing analysis software uses one or the other of these values depending on the context of the timing test being evaluated.

Timing Assertions for Synopsys Software Tools

Prior sections developed an example of timing assertions for the TXxDCLK or RXxDCLK pins which used the IBM EinsTimer software to perform static timing analysis. In this section, the equivalent timing assertions for analysis using Synopsys timing analysis software is presented.

The following timing assertion defines the clock waveform on the HSS core pin:

```
create_clock -name DCLK -period 1.882 -waveform [ list 0.0 0.941 ] \
                HSS_TX_CORE/TXADCLK
```

Although this timing assertion is similar to the EinsTimer phase definition assertion in "Timing Adjusts" under, the above assertion does more than just define the clock phase. It additionally sets the named HSS core pin as the source of this clock, and sets the signal waveform on this pin with the specified ideal arrival times. (Unlike EinsTimer, any signal propagation through the Synopsys timing model for the HSS core is ignored once this timing assertion is applied to the clock output pin.) The Synopsys *create_clock* assertion is therefore equivalent to the *et::create_clock*, *et::adjust_signal*, and *et::rename_phase* EinsTimer assertions in Sect. 10.3.1.2.

The Synopsys *set_clock_latency* timing assertion adjusts the timing of a signal in a similar manner to the EinsTimer *et::adjust_signal* assertion. This assertion is used to adjust the HSS clock output to account for the effects of both jitter and duty cycle variation. The following Synopsys timing assertions are the equivalent of the EinsTimer assertions in "Duty Cycle Variation" under Sect. 10.3.1.2:

```
create_clock -name DCLK -period 1.882 -waveform [ list 0.0 0.941 ] \
              HSS_TX_CORE/TXADCLK

set_clock_latency -source -rise -late 0.071 [ get_clocks DCLK ]
set_clock_latency -source -rise -early 0.0 [ get_clocks DCLK ]
set_clock_latency -source -fall -late 0.0941 [ get_clocks DCLK ]
set_clock_latency -source -fall -early -0.0941 [ get_clocks DCLK ]
```

The first two *set_clock_latency* assertions adjust the timing of the rising edge of the clock used in setup tests to account for jitter. Remaining assertions adjust the timing of the falling edge of the clock to account for duty cycle.

10.3.2 Receiver Parallel Data Outputs

The timing relationship between *RXxD*[19:0] and *RXxDCLK* is best explained by looking at the expected timing in a functional environment. Figure 10.12 shows HSS receive data connected to downstream flip-flops clocked by RXxDCLK. While the fanout of the data signals is limited to a few flip-flops, RXxDCLK is driven to a larger fanout through a clock tree.

Figure 10.12 also shows the timing for the RXxD signals and for the RXxDCLK at points A and B. The rising edge of RXxDCLK launches data at the HSS core, and the *same* clock edge (delayed by the clock tree) captures the data at the flip-flops in the chip logic. This is possible because the delay with which RXxDCLK propagates through the clock tree is typically greater than the delay incurred by the RXxD signals.

There are two issues with this from a static timing analysis perspective. The first of these issues results because static timing analysis software assumes that if a clock edge is used to launch data, then the data should be captured by the *next* clock edge. However, the interface has in this case been designed such that data should be captured by the *same* clock edge. As can be seen in the figure, the data is not inherently held long enough to be captured by the next clock edge, especially when the delay of the clock tree is considered. It is therefore necessary to adjust the data arrival time to arrive one clock cycle later.

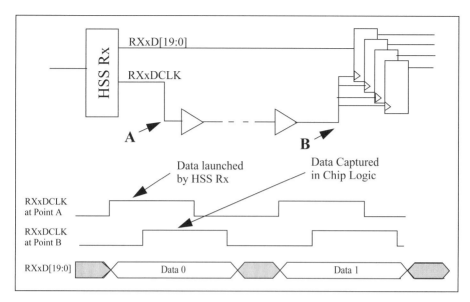

Fig. 10.12 HSS receive data timing

Assume an HSS EX10 core with the instance name *HSS_RX_CORE* is running at a data rate of 8.5 Gbps with a 16-bit data bus width, as was the case in prior examples. The cycle time is therefore 1.882 ns. The following EinsTimer timing assertions perform the necessary adjustment:

```
et::adjust_signal -pins { "HSS_RX_CORE/RXAD*" } -rise -fall \
                -time 1.882 -phase * -late
et::adjust_signal -pins { "HSS_RX_CORE/RXAD*" } -rise -fall \
                -time 1.729 -phase * -early
```

Note that separate adjust times have specified for *late* and *early* signal arrival. While the late timing value is usually adjusted by the full clock cycle time, the early timing value must account for period clock jitter, and is therefore, cycle time − jitter = 1.882 − 0.153 ns = 1.729 ns.

Without these adjustments, static timing analysis will report hold time violations on the RXxD signals. This could be corrected by adding buffers to delay these signals, however, such logic would be completely unnecessary.

The following Synopsys timing assertions perform the equivalent adjustments:

```
set_input_delay 1.882 -max [ get_pin HSS_RX_CORE/RXAD* ]
set_input_delay 1.729 -min [ get_pin HSS_RX_CORE/RXAD* ]
```

The second issue from a static timing analysis perspective occurs when timing analysis is performed on early versions of the chip layout. The chip physical design process often performs the early stages of layout and timing closure using ideal clock timing. When such analysis is performed, the RXxD signals do incur some propagation delay, while there is no delay incurred by RXxDCLK. Under these conditions, static timing analysis reports setup time violations on most or all of the sinks for the RXxD signals.

This issue is fixed by inserting an arbitrary delay on the RXxDCLK nets. Optimal methods for doing this vary somewhat based on static timing analysis software being used. For example, the solution for Synopsys software is to modify the timing assertions for the RXxDCLK pins that were described in "Timing Assertions for Synopsys Software Tools" under Sect. 10.3.1.2. The delay values associated with the set_clock_latency timing assertions are modified to include an arbitrary delay that is representative of the delay of the postlayout clock tree. This delay is used during prelayout timing analysis, and is removed for postlayout timing analysis.

10.3.3 Register Interface

The HSS EX10 core defined in Chap. 2 has an interface for reading and writing registers in the core. Although many variations exist for register interfaces, the example covered here for the HSS EX10 is representative.

Pins associated with this interface were defined in Table 2.1. Register read and write cycle timing is shown in Fig. 10.13, with the corresponding timing parameter values defined in Table 10.6. This interface consists of an address enable (*PRTAEN*), address bus (*PRTADDR*), write strobe signal *(PRTWRITE)*, and input/output data busses (*PRTDATAIN, PRTDATAOUT*). Although signals are synchronized to various clock domains internal to the HSS EX10 core, the interface appears as an asynchronous interface to the chip logic outside of the core.

Read cycles occur whenever *PRTAEN* is held high and *PRTWRITE* remains low as shown in Fig. 10.13. Register read data on *PRTDATAOUT* is valid within the output valid time (T_{ov}) after *PRTAEN* and *PRTADDR* are stable. This is a combinatorial propagation delay path.

Write data is strobed into registers by *PRTWRITE* = 1 as shown in Fig. 10.13. *PRTAEN, PRTADDR,* and *PRTDATAIN* must be stable before the rising edge of *PRTWRITE*, as defined by setup time T_{su}, and must be held after the falling edge of *PRTWRITE*, as defined by hold time T_{hd}. The *PRTWRITE* signal must also comply with minimum pulse width T_{pw}, and the minimum time between write cycle pulses T_{ipw}.

The timing model for the HSS EX10 core implements checks for each of the timing parameters described above. Depending on the design of the logic which drives this interface, timing assertions may be needed for static timing analysis software to properly evaluate this interface.

First, consider that *PRTWRITE* is a clock signal for purposes of timing analysis. On the timing model, this pin is the clock side of setup and hold tests, and is a termination point for pulse width tests. Chip logic may drive this signal from control logic which propagates a data phase to this pin. If this is the case, timing assertions must be written to redefine the arrival times of signals at this pin as clock signals. This is performed in a similar manner to the description in "Timing Adjusts" under Sect. 10.3.1.2 for *DCLK*.

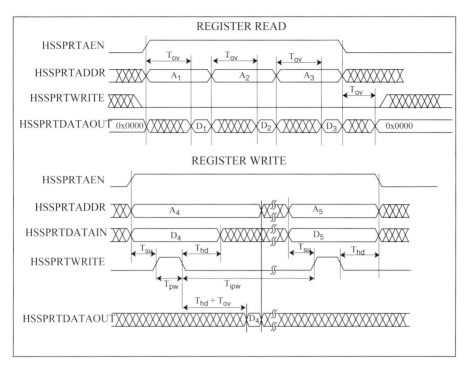

Fig. 10.13 HSS receive data timing

Table 10.6 Register access timing parameters

Symbol	Description	Limit	High range (ns)	Low range (ns)
T_{ov}	Output valid delay of HSSPRTDATAOUT[15:0], relative to HSSPRTADDR[8:0] changing and HSSPRTAEN=1	Max	10	10
T_{su}	Setup time of HSSPRTAEN, HSSPRTADDR[8:0] and HSSPRTDATAIN[15:0] relative to the rise of HSSPRTWRITE	Min.	15	22
T_{pw}	Pulse width of HSSPRTWRITE	Min.	15	22
T_{hd}	Hold time of HSSPRTADDR[8:0] and HSSPRTDATAIN[15:0] relative to the fall of HSSPRTWRITE	Min.	25	38
T_{ipw}	Write InActive pulse width of HSSPRTWRITE	Min.	200	200

Second, note that the timing parameters defined in Table 10.6 are relatively slow compared to the cycle times of other interfaces on the HSS core. If these signals are driven from control logic which is clocked at higher clock rates, the T_{ov} cycle time may be greater than the clock cycle time, and this path must be treated as a multicycle path. Timing assertions may use any of several coding styles to accomplish this.

10.3.4 Transmitter Synchronization

The HSS EX10 core supports transmitter synchronization using the *HSSRESYNCCLKIN* and *HSSRESYNCCLKOUT* pins as was described in Sect. 2.4.7. This scheme requires logic interconnecting these pins as shown in Fig. 2.15 (for synchronizing multiple HSS cores).

10.3.4.1 Critical Timing

The timing for the *Resync* net in Fig. 2.15 is a critical path which is launched by a rising edge clock and captured by a falling edge clock. This half cycle path must meet a tight timing budget. The falling edge of the clock is used to solve hold time issues and provides excess margin, however, this creates issues meeting the setup time. The following factors contribute to the timing budget for this net:

- Must support the minimum *HSSREFCLKx* period
- Must tolerate skew in the clock tree for the *HSSRESYNCCLKOUT* which clocks the flip-flop driving the net
- Must tolerate HSS PLL static phase error variation. This is the core-to-core variation in the skew between the *HSSREFCLKx* input and the *HSSRESYNCCLKOUT* output when the HSS PLL is locked
- Must tolerate skew in the clock tree used for *HSSRESYNCCLKOUT* in each of the staged pipelines
- Must tolerate duty cycle of *HSSRESYNCCLKOUT* signal when using both rising and falling edges

Based on these considerations, an example of a timing budget for the *RESYNC* signal to be sampled in each of the pipeline clock domains is

1,250-ps period * (0.40) = 500 ps (penalty @ 40% duty cycle)
- 100-ps REFCLK skew
- 100-ps PLL static phase error variation
- 100-ps RESYNCCLKOUT skew

= 200 ps (clock-to-output of resync latch, plus setup time of pipeline input latch).

This budget is insufficient for most "slow-chip" cases. It is therefore necessary to "slack steal" by taking the negative clock edge from an earlier point in the clock tree. With reasonable effort, excess hold slack can be shifted to improve setup slacks, and thus close timing on this path.

10.3.4.2 Timing Assertions

In order to time this transmitter synchronization scheme properly, appropriate timing assertions must be applied to the *HSSRESYNCCLKOUT* output. These are clock outputs, and must be defined as such in a similar manner to how the *DCLK* clock outputs were handled in "Timing Adjusts" under Sect. 10.3.1.2. It is necessary to check cases for each reference clock frequency which may be used by the application. These different clock frequencies can be checked as part of the same analysis run.

The first step is to define the various clocks. This can be performed using the following timing assertions for the IBM EinsTimer tool

```
et::create_clock -period $clock_period -waveform { 0.000 $falling_edge } \
        -name RESYNCCLK${ref_freq}_${core}
```

or the corresponding Synopsys timing assertions:

```
create_clock -name RESYNCCLK${ref_freq}_${core} -period $clock_period \
        -waveform [ list 0.000 $falling_edge ] \
        CORE${core}/HSSRESYNCCLKOUT
```

where a separate clock must be defined for the *HSSRESYNCCLKOUT* pin of each core instance (*$core*) in the group. If multiple reference clock frequencies may be used by the application, then separate clocks must be defined on each of these output pins and for each frequency case (*$ref_freq*). The ideal time of the falling edge (*$falling_edge*) and the clock period (*$period*) are determined for each reference clock frequency.

Next, it is necessary to define that clocks of different frequencies should not be compared. This is performed using the EinsTimer timing assertion:

```
et::set_phase_pair_exclusion -clock1 RESYNCCLK${ref_freq_1}_${core_1} \
        -clock2 RESYNCCLK${ref_freq_2}_${core2}
```

or the Synopsys timing assertions:

```
set_false_path -from RESYNCCLK${ref_freq_1}_${core_1} \
        -to RESYNCCLK${ref_freq_2}_${core_2}
set_false_path -from RESYNCCLK${ref_freq_2}_${core_2} \
        -to RESYNCCLK${ref_freq_1}_${core_1}
```

where these assertions must be iterated such that each pair of clocks that have different frequencies is defined as a false path.

Assume the core-to-core skew for the HSS EX10 core has been specified to be 200 ps. This skew affects the arrival times of clocks coming from different cores. In order to account for this skew value in the timing analysis, the arrival times of the clocks must be adjusted by the skew amount. This is performed using the following EinsTimer timing assertions:

```
et::set_user_delta_adjust -data RESYNCCLK${ref_freq}_${base_core} \
        -clock RESYNCCLK${ref_freq}_${core_N} \
        -late -adjust -0.200
```

```
et::set_user_delta_adjust -data RESYNCCLK${ref_freq}_${base_core} \
                -clock RESYNCCLK${ref_freq}_${core_N} \
                -early -adjust -0.200
```

or the Synopsys timing assertion:

```
set_clock_uncertainty 0.200 [get_clocks RESYNCCLK${ref_freq}_${core_N}]
```

where *$base_core* corresponds to the *HSSRESYNCCLKOUT* which launches the *RESYNC* signal in Fig. 2.15, and *$core_N* corresponds to all other HSS cores. The above commands must be iterated for each value of *$core_N* and *$ref_freq*. The result of these commands is that the arrival times of clocks from all other HSS cores have been adjusted to arrive from 200 ps early to 200 ps late relative to the clocks from the first core. This skew value is independent of the clock frequency.

Finally, in a similar manner to how the *DCLK* clock outputs were handled in "Timing Adjusts" under Sect. 10.3.1.2, EinsTimer assertions are required to adjust timing on the *HSSRESYNCCLKOUT* pins, and to define these pins as clocks:

```
et:::set_arrival -ports CORE${i}/HSSRESYNCCLKOUT \
                -phase RESYNCCLK${ref_freq}_${core}+ -time 0.000 -rise
et:::set_arrival -ports CORE${i}/HSSRESYNCCLKOUT \
                -phase RESYNCCLK${ref_freq}_${core}+ \
                -time [ expr 0.55 * $clock_period ] -fall -late
et:::set_arrival -ports CORE${i}/HSSRESYNCCLKOUT \
                -phase RESYNCCLK${ref_freq}_${core}+ \
                -time [ expr 0.45 * $clock_period ] -fall -early
et::set_clock_jitter -clocks RESYNCCLK${ref_freq}_${core}+ \
                -jitter_late $jitter -jitter_early 0.0 \
                -half_cycle_jitter 0.0 -pll_gate CORES${core}
```

Corresponding Synopsys timing assertions are:

```
set_clock_latency -source -rise -late $jitter \
                [ get_clocks RESYNCCLK${ref_freq}_${core} ]
set_clock_latency -source -rise -early 0.0 \
                [ get_clocks RESYNCCLK${ref_freq}_${core} ]
set_clock_latency -source -fall -late [expr $dc_hi_limit * $clock_period ] \
                [ get_clocks RESYNCCLK${ref_freq}_${core} ]
set_clock_latency -source -fall -early [expr $dc_lo_limit * $clock_period ] \
                [ get_clocks RESYNCCLK${ref_freq}_${core} ]
```

where these commands must be iterated for each *$ref_clock* and *$core*, and the values of clock period (*$clock_period*) is set based on the clock frequency. Parameters for period jitter (*$jitter*) and duty cycle high and low limits (*$dc_hi_limit* and *$dc_lo_limit*) are set based on the clock specification for the HSS core.

10.3.5 Serial Data Timing

The transmit and receive serial data signals cannot be analyzed using static timing analysis, but rather should be analyzed using Spice as was defined in Sect. 8.3. The timing model for the HSS EX10 core does not model any propagation delay timing arcs to the transmit serial data pins, nor does it implement any timing checks on the receive serial data pins.

10.3.6 Skew Management

Two kinds of skew are discussed below: lane-to-lane skew as defined in Sect. 4.1.2.5, and skew within the differential signal pair.

10.3.6.1 Lane-to-Lane Skew

As defined in Sect. 4.1.2.5, lane-to-lane *skew* is the constant portion of the arrival time difference between any two data signals of a multilane interface. Skew results from differences in reference clock routing to the Serdes cores, differences in clock routing within the Serdes cores, and differences in routing of serial data signals in the package and circuit boards. Because skew impacts the design of deskew logic at the receiver, many multilane protocol standards constrain the amount of skew that may be present at the package output pins of the transmitter chip.

Skew is not usually verified through timing analysis. Rather, skew is managed in the chip design process by determining the transmit skew requirements, and then defining a skew budget which allocates how much of this skew may be consumed by each contributor. Contributors to skew at the transmitter include:

- Clock skew in the reference clock distribution network connections to the *HSSREFCLK* pins. This contributor is within the control of the chip physical designer, and usually can be constrained to a small value.

- Skew due to data and clock routing trace length differences within the Serdes core. This is specified as lane-to-lane skew for the HSS core.

- Skew due to tolerances resulting from the resynchronization scheme used to synchronize transmitters in different cores. These tolerances result from PLL design tolerances that become a factor when the various transmitters being synchronized are in different cores and therefore are clocked by different PLL slices. This skew contributor may be specified as a core-to-core skew for the HSS core. Sometimes this core-to-core skew is specified such that it incorporates both the core-to-core and lane-to-lane contributors (and thus is used instead of the lane-to-lane value when multiple cores are involved).

- Skew due to signal routing differences in the package design result in time-of-flight variation. The extent to which skew can be minimized in the package design depends on how many signals are in the synchronized group, signal density, pin assignment constraints, etc. Also, skew management may conflict with other signal integrity requirements such as minimizing impedance discontinuities. As described in Table 10.7, a reasonable target for skew "Matching within a Bundle" is around 29 ps given current package technologies.

Assuming the clock tree, HSS core, and package each meet the skew requirements defined by the specified skew budget, then the skew requirement at the transmitter output is met by design. Note that the above assumes that *HSSRESYNCCLKIN*, or an equivalent function, is being used to resynchronize the transmitters. If this is not the case, then the lane-to-lane or core-to-core skew contributed by the HSS core is not defined.

10.3.6.2 Differential Pair Skew

Skew between the true and complement legs of the differential signal must be controlled to minimize duty cycle distortion (DCD), a key component of deterministic jitter (DJ). Typical package contributions to this skew are specified as the "Matching within Pair" parameter in Table 10.7. The specification for the skew contribution of the transmitter device is typically supplied by the vendor, and is generally in the range of 5 ps or less at higher baud rates. Skew contribution due to the channel can be analyzed using the analysis methods described in Chap. 8 to determine the effects of any DCD on link performance.

10.3.7 Timing Backannotation for Simulation

Fig. 10.1 implied support for determining propagation delays and timing checks associated with a chip design and *backannotating* this timing information into a simulation environment in order to perform simulations of the chip with actual timing. This is generally implemented by performing static timing analysis on the design and generating a *Standard Delay Format* (SDF) file. This file can be read by most event-driven simulators. Simulation models for ASIC library cells and cores on the chip are parameterized; the event-driven simulator sets delay parameters and timing check parameters based on the contents of the SDF file. The timing used in the simulation thereby reflects the actual timing of the postlayout chip.

In the case of HSS cores, several clocks are generated by the core. These clock outputs required timing assertions, as defined previously, in order to correctly evaluate chip timing. These clock outputs also cause problems for SDF generation. The parameters for the timing arc from *HSSREFCLK* to clock outputs are not set correctly. Also, output-to-output timing arcs such as the timing arc from the *RXxDCLK* outputs to the corresponding *RXxD* parallel data outputs are not set correctly. The following approach is a workaround for these issues:

- Timing analysis for SDF generation must be performed using a special timing model for the HSS core which does not contain any timing arcs to clock outputs, or any timing arcs for output-to-output paths.

- Simulation timing parameters (for arrival times, slews, etc.) for output clocks must be either set manually by the user or permitted to use default values.

- Simulation timing parameters (for arrival times, slews, etc.) for parallel data outputs (or any other output-to-output timing paths) must be either set manually by the user or permitted to use default values.

Of course, the user may be able to write scripts to extract the necessary parameter values from timing reports, perform any necessary manipulations of these values, and generate commands for the target event-driven simulation software. Alternatively, while the default delay values may not reflect the actual chip design, simulation with the defaults may be sufficient to meet the user's intended purpose. (Keep in mind that static timing analysis is generally used as the primary method of verifying chip timing. Full-timing event-driven simulation is computationally intensive, and is used only to a limited extent.)

10.4 Chip Floorplan and Package Considerations

This section focuses on chip physical floorplan and package design constraints. Connecting high-speed signals between silicon chips and circuit boards through chip packages requires certain electrical parameters be constrained in order to avoid signal integrity concerns.

10.4.1 Packages

Fig. 10.14 illustrates common methods of connecting silicon chips to packages. *Wirebond* packages include both *Fine Pitch Plastic Ball Grid Array* (FBGA) and *Electrically Enhanced Plastic Ball Grid Array* (EPBGA) package types. FBGA packages use wires to connect wirebond pads on the chip periphery to package balls. This type of package is relatively inexpensive, but does not have good signal integrity qualities. EPBGA packages are a newer technology which improves signal integrity by using short wires to connect the chip to substrate; substrate wiring then connects the signal to the package balls. EPBGA packages are attractive from a cost perspective relative to Flip-Chip packages, and keeping wires short significantly improves signal integrity.

At higher baud rates, *Flip-Chip* packages are commonly used to meet stringent electrical parameters. These packages bond C4 pads on the silicon directly to pads on the chip substrate, and substrate wiring then connects signals to package balls. (*C4* is short for *Controlled Collapse Chip Connections.*) FC-PBGA packages tend to be more expensive since substrates must usually be customized based on the chip I/O assignments.

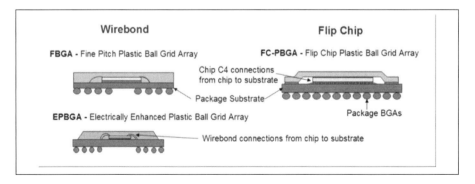

Fig. 10.14 Cross-sectional view of wirebond and flip-chip packages

Electrical parameters of the packaged die must achieve stringent requirements in order to support high-speed signals. Table 10.7 provides examples of package electrical parameters for the HSS EX10 serial data signals. Parameters in this table are readily achievable with an FC-PBGA.

The first portion of this table species resistance and inductance parameters for various HSS power supply inputs. These specifications represent limits assumed for HSS circuit design, and are constraints imposed on the package substrate design.

The lower portion of Table 10.7 specifies package requirements for differential signal I/O, including specification of skew parameters and signal integrity parameters.

"Matching within a pair" is one of the skew parameters specified in the table, and specifies the maximum time-of-flight skew between the true and complement legs of the differential signal pair; excessive skew between these signal legs contributes to DCD. "Matching within a bundle" specifies the time-of-flight skew requirement between different serial lanes of a multilane interface. This skew contributes to the protocol skew budget as was discussed in Sect. 4.2.6.3.

The signal integrity and achievable baud rate of Serdes signals are a function of the S_{DD21} (insertion loss) and S_{DD11} or S_{DD22} (return loss) of the package. Channel response and S-Parameters were discussed in "Channel Response" under Sect. 8.4.1.2 in Chap. 8. Figure 10.15 illustrates the Rx and Tx insertion loss requirements for the package side of a Serdes application, assuming the HSS signal termination is a 100-Ω resistive load. Table 10.7 specifies the insertion loss at $f_b / 2$ must be -2 dB or less. Figure 10.16 plots return loss specifications for various protocol standards as a function of baud rate. Package return loss must be less than whichever of these curves are relevant to the application. The insertion loss and return loss contributions of the transmitter package were considered in (8.30); the contributions of the receiver package were considered in (8.31).

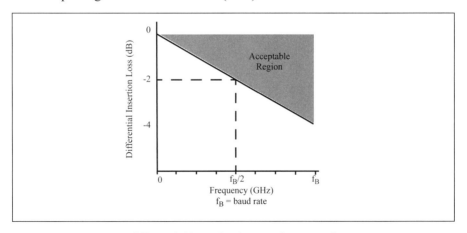

Fig. 10.15 Rx and Tx differential insertion loss package requirements

Table 10.7 Package electrical parameters for serdes

Parameter		Min.	Nom.	Max.	Units
Core power supplies					
AVDD25	Resistance			0.8 at 1.8V nom 2.0 at 2.5V nom	ohm
	Inductance			5	nH
ATST	Resistance			4	ohm
	Inductance			11	nH
AVTT, AVTR (per C4)	Resistance			0.357	ohm
	Inductance			4	nH
VDD/GND Compression				25 at 1.2V	mV
Differential signal I/O					
Rx & Tx serial I/O (including vias)	Resistance			2.6	ohm
	Differential impedance	85	100	115	ohm
Matching within pair				2.9	ps
Matching within bundle				29.1	ps
BGA adjacency			TNSEWD Swappable		
Rx trace and via isolation: Tx link to Rx link				-40 at 0 to $f_b/2$	dB
Rx trace and via isolation: Rx link to Rx link				-35 at 0 to $f_b/2$	dB
Tx trace and via isolation				-30 at 0 to $f_b/2$	dB
Rx differential return loss S_{DD11}					
Rx differential insertion loss S_{DD21}		-2 at $f_b/2$			dB
Tx differential return loss S_{DD22}					
Tx differential insertion loss S_{DD21}		-2 at $f_b/2$			dB
Tx transition time		40			ps
Tx voltage swing				0.7	V

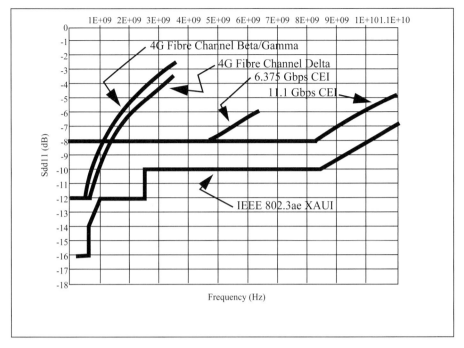

Fig. 10.16 Selected differential return loss package requirements

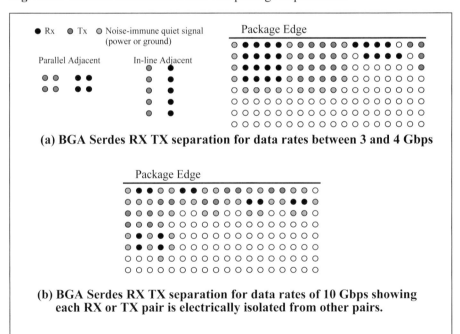

Fig. 10.17 BGA serdes RX TX separation

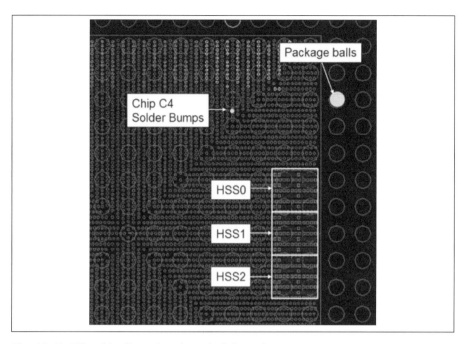

Fig. 10.18 Flip-chip die and package ball footprint

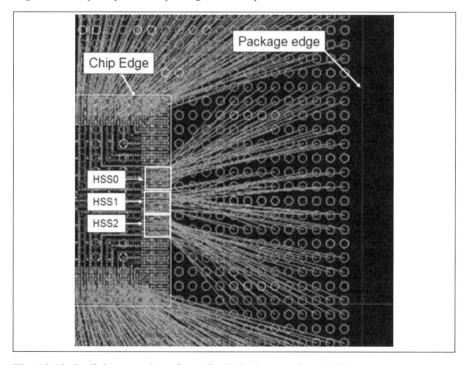

Fig. 10.19 Radial connections from die C4 balls to package BGAs

The "BGA Adjacency" entry in Table 10.7 specifies TNSEWD, which indicates that the module I/O for the true and complement legs of the differential signal must be adjacent in either the north, south, east, west, or diagonal direction on the chip.

In addition, the package design must also consider BGA assignments relative to other Tx/Rx signal pairs in order to minimize coupled noise and resistance in the package design. Power, Ground, or Test Signals may be positioned between high-speed signal pairs for noise isolation. Figure 10.17a provides examples of BGA assignments for 3–4 Gbps differential signals. Differential signals within a bundle are sometimes allowed to be adjacent to each other at these baud rates, with power and ground BGAs assigned between bundles. Figure 10.17b provides examples of BGA assignments for 10-Gbps differential signals. At higher baud rates each Tx/Rx signal pair is electrically isolated from other signal pairs.

Figure 10.18 illustrates HSS cores on the edge of a die. The HSS cores are designed to be placed directly underneath the chip C4 solder balls that connect to the differential signals. On the Flip-Chip package, the substrate wires signals radially from the C4 pads to the package balls as is shown in Fig. 10.19. To minimize noise, inner C4s are routed to inner BGAs, and outer C4s are routed to outer BGAs. Figure 10.19 assumes a 4-Gbps baud rate with BGA adjacency rules similar to the examples in Fig. 10.17a.

At lower baud rates, HSS cores can be packaged in either flip-chip or carefully designed wirebond substrates. The benefits of HSS cores are just as attractive for low cost applications. Although the less expensive wirebond packaging does not provide an optimal electrical environment for high-speed applications, it can generally support up to 3.2 Gbps assuming care is taken to ensure the package design meets electrical performance targets. Higher baud rates are also possible with new developments in wirebond packaging technology that specifically target Serdes support.

The insertion loss specifications in Fig. 10.15 and return loss specifications in Fig. 10.16 also can be applied to package designs for applications in this lower baud rate range. The frequency-based "Trace and Via Isolation" requirements in Table 10.7 also apply. Differential signal matching for lower baud rate applications can be relaxed somewhat relative to the specifications in Table 10.7; "Matching Within a Pair" is typically specified as 5 ps, and "Matching with a Bundle" is typically 50 ps.

For HSS applications, FBGA wirebond substrates typically use two signal and two power layers. HSS signals and analog power/ground are routed as microstrips on the top layer of the laminate. An analog ground plane in the region of the HSS BGA and bondfingers is recommended to enable the use of vias to supply analog ground in cases where there is not sufficient space for a dedicated solder ball assignment. Additionally, a plane for the analog power supply (AVdd) in the HSS region is recommended. Figure 10.20 illustrates the signal pair matching and shielding of the Rx and Tx signal microstrip signals on the top layer of the laminate.

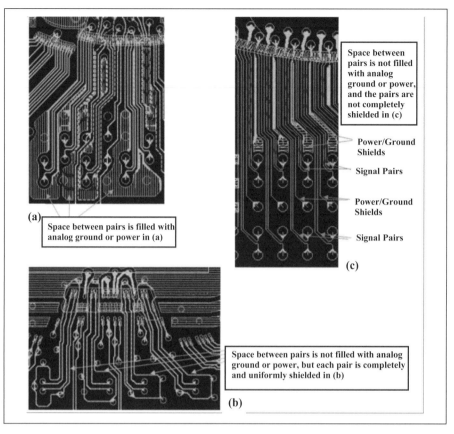

Fig. 10.20 Shielding and matched signal routing in wirebond package

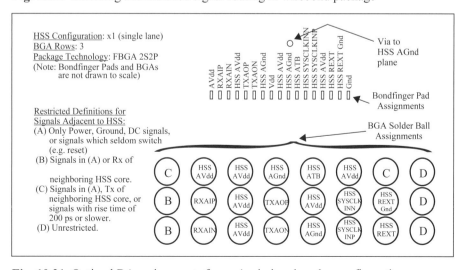

Fig. 10.21 Optimal BA assignments for a x1 wirebond serdes configuration

High-speed performance of bondwires in the package is strongly dependent upon bondwire length, with shorter bondwires performing better than longer ones. As the bondwire length increases, bandwidth degrades rapidly and coupling increases. Bondwire lengths for HSS signals are recommended to be 4 mm or less in order to meet insertion loss and isolation targets.

Signal assignment to the package solder ball (BGA) locations in the wirebond FBGA region has a large impact on both wireability and electrical performance. Assignment of neighboring signals results in coupling not only from solder balls, but also from vias and microstrips that are in close proximity. Table 10.8 summarizes adjacency rules for BGA assignments for wirebond HSS cores. Figure 10.21 shows an example of optimal package BGA and bondfinger assignments for a single lane Serdes that meets the electrical performance targets at the package solder balls and pin-through-hole (PTH) vias.

Table 10.8 Permissible BGA assignment adjacencies for wirebond HSS

HSS signal	Neighboring signals					
	Power/ ground	DC signals	Low activity factor signals	RXxIP/N of neighbor HSS	TXxOP/N of neighbor HSS	Switching signal with $t_{rise} < 200$ ps
HSSREFCLKT/C	Yes	Yes	Yes	No	No	No
RXxIP/N	Yes	Yes	Yes	Yes	No	No
TXxOP/N	Yes	Yes	Yes	No	Yes	Yes
Analog Vdd/Gnd	Yes	Yes	Yes	Yes	Yes	Yes

10.4.2 Chip Physical Design

10.4.2.1 HSS Core Footprint

The HSS EX10 core described in Chap. 2 supports 10-Gbps baud rates, and therefore must meet stringent electrical specifications. This core is therefore designed for use in FC-PBGA packages. The core should be placed under or adjacent to the C4 solder balls on the die, which connect the differential signals to the package substrate once the die is packaged. Figure 10.22 illustrates the footprint of this core when placed on the south side of the die. C4 solder balls are shown, and both Tx and Rx signal pins are located adjacent to C4s. This arrangement minimizes I/O wiring. A very short, fat metal wire and a stacked via is sufficient to wire the connection.

HSS cores designed for wirebond packages are placed around the edge of the chip in the region normally occupied by peripheral wirebond I/O. Figure 10.23a shows the chip edge for a single row wirebond chip layout with a single lane HSS core positioned next to the edge. The layout uses wide-wire, low-resistance connections to the bondfinger pads, seen at the die edge, and follows the restrictions for adjacency assignments described in Fig. 10.21.

Figure 10.23b extends this to a layout containing eight-lane wirebond HSS core. Tx and Rx slice circuitry and corresponding bondfingers are located to the right and left of the PLL slice.

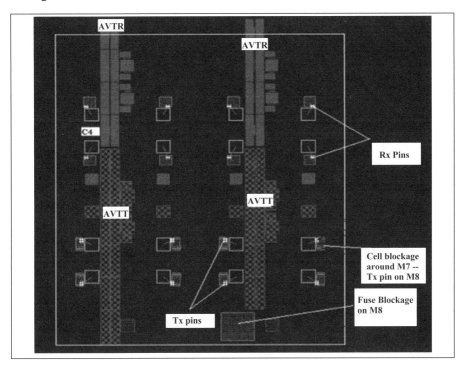

Fig. 10.22 HSS EX10 4 port full duplex core footprint

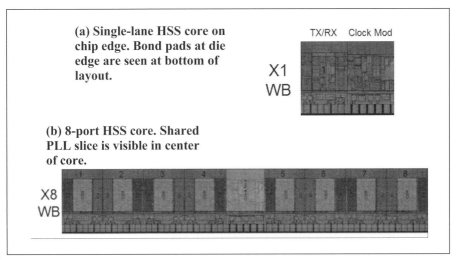

Fig. 10.23 HSS cores on wirebond chip edge

10.4.2.2 Chip Layout

Ideally, HSS cores are positioned during chip layout such that the Rx and Tx pins of the core are as close as possible to their assigned C4 pads as shown in Fig. 10.22. The core should always be oriented so that HSS internal pins are along the side of the core facing away from the edge of the die. The HSS core should be placed as close as possible to the die edge to maximize the die area available for chip logic placement. (Logic gates are generally not placed in any space between the HSS core and the edge of the die.) Placing the HSS core toward the edge of the die also has advantages from a noise perspective.

On wirebond chips, HSS cores are positioned along the edge of the die near bondfingers as shown in Fig. 10.23. The Rx and Tx pins of the core are as close as possible to their assigned bondfingers.

Figure 10.24 shows the recommended die locations for the HSS EX10 core assuming flip-chip packaging. The HSS cores can be positioned in one or two concentric rings around the die. Ideally, all cores would be located in the outer ring, as close to the die edge as possible. If more cores are required than can be placed in this ring, then cores can also be placed in the inner ring. The gap between the rings accommodates wiring tracks for the on-chip logic connections to the outer ring of HSS cores.

Once HSS cores have been placed, power routing is performed to connect power supplies to the cores, and the serial I/O signals are wired. Figure 10.25 illustrates power routing of the AVTT and AVTR power supplies for the HSS EX10 core. Fig. 10.26 illustrates an example of a short, fat wire connection between a serial data signal and the adjacent C4 pad.

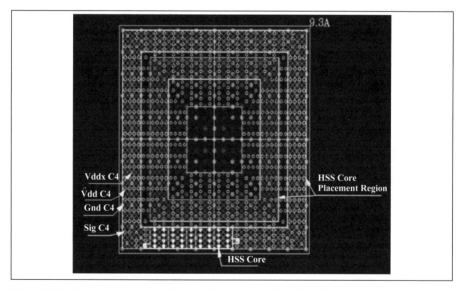

Fig. 10.24 Recommended serdes on-die locations for flip chip packages

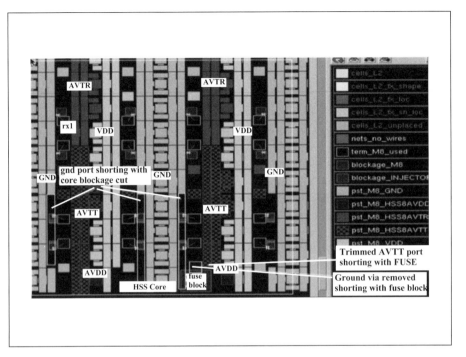

Fig. 10.25 HSS EX10 core power routing (AVTT and AVTR)

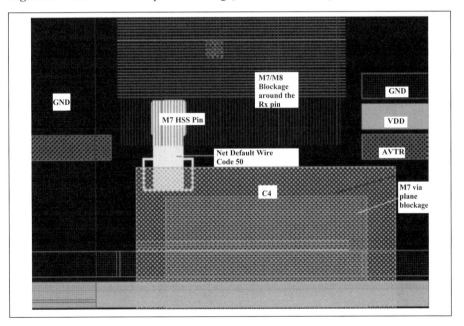

Fig. 10.26 RX pin to C4 wiring

10.4.2.3 Floorplan Considerations

Whether packaged in a flip-chip or wirebond package, HSS cores are susceptible to on-chip noise, and should not be placed near banks of noisy switching I/O or embedded memories. HSS cores are best placed on separate chip edges away from any noisy switching I/O. Layout engineers often use the "1-mm rule" for separating HSS cores from any embedded memories.

HSS cores are typically surrounded by on-chip decoupling capacitors in order to minimize the power supply voltage compression and transient noise at the core. Figure 10.27 illustrates a wirebond chip floorplan with a single lane HSS core on the south-side die edge surrounded by decoupling capacitors.

Figure 10.28 illustrates a chip floorplan with an IF PLL in the south center of the die edge, and two pairs of eight-port Serdes cores on either side of this PLL. Another IF PLL is located in the northeast corner of the die, along with two four-port Serdes cores. On-chip decoupling capacitors have been placed around the Serdes and the PLLs, as well as around the numerous embedded memories in the center of the die. The outlining of these blocks in the left view shown in Fig. 10.28 actually results from the display of these decoupling capacitors.

Legacy "IR Drop" analysis calculates power supply compression as a static value resulting from average supply current. This static analysis is generally insufficient and does not take into account transient currents which cause voltage noise on the power supply. Excessive compression on the power supply input of the HSS core could result in the core not being able to operate at the specified data rate. Even more modest transient noise can impact jitter performance of the core. An example of a typical specification for voltage compression due to static and transient effects is no more than 25 mV Vdd/Gnd compression for a 1.2V supply as listed in Table 10.7.

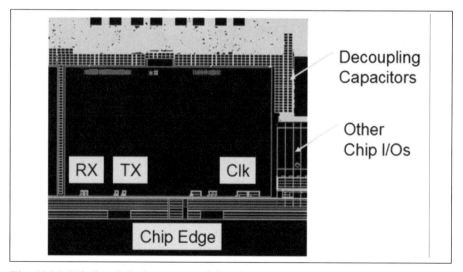

Fig. 10.27 Wirebond die layout containing single-lane HSS core

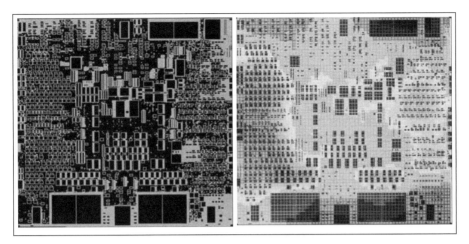

Fig. 10.28 Flip-chip die layout containing HSS cores

The east and west sides of the die in Fig. 10.28 contain banks of switching I/Os. The die view on the right is a heat graph of the transient power supply analysis performed on this die. The shading indicates the largest transient voltage compression is around the switching I/Os, while the compression around the PLLs and HSS cores is minimal.

Chip layout and package designs for HSS cores must address all of the topics discussed in this section in order to avoid impacting signal integrity. These topics included flip-chip C4 ball or wirebond bondfinger pad adjacency considerations, package BGA adjacency considerations, differential signal routing and shielding in the package, fat-wire routing and shielding of HSS signals and analog supplies on chip, and noise mitigation in the chip layout. Chip layout noise mitigation requires performing layout to isolate the HSS cores from noise sources, and inclusion of sufficient decoupling capacitance.

10.5 References

For more detail on the ASIC chip design methodology described in this chapter, see:

1. "Design Methodology for IBM ASIC Products", J.J. Engel, T.S. Guzowski, A. Hunt, D.E. LAckey, L.D. Pickup, R.A. Proctor, K. Reynolds, A.M. Rincon, D.R. Stauffer, IBM Journal of Research and Development, July 1996.

Interested IBM employees and IBM ASIC customers may wish to consult the following IBM HSS databooks for more detailed descriptions of timing assertions, test wrappers, and HSS footprints.

2. "High Speed Serdes (HSS) – 10G Optimized in Cu-65 Core Databook", SA15-6164-00, IBM.

10.6 Exercises

1. Specify the truth table for a 2-to-1 multiplexor assuming that inputs may have a value of "0," "1," or "X."

2. Specify the state table for a positive-edge triggered flip-flop with an asynchronous reset input assuming inputs have a value of "0," "1," or "X."

3. Design logic to use the signal driving the HSSRESET input of an HSS EX10 core to force core inputs to values specified in Table 10.1 during the reset. (HSSREFCLK[T,C] is not gated.)

4. For each of the following functions of the transmitter slice of the HSS EX10 core, explain to extent to which you would expect these functions to be modeled in a digital simulation model:

   (a) Function of the *Transmit Power Register*

   (b) Functions of the *Transmit Test Control Register*

   (c) Function of the *Slow Slew Control* in the *Transmit Driver Mode Control Register*

   (d) Function of the *Rate Select* in the *Transmit Configuration Mode Register*

   (e) Function of the *TXxELECIDLE* pins

   (f) Function of the *TXxRCVRDETEN* pins and corresponding *TXxRCVRDETTRUE* and *TXxRECVRDETFALSE* output pins

   (g) Function of the *TXxOBS* pins

   (h) Function of the *TXxBEACONEN* pins

5. For each of the following functions of the receiver slice of the HSS EX10 core, explain to extent to which you would expect these functions to be modeled in a digital simulation model:

   (a) Function of the *DFE Data and Edge Sample Register* and the *DFE Amplitude Sample Register*

   (b) Functions of the control signals in the *Receive Test Control Register*

   (c) Functions of the *Signal Detect Level* and *Signal Detect Power Down* in the *Signal Detect Control Register*

   (d) Function of the *Parallel Data Bus Width* in the *Receive Configuration Mode Register*

   (e) Function of the *Sonet Clock Control Register*

   (f) Function of the *DFE Tap X Registers* (for taps 1–5)

   (g) Function of the *RXxDATASYNC* pins

   (h) Function of the *RXxSIGDET* pins

6. The macro test controllability and observability requirements for the HSS EX10 core are specified in Table 10.3, and the corresponding requirements for JTAG Test are specified in Table 10.4.

   (a) Design logic which implements these requirements for the PLL slice of an HSS EX10 core.

   (b) Design logic which implements these requirements for the Transmitter slice of an HSS EX10 core.

   (c) Design logic which implements these requirements for the Receiver slice of an HSS EX10 core.

7. Input and output pins in Table 10.3 which must controlled or observed during the test sequence must be accessible through I/O pins during the test. Assuming input pins can be shared, but output pins cannot be shared, how many total I/O pins are required to support Macro Test for a single HSS EX10 core with four transmit and four receive slices?

8. Making the same assumptions as stated in Exercise 7, how many total I/O pins are required to support Macro Test for four HSS EX10 cores on the same chip?

9. Write EinsTimer and Synopsys timing assertions for the *TXADCLK* pin of an HSS EX10 core which account for jitter and duty cycle variation of this clock. Assume the clock period is 2.10 ns, the data width is set for 10 bits, and the duty cycle variation is $\pm 5\%$ of the cycle time.

10. Write EinsTimer and Synopsys timing assertions for the *RXADCLK* pin of an HSS EX10 core which account for jitter and duty cycle variation of this clock. Assume the clock period is 1.90 ns, the data width is set for 16 bits, and the duty cycle variation is $\pm 7\%$ of the cycle time.

11. Sect. 10.3.2 describes adjustments to the timing of the *RXADCLK* and associated *RXAD** pins to avoid false setup and hold violations. Fig. 10.12 illustrated the intended launch and capture timing for the *RXAD** pins. Assuming analysis is being performed on a postlayout netlist, use a timing diagram to illustrate the various timing adjustments to *RXADCLK* (per Sect. 10.3.1) and *RXAD** data signals (per Sect. 10.3.2) that contribute to performing setup tests at the flip-flops capturing the data. Repeat this illustration for hold tests.

12. Fig. 10.12 illustrated the intended launch and capture timing for the *RXAD** pins. Looking at this figure, it appears that rather than counting on clock tree delay to meet timing requirements for this interface, it would be possible to use the falling edge of *RXADCLK* to capture the data. Explain why this is not a good idea. (Hint: What if the data width and rate select are changed by the application?)

13. Using Fig. 10.12, explain why setup time violations occur when performing timing analysis on a prelayout netlist with an ideal clock tree (i.e., clock tree delay is zero).

14. Why must skew between the true and complement legs of the differential signal be minimized? What are the impacts as this skew increases?

15. Write EinsTimer timing assertions for the *HSSRESYNCCLKOUT* pins of two HSS EX10 cores which must be synchronized together which only use the divider ratios 16 and 20. Assume the frequency of *HSSRESYNCCLKOUT* is 2.5 ns when the divider ratio of 16 is used, that clock jitter is 0.080 ns, duty cycle variation is ±6% of the clock period, and core-to-core skew is specified to be 150 ps.

16. Write Synopsys timing assertions for the *HSSRESYNCCLKOUT* pins of two HSS EX10 cores with similar requirements to those specified in Exercise 15.

17. Table 5.5 describes the lane-to-lane skew requirements for the transmitter output of an OIF SFI-5.2 interface. The baud rate for SFI-5.2 is 9.95328 Gbps. Assume the HSS EX10 core-to-core skew is 450 ps. Suggest a skew budget for all skew contributors on the chip which complies with the SFI-5.2 specification for transmit skew.

18. Does Table 5.5 impose skew requirements on the receiving chip for the OIF SFI-5.2 interface? Explain.

19. Applying your knowledge of signal integrity for differential signals, why do you think FC-PBGA packages are better suited for higher baud rates as compared to wirebond packages?

20. The placement of one example of an HSS core is shown on the south side of the die, left of center, in Fig. 10.24. How many of these HSS cores can you fit on the die, utilizing both the outer and inner rings?

Index

Numerics

10 Gigabit Attachment Unit Interface (XAUI) 57, 167, 198, 204, 206, 218–220, 236
10 Gigabit Serial Electrical Interface (XFI) 207–213
64B/66B Block Code 138, 368
  CEI-P 155, 166, 184, 186, 187
  Fibre Channel 236
  run length 15
  XFI (Ethernet) 167, 198, 208, 210
8B/10B Block Code 9, 137, 138, 368
  Fibre Channel 168, 222, 225, 236
  PCI Express 241, 242, 244, 245, 250
  run length 15
  signal connections 54
  XGXS (Ethernet) 154, 167, 198, 203–207

A

AC JTAG. *See* JTAG 1149.6 (AC JTAG)
active (AC) power 321, 395–400, 408, 409, 411, 412, 414–417
activity factor (AF) 321, 395, 398, 399, 412, 416, 417
aliasing 154, 171
align symbol 154, 207
American National Standards Institute (ANSI) 169, 220
application layer 125
At-Speed Structural Test 318, 320
Automatic Gain Control (AGC) 312, 333, 334
Automatic Test Pattern Generator (ATPG) 432
AVdd power supply 104, 261, 411, 414, 459, 462
AVtt/AVtr power supply 36, 71, 109, 411, 414, 459, 466

B

backannotation (timing) 456
bang-bang phase detector 116
barrel shifter 154–156, 245
bathtub curve 24, 359, 360, 384, 389
beacon signalling 37, 66, 73, 80, 248, 249, 412
bit alignment 125, 134, 152, 153, 155, 156, 205, 206
Bit Error Rate (BER) 4, 369–371, 384, 389
  CEI 195
  clock jitter 268, 274
  error correction 143–146, 201
  Ethernet 220
  Fibre Channel 233
  relation to HSS circuits 66, 71, 100, 118
  relation to run length 15
  signal integrity 19–24, 345–361
  Sonet/SDH 15, 140, 171
  test 305, 328–333
Bit Interleaved Parity (BIP) 144, 171–173
bit interleaving 136, 156, 180, 183, 333
bit order
  HSS core 55, 70
  interface 125, 134, 142, 202, 225, 227
Bounded Uncorrelated Jitter (BUJ) 333, 335, 348, 349, 354–356, 360
Built-in Self-Test (BIST) 36, 40, 78, 99, 365, 442
  Logic BIST (LBIST) 295, 315–318
BYPASS_CAL 426, 430
byte order (of interface) 125, 134, 225
byte striping 136, 155, 156, 236, 244–246, 250, 333

C

C4 pads 457, 462, 465, 466, 469

channel (interconnect) 10, 19–21, 71, 74, 337, 368, 369, 376, 391
   analysis 23, 24, 79, 196, 365–367, 369, 371, 387, 456
   channel compliance 127, 128, 190, 191, 193, 196, 371
   crosstalk 21, 23, 377, 378, 380, 383
   equalization 54, 56, 58, 70, 79, 109, 110, 127, 247, 374, 381, 382
   jitter contribution 348, 351, 352, 361, 362, 370, 380
   response 10, 13, 19–21, 23, 26, 196, 343, 370–380
   skew 153, 156
   S-Parameters 377, 379, 384, 387
channel length (Leff) 326, 401, 402
characterization test 325–337
Clock and Data Recovery (CDR) 8, 14, 15, 66, 68, 78, 87, 90, 93, 114, 116–118, 156, 157, 189, 227, 246, 248, 274, 276–279, 311, 333, 347, 352, 353, 362, 378, 380, 383, 388, 427, 446
clock buffer (differential) 265, 279, 280, 444
Clock Forwarding 3
comma symbol (COM) 154, 242–245
Common Electrical I/O (CEI) 146, 166, 179, 180, 183, 184, 188, 190–197, 277, 278, 360–362, 371, 376
Common Electrical I/O Protocol (CEI-P) 146, 155, 166, 167, 179, 184, 186–190
common mode noise 6, 263, 280
Common Mode Voltage (Vcm) 17, 36, 65, 66, 71, 72, 104, 191–194, 219, 246–248, 263, 281–283, 302, 334
compliance channel 327, 330, 333, 335
compliance point 126–128, 326
   Fibre Channel 230, 232, 233, 329
   SFI Reference Model 181, 184, 277
Content Addressable Memories (CAMs) 126
continuous-time equalizer 100, 194
Correlated Bounded Gaussian Jitter (CBGJ) 349, 354–356, 360–362

Correlated Bounded High Probability Jitter (CBHPJ) 360
coupling (AC or DC) 34, 71, 114, 191, 193, 247, 248, 334, 368, 369
crosstalk 9, 19, 21–23, 56, 57, 73, 216, 348, 349, 353–355, 369, 370, 377, 378, 380, 383, 384, 391
   FEXT 377, 378
   NEXT 9, 378
Cumulative Distribution Function (defined) 344
current-voltage (IV) curve 400–402
cursors (of pulse response) 372, 374, 381, 382
cycle-to-cycle jitter 271, 272, 275
Cyclic Redundancy Check (CRC) 133, 144, 145, 186, 224, 225, 227, 241, 249

D

Data Dependent Jitter (DDJ) 20, 348–352, 360, 370, 371, 380
data eye 4
   amplitude 57, 79, 111, 128
   characterization 326, 328–332, 348
   duobinary 24, 25
   equalization 10, 13, 110, 111, 216
   eye mask (See eye diagram)
   NRZ 24
   PAM-4 24–26
   relation to BER 4, 24, 332, 348, 359, 360, 370, 389
   sampling 7, 14, 15, 137, 446
   signal impairments 5, 6, 21, 22, 274, 347, 356, 362, 370, 372
   signal integrity 79, 196
   simulations 365, 369
   statistical analysis 357, 359, 360, 371–373, 380–385
   training 87, 427
   virtual eye (See statistical eye)
data link layer 125, 126
   PCI Express 240–242, 245, 249
data packet 125, 133, 134, 140, 144, 145, 150, 153, 154, 168, 176
   Ethernet 134, 154, 167, 197, 198, 201, 204, 208, 211, 212

data packet (*Cont'd*)
  Fibre Channel 167
  PCI Express 154, 240–245
  SPI 178, 184, 186
DC Stuck Fault Testing. *See* scan test
Decision Feedback Equalizer (DFE), 13,
  14, 17, 23, 50, 100, 114, 116, 129, 193,
  233, 312, 333
  circuit architecture 118–121
  digital eye 79
  HSS EX10 DFE 46, 48–52, 68, 70,
    71
  jitter budget 362
  macro test 323
  multilevel signalling 26
  power dissipation 337, 412, 417,
    419
  simulation and analysis 23, 365,
    372, 379–382, 388, 389, 431
  training 87, 90, 93, 189, 427
  use of signal detect 73
decoupling capacitors
  on signal 18, 19, 74, 247, 248, 301,
    303, 332
  power supply 261, 468
deemphasis 10, 128, 247
deserializer stage 9, 70, 116, 121
Design-For-Test (DFT) design
  automation 440
deskew channel 155, 180–183
deskew logic 151, 153–156, 183, 184,
  187, 455
Deterministic Jitter (DJ) 14, 22, 23, 184,
  193, 219, 233–235, 247, 274, 328,
  346, 348–354, 356, 357, 360, 373,
  378, 384, 456
DFE Amplitude Sample Register 50, 79
DFE Control Register 49, 79, 93
DFE Data and Edge Sample Register 50,
  79
DFE TapX Register 51, 52
DI 433, 434
differential clock analysis 286–290
differential driver 15–17, 415
differential receiver 17, 264, 268, 282,
  283, 325, 415

differential signal 5, 6, 8, 54, 71, 81, 281,
  430
  amplitude 46, 56, 72, 91, 281, 282,
    323, 325, 329–332, 335, 411, 419
  crosstalk 21, 56, 378
  footprint 462, 464
  input level 282
  JTAG 18, 301–303
  noise rejection 263
  power supply compression 263
  reference clock 261–263
  skew 455, 456, 458
  S-Parameters 377
  timing analysis 444
  wiring 280, 281
Differential Voltage (Vdiff) 17, 46, 281,
  282, 419
digital eye 50, 52, 79, 332
Digital Eye Control Register 50, 79
double data rate (DDR) 3, 4, 57, 202
Drain Induced Barrier Lowering (DIBL)
  402
Dual-Dirac probability density function
  24, 343, 346, 350, 351, 353, 357, 360,
  371, 380, 383
Duty Cycle Distortion (DCD) 102, 103,
  192–194, 219, 220, 279, 280, 282,
  284, 328, 348–350, 357, 360, 380,
  456, 458

E

elastic FIFO 125, 134, 147–151, 153,
  156, 243, 244, 250, 304, 306, 446
Electrical Idle State 38, 65, 243, 247,
  248, 413
Electrically Enhanced Plastic Ball Grid
  Array (EPBGA) package 457
Electromagnetic Interference (EMI) 3,
  78, 247, 264, 265, 267, 352,
  377
error correction 125, 134, 143, 145,
  146, 155, 165, 166, 178, 180,
  188–190, 200, 201, 213, 217
Ethernet. *See* IEEE 802.3
evaluation board 326
External Serial Loopback 305, 306

eye (closure, opening, width) *See* data eye.

eye diagram 107, 126, 128, 129, 195, 196, 218, 219, 233–235, 246, 247, 328, 330–332, 369, 389, 391

F

fall time. *See* slew rate

Feed Forward Equalizer (FFE) 8, 10, 11, 44–46, 58–63, 79, 87, 88, 91, 92, 99, 107–111, 128, 191, 200, 213–216, 218, 247, 325, 327, 328, 337, 368, 372–375, 386, 389–390, 431

FFE Coefficient Negotiation 59, 216

Fibre Channel 57, 65, 72, 133, 140, 144, 167, 168, 220–237

Fine Pitch Plastic Ball Grid Array (FBGA) package 457

Flip-Chip packages 457, 458, 462, 466, 468, 469

floorplanning
    clock 279, 280
    core placement on die 464–466, 468
    decoupling capacitors 468
    I/O wiring 464, 466
    pad assignments 462, 464
    power routing 466
    transient power supply analysis 469

Flywheel 48, 66, 78

Forward Error Correction (FEC). *See* error correction

frame alignment 173, 183

full duplex core 8, 17, 31, 99, 100, 134, 304, 305, 414

G

Gate Dielectric Current (Igate) 400, 404, 405

Gate Induced Drain Leakage Current (Igidl) 400, 406, 407

Gaussian distribution 22, 137, 140, 343–346, 349, 355, 371
    bounded 345, 346, 349, 354, 356

Gaussian Jitter (GJ) 344, 345, 349, 356, 360, 380, 384

Golden PLL 331

H

High Probability Jitter (HPJ) 348

High Speed Serdes (definition) 8

HSSACJAC 34, 439

HSSACJPC 34, 75, 439

HSSCDR software 24, 328, 371, 386–391

HSSDIVSEL 33, 82, 264–266, 429, 437

HSSEYEQUALITY 35, 52, 79

HSSJTAGCE 34, 36, 40, 63, 64, 74, 439

HSSPDWNPLL 33, 60, 73, 429, 437, 439

HSSPLLLOCK 33, 83, 107, 426, 437, 438

HSSPRTADDR 34, 41, 437, 443, 450, 451

HSSPRTAEN 34, 437, 443, 450, 451

HSSPRTDATAIN 34, 437, 443, 450, 451

HSSPRTDATAOUT 34, 443, 450, 451

HSSPRTREADY 34, 83, 84, 88, 90

HSSPRTWRITE 34, 437, 443, 450, 451

HSSRECCAL 33, 83, 84, 437

HSSREFCLKT/C 33, 80–83, 88, 186, 267, 429, 443–445, 452, 455, 456

HSSREFDIV 33, 82, 264–266, 429, 437

HSSRESET 33, 82, 84, 87, 88, 425, 426, 429, 437

HSSRESETOUT 33, 82

HSSRESYNCCLKIN 33, 84, 429, 437, 443, 446, 452, 456

HSSRESYNCCLKOUT 34, 443, 452–454

HSSRXACMODE 34, 71, 369, 439

HSSSTATEL2 35, 87, 429, 437

HSSTXACMODE 34, 368, 439

I

I/Q clock generator 102

IDDQ Test. *See* leakage test

idle symbol (IDL) 134, 150, 153, 203–205, 207, 211, 212, 224, 243, 244

IEEE 802.3 (Ethernet) 133, 134, 138, 140, 154, 167, 168, 178, 197–220, 224, 236–238, 240

IEEE 802.3ap Backplane Ethernet 59,
   91, 146, 167, 200, 213–218
Infiniband 57, 72, 168
Input Offset Compensation 114
insertion loss 19–21, 26, 351, 376, 377,
   389, 458, 462
Institute of Electrical and Electronics
   Engineers (IEEE) 197
Intermediate Frequency (IF) PLL 19,
   261–268, 275, 279–287, 325, 468
Internal Status Register 52, 79, 90, 93
International Committee for Information
   Technology Standards (INCITS) 168
International Telecommunications
   Union (ITU-T) 146, 169
interoperability points. *See* compliance
   point
Intersymbol Interference (ISI) 14, 58,
   118, 247, 328, 333, 335, 348, 351,
   370, 371, 382, 383
IR Drop analysis 468

J

jitter 1–3, 15, 21–24, 66, 80–82, 99, 103,
   127, 130, 177, 181, 183, 196, 218,
   219, 221, 261, 262, 264–267, 284,
   286, 325–327, 327, 330, 343,
   345–363, 373, 378, 383, 384, 415,
   448, 468
jitter (clock) 268–279, 446, 447
jitter amplification 277
jitter budget 117, 181, 184, 227,
   360–362
jitter gain See jitter transfer
jitter generation 21, 268, 371, 378, 380,
   383
jitter peaking 276–278
jitter tolerance 22, 36, 66, 78, 218,
   220, 246, 247, 264, 277, 321, 335,
   350–363
jitter transfer 22, 278
jitter transparent application 278
Joint Test Action Group (JTAG) 296,
   303
JTAG 1149.6 (AC JTAG) 6, 56, 61–76,
   104, 289, 295, 303, 324, 432, 438

JTAG Test 6, 295–303, 353, 432, 433,
   438–441
   HSS EX10 implementation 61, 62,
      64–76
   JTAG Boundary Scan Cell 300,
      438, 440
   JTAG Boundary Scan Register 61,
      62, 299–300, 438
   JTAG Bypass Register 299
   JTAG Compliance Enable signal
      438, 439
   JTAG ID Register 299, 300
   JTAG Instruction Register
      298–300, 442
   JTAG RUNBIST Instruction 299, 442
   JTAG TAP Controller 61, 62, 297,
      298, 438
   receiver 56, 324, 353
   signals 34, 36, 37, 39, 40, 297
Junction Leakage Current (Ijxn) 400, 406

L

leakage current (DC power) 321, 322,
   395, 400–409, 411, 412, 414–415,
   433, 434
leakage test 321, 322, 434
Linear Feedback Shift Register (LFSR)
   15, 65, 140, 141, 145, 207, 307, 308,
   311, 314, 317
Link Enable Register 42, 60, 72, 91, 92
Link Reset Register 43
Logic BIST (LBIST). *See* Built-in
   Self-Test (BIST)
loop timing 80, 132, 176, 186, 261, 266,
   267, 277, 278, 306
loopback test 15, 36, 40, 47, 64, 65, 77,
   78, 304–307, 322, 432
LT 433, 434

M

Macro Test
   HSS core 322–325, 433–438, 440,
      441
   Macro Test Complete signal 436
   Macro Test Enable signal 435, 436
   PLL 321, 322

manufacturing test (of chip) 15, 261, 268, 295, 304, 318–325, 432–434
Multiple-Input-Shift-Register (MISR) 317, 318

N

network layer 125, 126
Network Processing Elements (NPEs) 126, 166, 178, 184

O

Optical Internetworking Forum (OIF) 155, 157, 165, 166, 177–197, 266, 277, 371, 376
Out of Band Signalling (OBS) 36, 65

P

package model 21, 24, 373, 379, 386, 388
packet. *See* data packet
Parallel Data Bus Width 43, 46, 53–55, 68, 69, 91, 92, 109, 114, 446
Parallel Diagnostic Loopback 307
parity (on deskew channel) 155, 181, 183
parity error detection 143, 144
PCI Express 37, 65, 80, 134, 154, 168, 237, 306
    beacon 66, 73, 80
    link state 35, 38, 40, 66, 80, 87
peaking amplifier 8, 13, 114
period jitter 270–272, 275, 449, 454
Periodic Jitter (PJ) 348, 352–354, 356, 357, 360, 380
Peripheral Component Interconnect Special Interest Group (PCI-SIG) 168, 237
phase jitter 269–272, 274, 275
Phase Locked Loop (PLL) 14, 15, 19, 31, 99–107, 147, 156, 177, 266, 268, 337, 409, 410, 443, 455
    bandgap voltage 103, 104
    charge pump 101
    clock distribution macro 102

HSS EX10 PLL Slice 32–35, 41–43, 54, 55, 60, 66, 68, 71–73, 80–87, 264, 413
    jitter 275, 276, 446, 452
    JTAG 439
    Lock Detect 107
    lock detect 83, 105, 206, 209
    loop timing 186, 267
    macro test 321, 322, 437
    phase-frequency detector 101
    power dissipation 414
    power down 249
    reset and configuration 87, 88, 90, 91, 426, 429, 430
    test 317
    voltage and current references 103, 104
phase noise 272–274
Phase Rotator 48, 52, 78, 116–118
Phase Rotator Control Register 48, 78
Phase Rotator Position Register 48, 78
physical layer 125, 126
    CEI 146
    Ethernet 197, 198, 202, 204
    Fibre Channel 167, 168, 225, 232
    PCI Express 238, 240–242, 244–246, 248–250
plesiosynchronous clock 15, 133, 150, 306, 333, 388
post-cursor taps 10, 108–110, 214, 215
post-cursors (of pulse response) 121, 362, 372, 379–382, 389
power dissipation 3, 8, 56, 59, 60, 62, 66, 72, 77, 80, 87, 91, 100, 101, 103, 120, 121, 183, 238, 248, 321, 395–419
    testing 337
power management 59, 72, 248–250, 412
    power states 38, 40, 66, 80, 87, 243, 247, 248, 413, 419
power supply (general) 6, 56, 88, 103, 131, 261, 262, 321, 337, 365, 368, 369, 395, 397, 398, 402, 404, 405, 408–410, 416, 418, 425, 427, 431, 458, 468
power supply (Vdd) 414, 415
power supply compression 261–263, 468

power supply noise 261, 263, 328, 348, 352, 468, 469

Power Supply Rejection Ratio (PSRR) 104

pre-cursor taps 10, 108, 109, 214, 215

pre-cursors (of pulse response) 372, 382

preemphasis 10, 54, 58, 128, 365, 368

Probability Density Function (PDF)
bounded Gaussian distribution 345
definition 343
Dual-Dirac distribution 346
Gaussian distribution 343
Total Jitter 357

Process/Voltage/Temperature (PVT)
corners 1, 3, 4, 6, 7, 130, 147, 325, 368, 369, 386, 388

protocol logic 90–92, 125, 126, 130–132, 134–157, 304–306, 347, 444, 446

Pseudo-Random Bit Sequence (PRBS) 295, 304–315, 317, 322, 432
Bit Sequences (Patterns) 311, 312, 328, 330
Checker 17, 40, 47, 68, 77, 78, 121, 310, 311
Generator 36, 44, 54, 64, 65, 114, 308, 309, 335
Test Sequence 312–315
use in scrambling 140
use in training frame 214, 216

pulse width distortion. *See* Duty Cycle Distortion (DCD)

Q

Q (of circuit) 24, 345, 346, 359, 360, 384

quiescent current (power) 321, 322, 395, 408, 409, 411, 412, 414, 415, 419, 434

R

Random Jitter (RJ) 22, 23, 101, 105, 193, 219, 282, 328, 333, 335, 348, 349, 354–356, 373, 378

Rate Select 43, 46, 54, 61, 62, 68, 76, 91, 107, 228, 432

Receive Configuration Mode Register 46, 68–70, 76, 91, 121, 412, 417, 419

Receive Test Control Register 47, 77, 78, 312, 314, 315

Receiver (Rx) Slice 31, 38–41, 46–52, 66–80, 100, 114–121, 250, 365, 417
JTAG 439
macro test 437
power dissipation 412, 414
reset and configuration 90, 92, 93, 427, 429

receiver detection 37, 66, 248

reference channel 233

register definitions (HSS EX10) 41–52

register interface 32, 34, 431, 450–452

Remote Line Loopback 306

Remote Payload Loopback 307

resynchronization function 33, 34, 55, 60, 61, 84–87, 91, 92, 148, 151, 157, 206, 452–455

return loss 21, 22, 195, 196, 332, 337, 376, 377, 379, 458, 459, 462

RI 433, 434

rise time. *See* slew rate

Root-Mean-Square (RMS) values 345, 355, 360, 361

run length 14, 15, 54, 64–66, 78, 137, 138, 140, 171, 175, 183, 210, 281, 308

RXxACJPDP/N 39, 74, 75, 437, 439

RXxACJZTP/N 39, 74, 75, 325, 439

RXxBSOUT 40, 73, 74, 324, 439

RXxD 39, 68, 70, 76, 90, 92, 443, 448, 450, 456

RXxDATASYNC 39, 68, 70, 153, 155–157, 183, 206, 224, 245, 429, 430, 437

RXxDCLK 39, 68–70, 73, 87, 88, 92, 132, 150, 157, 228, 274, 430, 443, 444, 446–448, 450, 456

RXxIP/N 39, 75, 323, 324, 430

RXxPHSLOCK 437

RXxPRBSEN 40, 47, 312, 314, 437

RXxPRBSERR 40, 47, 315

RXxPRBSFRCERR 40, 47, 437

RXxPRBSRST 40, 314, 437

RXxPRBSSYNC 40, 47, 315

RXxPWRDWN 40, 72, 413, 429, 437, 439
RXxRCVC16T/C 39, 48, 267, 443, 444
RXxSIGDET 39, 72, 228, 248, 437, 438
RXxSIGDETEN 41, 73, 80, 92, 414, 437
RXxSTATEL1 40, 80, 413, 429, 437

S

sampling jitter 373, 378, 380, 383, 388
Scalable System Packet Interface (SPIS)
    155, 166, 177–180, 184–188, 266
scan chains 315–319, 433–435, 440
    JTAG 300, 301
scan mode 315, 316, 319
scan test 318–321, 432–434
SCANGATE 433, 434
SCANIN 433, 434
SCANOUT 433, 434
scrambling 9, 15, 65, 125, 134, 140–143,
    155, 166, 168, 169, 171–173, 175,
    183, 184, 187, 188, 208, 210, 217,
    225, 227, 244, 245, 250, 368, 374, 387
    self-synchronizing scrambler
        140–143, 210, 227
    sidestream scrambler 140–142, 171,
        244
Serdes-Framer Interface (SFI-5 or
    SFI5.2) 155–158, 165, 166, 177–184
Serial ATA (SATA) 65, 78, 168, 268
Serial Attached SCSI (SAS) 65, 72, 78, 168
Serial Diagnostic Loopback 304–306
serializer stage 9, 61, 112–114
Signal Detect 41, 49, 68, 71–73, 80, 87,
    92, 93, 116, 129, 130, 248, 249, 324,
    335, 336, 365, 414, 417
Signal Detect Control Register 49, 72,
    73, 324, 414, 417
signal integrity 9, 10, 18–24, 35, 79, 196, 197,
    238, 264, 268, 279, 300, 326–328,
    343–391, 411, 419, 444, 455, 457
    clocks 281–290
        Spice analysis 363–369
        statistical analysis 370–384
simplex core 8, 17, 31, 65, 78, 99, 100,
    322, 323
simulation models (event-driven
    simulation) 423, 425–432

single data rate (SDR) 3, 4, 57
single-ended signal 6, 17, 232, 261, 263,
    281, 331
Sinusoidal Jitter (SJ) 220, 233, 235, 277,
    333, 335, 348, 349, 353, 361, 362, 389
skew (on a multi-lane interface) 84, 91,
    130, 131, 135, 148, 149, 151, 153,
    155–157, 184, 206, 261, 266, 318,
    333, 347, 362, 446, 452–456, 458
skew budget 156, 157, 184, 266, 455, 458
skip symbol (SKP) 134, 154, 207,
    242–245
slew rate 4, 45, 56, 57, 91, 117, 279, 281,
    282, 284, 368, 430
Sonet Clock Control Register 48, 80
source synchronous interface 2–7, 165
S-Parameters 19, 24, 196, 326, 328, 376,
    377, 379, 383, 384, 387, 389, 458
Spice 23, 279, 281, 282, 286–290,
    363–370, 444, 454
    behavioral models 365–367
    extracted models 363–365
speed negotiation 91, 227, 228
Spread Spectrum Clocking (SSC) 48, 78,
    93, 247, 267
Standard Delay Format (SDF) 456
StatEye software 24, 197, 371, 386
static timing analysis. See timing analysis
statistical eye 14, 24, 79, 129, 196, 357,
    380–386, 389, 391
statistical signal integrity analysis
    371–390
status channel 184, 186, 187
Sub-Threshold Leakage Current (Isubvt)
    400, 402–404, 406, 407
synchronous clock 132, 133, 306, 333
Synchronous Digital Hierarchy (SDH).
    See Synchronous Optical NETwork
    (SONET)
Synchronous Optical NETwork
    (SONET) 15, 39, 48, 65, 132, 133,
    140, 165, 166, 168–178, 180, 183,
    198, 207, 208, 212, 213, 266, 307
    BIP 144
    Forward Error Correction 146
    scrambler 142

T

TCP/IP model 125
termination (of signal) 17, 34, 56, 60, 71,
    109, 114, 127, 219, 233, 324, 325,
    365, 368, 369, 411, 414
test synthesis (DFT) 423
test wrapper 433, 440, 441
TESTENABLE 433, 434
threshold voltage (Vt) 325, 395,
    400–402, 406, 417
timing analysis 423, 443–454, 456, 457
    clock duty cycle 447, 448
    clock jitter 275, 446–449, 454
    register interface 450, 452
    resynchronization function 452–454
    RXxD receive data 448, 449
    serial data 454
timing assertions
    HSSRESYNCCLKOUT 453, 454
    RXxD receive data 449
    TXxDCLK and RXxDCLK
        445–448
timing model (of HSS) 443, 444, 448,
    450, 454, 456
Total Jitter (TJ) 22, 23, 184, 192, 219,
    233–235, 246, 247, 328, 347, 349,
    356, 357, 359, 360, 362, 380, 383
Transmit Coefficient Control Register
    44, 59, 92
Transmit Configuration Mode Register
    43, 53–55, 61, 91, 107, 114
Transmit Driver Mode Control Register
    45, 57–60, 91, 109, 368
Transmit Polarity Register 46, 58, 59
Transmit Power Register 46, 56–59,
    64, 91, 108, 109, 325, 365, 367,
    412
Transmit TapX Coefficient Register 45,
    58, 59, 91, 108, 325, 365, 368
Transmit Test Control Register 44, 64,
    78, 314
Transmitter (Tx) Slice 31, 35–38, 43–46,
    53–66, 99, 107–114, 250
    JTAG 439
    macro test 437
    power dissipation 412, 414

    power down 431
    reset and configuration 90–92, 429
transport layer 125
TXxBEACONEN 37, 66
TXxBSIN 36, 62, 63, 439
TXxBSOUT 36, 439
TXxBYPASS 37, 62, 437
TXxD 36, 53–55, 61, 90, 91,
    136, 443
TXxDCLK 31, 36, 53–55, 60, 61, 65, 84,
    88, 91, 148, 157, 228, 232, 274, 443,
    444, 446, 447
TXxELECIDLE 38, 65, 413, 437
TXxJTAGAMPL 37, 63, 64, 439
TXxJTAGTS 36, 63, 64, 439
TXxOBS 36, 65
TXxOP/N 36, 62, 108, 325
TXxPRBSEN 36, 44, 437
TXxPRBSRST 36, 437
TXxPWRDWN 37, 60, 413, 429, 437,
    439
TXxRCVRDETEN 37, 66
TXxRCVRDETFALSE 37, 66
TXxRCVRDETTRUE 37, 66
TXxSTATEL1 38, 66, 413, 429, 437
TXxTS 36, 60, 64, 91, 92, 437

U

Uncorrelated Bounded High Probability
    Jitter (UBHPJ) 192–194, 354
Uncorrelated High Probability Jitter
    (UHPJ) 192, 193
Uncorrelated Unbounded Gaussian Jitter
    (UUGJ) 192–194, 349, 354–356, 360,
    361
Unit Interval (defined) 22

V

Variable Gain Amplifier (VGA) 52, 70,
    71, 114, 365
VCO calibration 42, 83, 84, 88, 90, 105,
    426, 430
VCO Coarse Calibration Control
    Register 42, 83, 84
VCO Coarse Calibration Status Register
    42, 83

Vdd power supply 288, 410, 411, 418

Vector Network Analyzer (VNA) 19, 376, 377

Voltage Controlled Oscillator (VCO) 83, 101, 105, 275, 426

Voltage Regulator (PLL) 100, 103, 104, 322

voltage screen test 320, 321

W

WAN Interface Sublayer (WIS) 198, 200, 208, 212, 213

wander (on a multi-lane interface) 130, 131, 147, 149, 151, 156, 157, 184, 333, 347, 362

Wirebond packages 297, 457, 462, 464, 466, 468, 469

wrap back. *See* loopback test

X

XAUI. *See* 10 Gigabit Attachment Unit Interface (XAUI)

XFI. *See* 10 Gigabit Serial Electrical Interface (XFI)

XGMII Extended Sublayer (XGXS) 154, 167, 198, 204–207

Z

ZDI 433, 434

ZRI 433, 434